Introduction to Geospatial Technologies

Third Edition

Bradley A. Shellito

Youngstown State University

W. H. FREEMAN & COMPANY
A Macmillan Education Imprint

Publisher: Kate Ahr Parker
Executive Editor: Bill Minick
Senior Development Editor: Blythe Robbins
Editorial Assistant: Abigail Fagan
Marketing Manager: Maureen Rachford
Photo Editor: Sheena Goldstein and Christine Buese
Design Manager: Vicki Tomaselli
Interior and Cover Designer: Patrice Sheridan
Production Manager: Paul Rohloff
Project Management: MPS North America LLC
Composition: MPS Ltd.
Printing and Binding: RR Donnelley
Cover Photos: kropic1/Shutterstock (main photo);
NASA (top inset); Tshooter/Shutterstock (bottom inset)

ISBN-13: 978-1-4641-8872-5
ISBN-10: 1-4641-8872-6

Library of Congress Control Number: 2015946168

Printed in the United States of America

First printing

W. H. Freeman and Company
One New York Plaza
Suite 4500
New York, NY 10004-1562

www.whfreeman.com/geography

Dedication

This book is dedicated to my parents, David and Elizabeth, who taught me that with a lot of hard work—and a little luck—all things are possible.

About the Author

[Source: Neal P. McNally]

Bradley A. Shellito is a geographer whose work focuses on the application of geospatial technologies. Dr. Shellito has been a professor at Youngstown State University (YSU) since 2004, and was previously a faculty member at Old Dominion University. He teaches classes in GIS, remote sensing, GPS, and 3D visualization, and his research interests involve using these concepts within a variety of real-world issues. His second book, *Discovering GIS and ArcGIS*, was also published by Macmillan Education. He also serves as YSU's Principal Investigator in OhioView, a statewide geospatial consortium. A native of the Youngstown area, Dr. Shellito received his bachelor's degree from YSU, his master's degree from the Ohio State University, and his doctorate from Michigan State University.

Contents in Brief

Contents

CHAPTER 6

Using GIS for Spatial Analysis

Database Query and Selection, Buffers, Overlay Operations, Geoprocessing Concepts, and Modeling with GIS 177

CHAPTER 7

Using GIS to Make a Map

Scale, Map Elements, Map Layouts, Type, Thematic Maps, Data Classification Methods, Color Choices, and Digital Map Distribution Formats 223

CHAPTER 8

Getting There Quicker with Geospatial Technology

Satellite Navigation Systems, Road Maps in a Digital World, Creating a Street Network, Geocoding, Shortest Paths, and Street Networks Online   275

PART 3 Remote Sensing   309

CHAPTER 9

Remotely Sensed Images from Above

Where Aerial Photography Came From, UAS, Color Infrared Photos, Orthophotos, Oblique Photos, Visual Image Interpretation, and Photogrammetric Measurements   309

Preface

Why I Wrote *Introduction to Geospatial Technologies*

When people ask me what I teach, I say "geospatial technology." The usual response to this statement is a blank stare, a baffled "What?" or a variation on "I've never heard of that." However, if I say I teach "technologies like GPS, taking images from satellites, and using online tools like Google Earth or MapQuest," the response generally improves to: "GPS is great," or "Google Earth is so cool," or even "Why do I get the wrong directions from that thing in my car?" Although geospatial technologies are everywhere these days—from software to Websites to cell phones—it seems that the phrase "geospatial technology" hasn't really permeated into common parlance.

I hope that this book will help remedy this situation. As its title implies, the goal of this book is to introduce several aspects of geospatial technologies—not only what they are and how they operate, but also how they are used in hands-on applications. In other words, the book covers a little bit of everything, from theory to application.

In a sense, the book's goal is to offer students an overview of several different fields and techniques and to provide a solid foundation on which further knowledge in more specialized classes can be built, such as those delving further into geographic information systems (GIS) or remote sensing. Whether the book is used for a basic introductory course, a class for non-majors, or as an introduction to widely used geospatial software packages, this book is aimed at beginners who are just starting out. At Youngstown State University (YSU), I teach an introductory class titled "Geospatial Foundations," but similar classes at other universities may have names like "The Digital Earth," "Introduction to Geospatial Analysis," "Survey of Geospatial Technologies," "Introduction to GIS," or "Computer Applications in Geography." All of these courses seem aimed at the audience for which *Introduction to Geospatial Technologies* was written.

Organization of the Book

This book is divided into four main parts.

Part 1: Geospatial Data and GPS focuses on geospatial technology as it relates to spatial measurements and data.

▶ Chapter 1, "It's a Geospatial World Out There," introduces some basic concepts and provides an overview of jobs, careers, and some key technologies and applications (such as Google Earth).

▶ Chapter 2, "Where in the Geospatial World Are You?," explains how coordinates for location-based data and measurements from a three-dimensional (3D) world are translated into a two-dimensional (2D) map on a computer screen.

▶ Chapter 3, "Getting Your Data to Match the Map," discusses reprojection and georeferencing, important information when you're using any sort of geospatial data.

▶ Chapter 4, "Finding Your Location with the Global Positioning System," introduces GPS concepts. Taking a hand-held receiver outside, pressing a button, and then having the device specify your precise location and plot it on a map sounds almost like magic. This chapter demystifies GPS by explaining how the system works, why it's not always perfectly accurate, and how to get better location accuracy.

Part 2: Geographic Information Systems focuses on geographic information systems (GIS).

▶ Chapter 5, "Working with Digital Geospatial Data and GIS," serves as an introduction to GIS, examining how real-world data can be modeled and how GIS data can be created and used.

▶ Chapter 6, "Using GIS for Spatial Analysis," covers additional uses of GIS, including querying a database, creating buffers, and geoprocessing.

▶ Chapter 7, "Using GIS to Make a Map," offers instruction on how to classify your data and how to transform GIS data into a professional-looking map.

▶ Chapter 8, "Getting There Quicker with Geospatial Technology," discusses concepts related to road networks, such as: How are streets, highways, and interstates set up and used in geospatial technology? How does the computer translate a set of letters and numbers into a map of an actual street address? How do programs determine the shortest route from point *a* to point *b*?

Part 3: Remote Sensing examines issues related to remote sensing.

▶ Chapter 9, "Remotely Sensed Images from Above," focuses on aerial photography. It explains how the field started over 150 years ago with a man, a balloon, and a camera, and how it continues today with unmanned aircraft systems flying over Iraq and Afghanistan. This chapter also describes how to visually interpret features in aerial imagery and how to make accurate measurements from items present in an image.

▶ Chapter 10, "How Remote Sensing Works," delves into just what remote sensing is and how it works, and how exactly that image of a house is acquired by a sensor 500 miles away. This chapter also discusses all of the things that a remote sensing device can see that are invisible to the human eye.

▶ Chapter 11, "Images from Space," focuses on the field of satellite remote sensing and how satellites in orbit around Earth acquire images of the ground below.

▶ Chapter 12, "Studying Earth's Climate and Environment from Space," discusses the Earth Observing System, a series of environmental observatories that orbit the planet and continuously transmit data back to Earth about the land, seas, and atmosphere.

Part 4: Geospatial Applications focuses on individual topics in geospatial technology that combine GIS and remote sensing themes and applications.

▶ Chapter 13, "Digital Landscaping," describes how geospatial technologies model and handle terrain and topographic features. Being able to set up realistic terrain, landscape features, and surfaces is essential in mapping and planning.

▶ Chapter 14, "See the World in 3D," delves into the realm of 3D modeling, shows how geospatial technologies create 3D structures and objects, and then explains how to view or interact with them in programs like Google Earth.

▶ Chapter 15, "Life in the Geospatial Cloud and Other Current Developments," wraps things up with a look at the influence and advantages of the cloud, information regarding organizations and educational opportunities within geospatial technologies, and a look ahead to the future of the field.

Geospatial Lab Applications

Each chapter of *Introduction to Geospatial Technologies* covers one aspect of geospatial technology with an accompanying Geospatial Lab Application. The goal of these lab applications is not to teach software, but to help students work directly with the chapter's concepts. Each lab application uses freely available software that can be downloaded from the Internet or accessed through a Web browser. These software packages include:

▶ ArcGIS Online
▶ Google Earth Pro
▶ MapCruncher
▶ MultiSpec
▶ QGIS
▶ SketchUp
▶ Trimble GNSS Planning Online

Three of the chapters in Part 2 (Geographic Information Systems) offer two versions of the lab application. Instructors can choose to use either the free QGIS or ArcGIS for Desktop. The labs provide hands-on application of the concepts and theories covered in each chapter—it's one thing to read about how 3D structures can be created and placed into Google Earth, but it's another thing entirely to use SketchUp and Google Earth to do exactly that. Each lab application has integrated questions that students must answer while working through the lab. These questions are designed both to explore the various topics presented in the lab and also to keep students moving through the lab application. Note that words or phrases highlighted in purple text in the labs indicate menu items or icons that are clicked on or specific items that are typed in during the lab.

Some labs use sample data that comes with the software when it's installed; others require students to download sample data for use in the lab. Each lab provides links to a Website from which you can download the software. (The Website will also provide information regarding the necessary hardware or system requirements. Not all computers or lab facilities work the same, so be sure to check the software's Internet resources for help on installing the software.) The Instructors' section of this book's catalog page also offers a "tech tips" section with some additional information related to installing or utilizing some of the software.

The lab applications for each chapter are set up as follows:

▶ Chapter 1: This lab introduces the free Google Earth Pro as a tool for examining many facets of geospatial technology.

▶ Chapter 2: Students continue using Google Earth Pro, investigating some other functions of the software as they relate to coordinates and measurements.

▶ Chapter 3: Students use Microsoft's MapCruncher program to match a graphic of a campus map with remotely sensed imagery and real-world coordinates.

▶ Chapter 4: This lab uses Trimble GNSS Planning Online and some other Web resources to examine GPS planning and locations. It also provides suggestions for expanding the lab activities if you have access to a GPS receiver and want to get outside with it.

▶ Chapter 5: This lab introduces basic GIS concepts using QGIS. An alternate version of the lab uses ArcGIS for Desktop.

▶ Chapter 6: This lab continues investigating the functions of QGIS (or ArcGIS for Desktop) by using GIS to answer some spatial analysis questions.

▶ Chapter 7: This lab uses QGIS (or ArcGIS for Desktop) to design and print a map.

▶ Chapter 8: This lab uses Google Maps and Google Earth Pro to match a set of addresses and investigate shortest paths between stops on a network.

▶ Chapter 9: This lab tests students' visual image interpretation skills by putting them in the role of high-tech detectives who are trying to figure out just what a set of aerial images are actually showing.

▶ Chapter 10: This lab is an introduction to MultiSpec, which allows users to examine various aspects of remotely sensed imagery obtained by a satellite.

▶ Chapter 11: This lab continues using MultiSpec by asking students to work with imagery from the Landsat 8 satellite and investigate its sensors' capabilities.

▶ Chapter 12: This lab uses NASA data and Google Earth Pro to examine phenomena such as hurricanes, fires, and pollution on a global scale.

▶ Chapter 13: This lab uses Google Earth Pro to examine how terrain is used in geospatial technology (and film a video of flying over 3D-style terrain). It also uses the various terrain functions of Google Earth Pro for work with several digital terrain modeling features.

▶ Chapter 14: This lab introduces 3D modeling. Starting from an aerial image of a building, students design a 3D version of it using SketchUp, and then look at it in Google Earth.

▶ Chapter 15: This lab utilizes Esri's free ArcGIS Online to stream data from the cloud, create Web maps, wrap things up, and look at many of the book's concepts combined in a single package.

Additional Features

In addition to the lab applications, each chapter contains several **Hands-On Applications,** which utilize free Internet resources to help students further explore the world of geospatial technologies and get directly involved with some of the chapter concepts. There's a lot of material out there on the Internet, ranging from interactive mapmaking to real-time satellite tracking, and these Hands-On Applications introduce students to it. In the third edition, each Hands-On Application has a set of Expansion Questions for students to answer while working with that Application's Web resources.

Each chapter also has one or more boxes titled **Thinking Critically with Geospatial Technology.** These boxes present questions to consider regarding potential societal, privacy, design, or ethical issues posed by geospatial technologies and their applications. The questions presented in these boxes are open-ended and are intended to stimulate discussion about geospatial technologies and how they affect (or could affect) human beings. For instance, how much privacy do you really have if anyone, anywhere, can obtain a clear image of your house or neighborhood and directions to drive there with just a few clicks of a mouse?

Lastly, each chapter ends with two boxes. The first of these, chapter **Apps,** presents some representative apps for a mobile device related to the chapter's content that you may wish to investigate further. For instance, Chapter 8's *Geocoding and Shortest Paths Apps* box showcases apps for your phone or your tablet. Note that at the time of writing, all of these apps were free to obtain and install.

The second section, **Social Media,** highlights some representative Facebook, Twitter, and Instagram accounts, as well as YouTube videos and blogs, that are relevant to the chapter's topics. For instance, Chapter 11's *Satellite Imagery in Social Media* box features Facebook and Twitter accounts from satellite imagery sources such as DigitalGlobe or the USGS updates on Landsat, as well as videos of satellite imagery applications. (Note that all of these apps and social media accounts are examples, not recommended products.)

New to This Edition

The third edition contains multiple key updates. Each chapter has something new within it, whether it's a newly added or revised text section, Hands-On Application, or Lab Application. At the end of each chapter, there is an updated section on available smartphone and tablet apps as well as resources for using geospatial technologies in social media. The Lab Applications have been updated to use current software and techniques, including all new Lab Applications that utilize ArcGIS Online (Chapter 15), Trimble GNSS Planning Online (Chapter 4), and the now-free Google Earth Pro (Chapters 1, 2, 8, 12, and 13). In addition, Landsat 8 imagery is now used with Multispec (Chapters 10 and 11), and the most recently available version of QGIS is used for the GIS Lab sections (Chapters 5, 6, and 7).

There are many other updates and revisions throughout each chapter. For instance, Chapter 1 showcases using geospatial technologies on mobile devices and how geolocation works. The remote sensing focused chapters (9, 10, 11, and 12) have been expanded to include many topics about the state of remote sensing today, including UAS, Landsat 8, Sentinel-2, Skysat, Suomi-NPP, small satellites, cubesats, and using remote sensing for disaster monitoring. Chapter 13 includes more information about US Topos as well as the change from the National Elevation Dataset to the new 3DEP elevation data used by the USGS. Chapter 15 has been expanded and revised for a focus on the use of the cloud with geospatial technologies, including new Hands-On Applications that utilize Esri Story Maps. Also, throughout the book there are new Hands-On Applications that use new Web resources, including Tomnod, Landsat Live, Indiemapper, Census Mapping tools, and CityEngine Web Scenes.

Ancillary Materials and Student and Instructor Resources

The catalog page **macmillanhighered.com/shellito/catalog** offers a set of valuable resources for both students and instructors.

For **students,** the catalog page offers a multiple-choice self-test for each chapter, as well as an extensive set of references, categorized by topic, to provide further information on a particular topic. There are also a set of links to the free software packages needed to complete the lab activities, as well as the datasets required for specific lab applications. A set of world links is also provided.

For **instructors,** the catalog page offers an instructor's manual, which provides teaching tips for each chapter on presenting the book's material, a set of "tech tips" related to software installation and usage, a set of key references for the chapter materials, and an answer key for all the lab activities. A test bank of questions is also provided.

Acknowledgments and Thanks

Books like this don't just spring out of thin air—I owe a great deal to the many people who have provided inspiration, help, and support for what would eventually become this book.

Bill Minick, my editor, has offered invaluable help, advice, patience, and guidance throughout this entire project. I would also very much like to thank Abigail Fagan, Sheena Goldstein, Paul Rohloff, and Christine Buese at Macmillan Education for their extensive "behind the curtain" work that shaped this book into a finished product.

I want to thank Sean Young for his very helpful comments and feedback on the previous editions and for his close reading and review of the labs in this third edition. Neil Salkind and Stacey Czarnowski gave great representation and advice. The students who contributed to the development of the YSU 3D Campus Model (Rob Carter, Ginger Cartright, Paul Crabtree, Jason Delisio, Sherif El Seuofi, Nicole Eve, Paul Gromen, Wook Rak Jung, Colin LaForme, Sam Mancino, Jeremy Mickler, Eric Ondrasik, Craig Strahler, Jaime Webber, Sean Welton, and Nate Wood) deserve my sincere thanks. Many of the 3D examples presented in Chapter 14 wouldn't have existed without them. Others who supported this book in various ways include: Jack Daugherty, for tech support, assistance with the labs, and help with the GeoWall applications; Lisa Curll, for assistance with the design and formatting of the lab applications; Grant Wilson, for his insightful technical reviews of the ArcGIS lab applications; Margaret Pearce, for using an earlier draft of the manuscript with her students at the University of Kansas and for her extremely useful comments and feedback; Hal Withrow, for invaluable computer tech assistance; and Mark Guizlo for assistance with data sources.

I offer very special thanks to all of my professors, instructors, colleagues, and mentors, past and present (who are too numerous to list), from Youngstown State University, the Ohio State University, Michigan State University, Old Dominion University, OhioView, and everywhere else, for the help, knowledge, notes, information, skills, and tools they've given me over the years. I am also deeply indebted to the work of Tom Allen, John Jensen, Mandy Munro-Stasiuk, and the members of SATELLITES for some methods used in some of the chapters and labs.

Finally, I owe a debt of gratitude to the colleagues who reviewed the original proposal and various stages of the manuscript for the first, second, and third editions. Thank you for your insightful and constructive comments, which have helped to shape the final product:

Robbyn Abbitt, *Miami University*
Amy Ballard, *Central New Mexico Community College*
Chris Baynard, *University of North Florida*
Robert Benson, *Adams State College*
Edward Bevilacqua, *SUNY College of Environmental Science and Forestry*
Julie Cidell, *University of Illinois*
W. B. Clapham Jr., *Cleveland State University*
Russell G. Congalton, *University of New Hampshire*
Jamison Conley, *West Virginia University*
Kevin Czajkowski, *University of Toledo*
Nathaniel Dede-Bamfo, *Texas State University*
Adrienne Domas, *Michigan State University*
Christine Drennon, *Trinity University*
Charles Emerson, *Western Michigan University*
Jennifer Fu, *Florida International University*
Nandhini Gulasingam, *DePaul University*
Victor Gutzler, *Tarrant County College Southeast*
Melanie Johnson, *Paul Smith's College*
Marilyne Jollineau, *Brock University*
Jessica K. Kelly, *Millersville University*
Sara Beth Keough, *Saginaw Valley State University*
James Kernan, *SUNY Geneseo*
Kimberley Britt Klinker, *University of Richmond*
Michael Konvicka, *Lone Star College*
James Lein, *Ohio University*
James Leonard, *Marshall University*
Russanne Low, *University of Nebraska–Lincoln*
Chris Lukinbeal, *Arizona State University*
Marcos Luna, *Salem State University*
John McGee, *Virginia Tech*
George M. McLeod, *Old Dominion University*
Bradley Miller, *Michigan State University*
Trent Morrell, *Laramie County Community College*
Nancy Obermeyer, *Indiana State University*
Tonny Oyana, *The University of Tennessee*
Margaret Pearce, *University of Kansas*
Hugh Semple, *Eastern Michigan University*
Thomas Sigler, *University of Queensland*
Anita Simic, *Bowling Green State University*
Brian Tomaszewski, *The Rochester Institute of Technology*

Shuang-Ye Wu, *University of Dayton*
Sean Young, *University of Iowa*
Donald Zeigler, *Old Dominion University*
Arthur Zygielbaum, *University of Nebraska–Lincoln*

A Rapidly Changing Field

As Chapter 15 points out, geospatial technology has become so widespread and prevalent that no book can cover every concept, program, or online mapping or visualization tool (as much as I'd like this one to). I hope that the students who use this book will view the concepts and applications presented herein as an introduction to the subject—and that this will motivate them to take more advanced courses on the various aspects of geospatial technology.

One thing to keep in mind: In such a rapidly advancing field as geospatial technology, things can change pretty quickly. New satellites are being launched and old ones are ending their mission lives. Websites get updated and new updates for software and tools are released on a regular basis. As of the writing of this book, all of the Web data, software, and satellites were current, but if something's name has changed, a Website works differently, or if a satellite isn't producing any more data, there's probably something newer and shinier to take its place.

I'd very much like to hear from you regarding any thoughts or suggestions you might have for the book. You can reach me via email at **bashellito@ysu.edu** or follow me on Twitter **@GeoBradShellito**.

Bradley Shellito
Youngstown State University

Accessing Data Sets for Geospatial Lab Applications

Some of the Geospatial Lab Applications in this book use data that comes with the software, sample data that gets installed on your computer when you install the software itself, or data that you'll create during the course of the lab. However, the lab applications for Chapters 3, 5, 6, 7, 9, 10, 11, and 12 require you to download a set of data that you'll use with those labs.

The lab applications will direct you to copy the dataset before beginning the lab. Each dataset is stored in its own folder online. To download these folders, please visit **macmillanhighered.com/shellito/catalog.** Under "Student Options," you'll find access to student resources, including "Lab Data Sets."

This book was not prepared, approved, or endorsed by the owners or creators of any of the software products discussed herein. The graphical user interfaces, emblems, trademarks, and associated materials discussed in this book remain the intellectual property of their respective owners.

ArcGIS 10.3 (Esri), ArcGIS Online (Esri), Google Earth Pro 7.1 (Google), SketchUp Make (Trimble), MapCruncher 3.2 (Microsoft), MultiSpec 3.4, QGIS 2.8.1, Trimble GNSS Planning Online.

1 It's a Geospatial World Out There

An Introduction to Geospatial Technologies, Geospatial Jobs, Geospatial Data, Volunteered Geographic Information, Geolocation, and Google Earth

Have you ever done any of the following?

▶ Used a smartphone, tablet, or other mobile device to find your location, co-ordinates, or directions, or to look for the nearest restaurant or gas station?

▶ Used an online mapping service like MapQuest, Google Maps, or Bing Maps to find directions (and the best route) to a destination or to print a map of an area?

▶ Used an in-car navigation system (say, one from Garmin, Magellan, or TomTom) to navigate to or from a destination?

▶ Used social media (such as Facebook or Twitter) to add your location information to a post or tweet?

▶ Used a Global Positioning System (GPS) receiver while hiking, jogging, hunting, fishing, golfing, or geocaching?

▶ Used a Web resource to find a map of your neighborhood so that you can compare nearby housing values or see exactly where your property ends and your neighbor's begins?

▶ Used a virtual globe program (like Google Earth) or an online map to look at photos or images of your home, street, school, or workplace?

If so, then congratulations—you've used geospatial technologies. Anytime you're using any sort of technology-assisted information (on a computer, smartphone, or tablet) concerning maps, locations, directions, imagery, or analysis, you're putting geospatial technology applications to use.

1

Geospatial technology has become extremely widespread in society, with a multitude of uses in both the private and public sectors. However, more often than not, if you tell someone you're using geospatial technology, you'll be asked, "What's that?"

 ## What Is Geospatial Technology?

Although geospatial technology is being used in numerous fields today, the term "geospatial technology" doesn't appear to have seeped into everyday usage. Words like "satellite images" and "Google Earth" and acronyms like "GIS" and "GPS" are growing increasingly commonplace, yet the phrase "geospatial technology" seems relatively unknown, though it incorporates all of these things and more. **Geospatial technology** describes the use of a number of different high-tech systems and tools that acquire, analyze, manage, store, or visualize various types of location-based data. The field of geospatial technology encompasses several fields and techniques, including:

▶ **Geographic information system (GIS)**: Computer-based mapping, analysis, and retrieval of location-based data

▶ **Remote sensing**: Acquisition of data and imagery from the use of satellites (**satellite imagery**) or aircraft (**aerial photography**)

▶ **Global Positioning System (GPS)**: Acquisition of real-time location information from a series of satellites in Earth's orbit

There are numerous related fields that utilize one or more of these types of technologies. For instance, an in-car navigation system already contains extensive road-network data, mapped out and ready to use, which includes information about address ranges, speed limits, road connections, and special features of roads (such as one-way streets). It also requires the mapping of points of interest (such as gas stations or restaurants), and should be capable of referencing new user-defined destinations. It also has to be able to plot the car's real-time position in relation to these maps and may even have a feature that shows a representation of the surrounding landscape as taken from an overhead viewpoint. Many of these types of systems combine different geospatial technologies to work together in one application.

 ## Who Uses Geospatial Technology?

Geospatial technology is used in a wide variety of fields (**Figure 1.1**), including federal, state, and local government, forestry, law enforcement, public health, biology, and environmental studies (see *Hands-On Application 1.1: Industries Using Geospatial Technology* for a look at industries employing

---

geospatial technology a number of different high-tech systems that acquire, analyze, manage, store, or visualize various types of location-based data

Geographic information system (GIS) computer-based mapping, analysis, and retrieval of location-based data

remote sensing acquisition of data and imagery from the use of satellites or aircraft

satellite imagery digital images of Earth acquired by sensors onboard orbiting spaceborne platforms

aerial photography acquisition of imagery of the ground taken from an airborne platform

Global Positioning System (GPS) acquisition of real-time location information from a series of satellites in Earth's orbit

FIGURE 1.1 Examples of geospatial technology in action on the job. *[Source: (top left) AP Photo/ Wilfredo Lee (top right) © Ron Nickel/Design Pics/Corbis (center left) Joe Raedle/Getty Images (center right) Bob Nichols/USDA NRCS (bottom left) AP Photo/U.S. Geological Survey, Dr. Dan Dzurisin (bottom right) Justin Sullivan/Getty Images]*

HANDS-ON APPLICATION 1.1

Industries Using Geospatial Technology

Geospatial technology is being used in a variety of applications in numerous different fields today. For a deeper look at some of these applications, visit **www.esri.com/industries .html**, which is run by Esri (we'll discuss more about Esri in Chapter 5, but the short version is that they're the market leader in GIS). This site lists dozens of different fields that are using geospatial technology, and describes how GIS (and other Esri products) are being utilized in them. Examine a few that are connected to your own fields of interest. For instance, if your interest is in law enforcement, examine some

of the "Public Safety" applications. If you're involved in public or community health, examine some of the "Health and Human Services" applications, then describe how GIS is being utilized in some real-world, on-the-job applications.

Expansion Questions

- How is GIS being utilized in some real-world, on-the-job applications in fields of interest to you?

- Who in these fields is using GIS with what kinds of applications and why are they using GIS?

HANDS-ON APPLICATION 1.2

Jobs in the Geospatial Field

Businesses are hiring in the geospatial field. For examples of current job openings, visit some of the following Websites:

1. Geography Jobs: **www.geographyjobs.com**
2. Geosearch: **http://jobs.geosearch.com /JobSeeker/Jobs.aspx**
3. GIS Careers: **http://giscareers.com**
4. GIS Jobs.com: **www.gisjobs.com**
5. The GIS Jobs Clearinghouse: **www.gjc.org**
6. GIS Lounge: **http://jobs.gislounge.com**

These are just a sampling of Websites where employers post job openings worldwide. Examine several jobs from areas near where you are (or where you'd like to go to).

Expansion Questions

- What kinds of jobs are being advertised?
- What kinds of job qualifications, training, and skill sets are employers looking for?
- What kinds of starting salary ranges are employers offering?

people in these fields). As long as the job field involves the utilization of some sort of information or data that has a location associated with it, chances are that some sort of geospatial technology is being used. Geospatial technology has been heralded by the U.S. Department of Labor as one of the main emerging and evolving job fields in the United States, with enormous growth potential. For instance, the Department of Labor's O*NET (Occupational Information Network) OnLine utility contains job descriptions in fields such as "Geospatial Information Scientists and Technologists," "Remote Sensing Scientists and Technologists," and "Geographic Information Systems Technicians" (see *Hands-On Application 1.2: Jobs in the Geospatial Field* for more information about types of jobs).

The following are just a handful of examples of fields that utilize geospatial technology.

Archeology

The ability to pinpoint the location of artifacts uncovered on a dig, construct a map of the area, and then search for patterns on the site are all archeological functions that can be rendered quickly and efficiently with geospatial technology (**Figure 1.2**). Archeologists can utilize historical maps, current aerial photography or satellite imagery, and location information obtained on the site throughout the course of their work.

City Planning

Utilities, wastewater, green space, traffic, roads, zoning, and housing are all matters of concern to urban planners. Geospatial technology provides a means of working with all of these entities together for planning purposes. Strategies for smart urban growth and the management and updating of city resources can be examined through a variety of different applications.

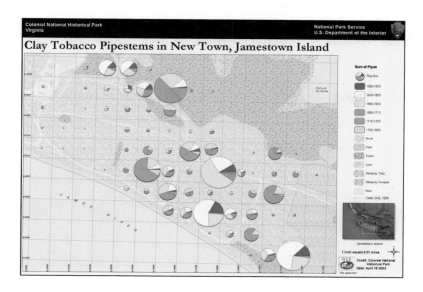

FIGURE 1.2 Using GIS for archeological mapping of Clay Tobacco Pipestems in Virginia. *[Source: Colonial National Historical Park, National Park Service]*

Environmental Monitoring

Processes that affect Earth's environment in many different ways can be tracked and assessed using geospatial technology. Information about land-use change, pollution, air quality, water quality, and global temperature levels is vital to environmental research, ranging from the monitoring of harmful algae blooms to studies in climate change (see **Figure 1.3**).

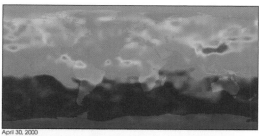

FIGURE 1.3 Global carbon monoxide concentrations as monitored by NASA remote sensing satellites. *[Source: NASA GSFC Scientific Visualization Studio, based on data from MOPITT (Canadian Space Agency and University of Toronto)]*

FIGURE 1.4 Geospatial technology used for assessing the risk of wildfires in Virginia.
[Source: Courtesy Virginia Department of Forestry]

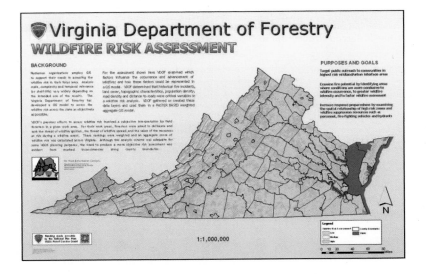

Forestry

All manner of forest monitoring, management, and protection can be aided through the use of geospatial technology. Modeling animal habitats and the pressures placed upon them, examining the spatial dimensions of forest fragmentation, and managing fires are among the many different ways that geospatial technology is utilized within the field of forestry (**Figure 1.4**).

Homeland Security

Geospatial technology is a key component in examining vulnerable areas with regard to homeland security. Risk assessment of everything from evacuation plans to smoke-plume modeling can be examined using geospatial technology. Disaster mitigation and recovery efforts can be greatly enhanced through the use of current satellite imagery and location-based capabilities.

Law Enforcement

The locations of various crimes can be plotted using GIS. Law enforcement officials can use this information to analyze patterns (**Figure 1.5**) and to determine potential new crime areas. Geospatial technology can be used in several other ways beyond mapping and analysis; for instance, police departments can gather high-resolution aerial photography of locations of patrols to provide further information about potentially dangerous areas.

Health and Human Services

Geospatial technology is used in a variety of health-related services. For example, monitoring of diseases, tracking sources of diseases, and mapping health-related issues (such as the spread of H1N1 or other influenza; see **Figure 1.6**) are all tasks that can be completed using geospatial technology applications.

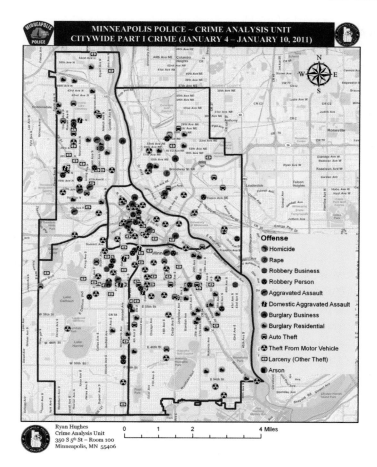

MINNEAPOLIS POLICE ~ CRIME ANALYSIS UNIT
CITYWIDE PART I CRIME (JANUARY 4 – JANUARY 10, 2011)

FIGURE 1.5 A GIS map showing crime locations in Minneapolis.
[Source: Minneapolis Police Crime Analysis Unit]

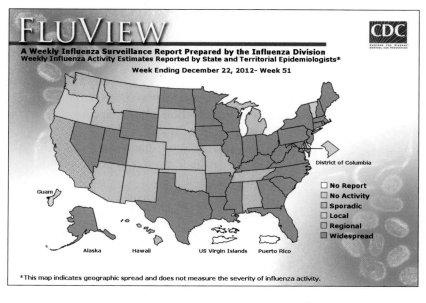

FIGURE 1.6 A CDC map examining the spatial distribution of influenza activity. *[Source: Centers for Disease Control and Prevention]*

Military and Intelligence

Geospatial technology plays a key role in today's military, defense, and intelligence worlds. Satellite imagery and aerial photography have long been used for intelligence gathering and reconnaissance, practices that continue today with an increased use of Unmanned Aerial Systems (UAS) and high-resolution satellite sensors. The field of Geospatial Intelligence (**GEOINT**) encompasses all types of remotely sensed imagery, GIS, and GPS data for information gathering and analysis related to national security and defense applications.

GEOINT Geospatial Intelligence

Real Estate

Through geospatial technology, realtors and appraisers (as well as home buyers and sellers) can create and examine maps of neighborhoods and quickly compare housing prices and values of nearby or similar properties. Other features of a property can be examined by viewing high-resolution aerial images, showing the topography, the terrain, and even if it's located on a floodplain. You can also examine where a property is located in relation to schools, highways, wastewater treatment plants, and other urban features.

geospatial data items that are tied to specific real-world locations

All of these areas and many, many more are reliant on the capabilities of geospatial technology. No matter the field, what distinguishes geospatial technology from other types of computer systems or technologies is that it explicitly handles **geospatial data**.

 ## What Is Geospatial Data?

Geospatial data (also often referred to as "spatial data") refers to location-based data, which is at the heart of geospatial technology applications. This ability to assign a location to data is what makes geospatial technology different from other systems. You're using geospatial concepts anytime you want to know "where" something is located. For instance, emergency dispatch systems can determine the location of a 911 call. They also have access to information that tells them where the local fire stations, ambulance facilities, and hospitals are. This type of information allows the closest emergency services to be routed to its location.

When the data you're using has a location that it can be tied to, you're working with geospatial data. This isn't just limited to point locations—the length and dimensions of a hiking trail (and the locations of comfort stations along the trail) are examples of real-world data with a spatial dimension. Other kinds of data, like the boundaries of housing parcels in a subdivision or a satellite image of the extent of an area impacted by a natural disaster, would fall under this category. Geospatial technology explicitly handles these types of location-based concepts.

However, not all of the data in the world is geospatial data. For instance, data regarding the location of a residential parcel of land containing a house and the parcel's dimensions on the ground would be geospatial information. However, other data, such as the names of the occupants, the assessed value of the house, and the value of the land is not geospatial information. A benefit of using geospatial technology is that this type of **non-spatial data** can be linked to a location. For instance, say you sell a lot of used books on eBay. You could make a map showing the locations of each of the purchasers of your online sales—this would be geospatial information. You could then link related non-spatial data (such as the name of the book that was sold or the price it was purchased for) to those locations, creating a basic database of sales.

Geospatial information can also be gathered about other characteristics, such as the landscape, terrain, or land use. Remote sensing provides images of the ground that serve as "snapshots" of particular locations at specific times, which can generate other forms of spatial information. For instance, a satellite image of Las Vegas, Nevada, can help ascertain the locations of new housing developments. Examining imagery from several dates can help track locations where new houses are appearing (**Figure 1.7**). Similarly,

non-spatial data data that is not directly linked to a geospatial location (such as tabular data)

FIGURE 1.7 Satellite imagery (from 1972, 1986, 1992, and 2000) showing the growth of Las Vegas, Nevada, over time. [Source: USGS]

1972

1986

1992

2000

geospatial technology can be used to determine where, spatially, new urban developments are most likely to occur, by analyzing different pressures of development on these areas.

At the national level, the USGS maintains a large geospatial data distribution program called **The National Map**. Users of The National Map can access multiple types of data, including National Elevation Datasets (see Chapter 13), hydrography (see Chapter 5), transportation (see Chapter 8), land-cover data (see Chapter 5), orthoimagery (see Chapter 9), boundary files, and many more, all in formats that GIS software can easily read. The National Map is just one part of the larger coordinated vision that comprises the National Geospatial Program, and it serves as a source of geospatial data for the United States (see **Figure 1.8** and *Hands-On Application 1.3: The National Map Viewer* for more information about The National Map). You can think of The National Map as a "one-stop-shopping" venue for free geospatial data—numerous types of datasets are available for a geographic area that you define, a very useful option that prevents you from having to find and download data from multiple sources, then clip or alter the data to make them fit together in a way that will provide the information you want about the region you're examining. If you're studying East Liverpool, Ohio, you can use The National Map to download land-cover data, elevation data, boundary files, water features, and transportation layers of only the East Liverpool area.

Many types of decision making are reliant on useful geospatial information. For instance, when a new library is proposed to be built in a particular area, the choice of location is going to be important—the new site should

The National Map an online basemap and data viewer with downloadable geospatial data maintained and operated by the USGS and part of the National Geospatial Program

FIGURE 1.8 The National Map Viewer. *[Source: U.S. Geological Survey]*

HANDS-ON APPLICATION 1.3

The National Map Viewer

You can access The National Map online at **http://nationalmap.gov/index .html**. Select the option for The National Map Viewer and Download Platform, then choose the option to go to the Viewer. A new Web page will open with the Viewer itself, the tool for accessing data via The National Map. The National Map can be used to view available data—from the Overlays options on the left side of map, you can select what layers to display. High-resolution imagery can be accessed as an additional visible layer from the Imagery option on the right side of the map.

The layers you're viewing in The National Map can also be downloaded for use in GIS or other software programs. First, search for an area of interest (go to Reno, Nevada). Zoom in to this area, and when the view of the map changes, press the Download Data button in the upper-right corner of the map. New download options will appear that allow you to select the extent of the area you want to download data for. Next, the data layers that are available for download will be shown (these will include structures, transportation, boundaries, hydrography, land cover, elevation, and more) as well as the data formats these layers are available in. Select the layers you want, choose Add to Cart, and then follow the steps to download your data (you will have to register with the USGS—at the time of writing, they will send you an e-mail with a link you can use to retrieve your data).

Expansion Questions

- What geographic areas are available as extents for download options?

- What types of data can be accessed via The National Map? Why does this make The National Map so versatile as a one-stop data destination?

maximize the number of people who can use it but minimize its impact on other nearby sites. The same type of geospatial thinking applies to the placement of new schools, fire stations, or fast-food chain restaurants. Geospatial information fuels other decisions that are location dependent. For instance, historical preservation efforts seek to defend historic Civil War battlefield sites from urban development. Knowing the location of the sections of land that are under the greatest development pressure helps in channeling preservation efforts to have their greatest impact. With more and more geospatial technology applications becoming available, these types of geospatial problems are being taken into consideration (see *Hands-On Application 1.4: Mapping Data Layers Online*; *Hands-On Application 1.5: Examining Real Estate Values Online*; and *Hands-On Application 1.6: Geospatial Technology and Emergency Preparedness* for examples).

What Does All This Have to Do with Geography?

The first sentence for the entry "geography" in *The Dictionary of Human Geography* describes geography as "the study of the Earth's surface as the space within which the human population lives." To simplify a definition

HANDS-ON APPLICATION 1.4

Mapping Data Layers Online

Online mapping applications that combine a number of different types of spatial data into a series of map layers are becoming increasingly common. A good, easy-to-use resource for viewing spatial information online is MapWV, a utility for mapping various layers of data about West Virginia. Go to **www.mapwv.gov /viewer/index.html**. Under the Layers tab on the right-hand side of the Map Viewer, you'll see two options: Reference Layers and Basemap Layers. Under the Basemap Layers option, you'll see a number of possible starting layers, including road maps and aerial photography (such as NAIP from different years; see Chapter 9 for more about NAIP imagery). Under the Reference Layers tab there are several other data layers for you to map—for instance, turn on the Hydrography, Public Land, and Populated Places layers (by placing checkmarks in their respective boxes), and bodies of water, publicly owned sites like parks and forests, and towns and cities will appear as locations on the map. You can zoom in on a major city (such as Morgantown), a specific county, or a particular place of interest (such as a state park). Tools for zooming in and out of the map and navigating are on the Map Viewer's left-hand side. Additional analytical tools available under the Tools tab will enable you to perform tasks like creating buffers or making measurements. Examine several of the layers for Morgantown, West Virginia, and see how they fit together in a geospatial context.

Expansion Questions

- How is the city of Morgantown represented differently using the Light Grey Canvas, Esri Imagery, USA Topo, and 1996 CIR Imagery basemap layers?

- How do the mapped water bodies of the hydrography layer match up with how they're shown on each of the four basemaps?

- How does the mapped political boundary layer (of Morgantown's boundaries) match up with how the city is shown on each of the four basemaps?

HANDS-ON APPLICATION 1.5

Examining Real Estate Values Online

Zillow is an online tool used to examine housing prices in cities across the United States. You can examine the prices of houses for sale in the neighborhood and see—spatially— where housing prices change. In other words, Zillow provides online mapping related to real estate prices and appraisal. To check it out, visit **www.zillow .com**. Starting with the "Find your way home" search option, look for houses in Virginia Beach, Virginia. Zoom in on some of the residential neighborhoods, and you'll see markers in blue (auction sites), purple (homes for rent), red (homes for sale), and yellow (recently sold). Selecting a marker will give you the sale (or sold) price, along with Zillow's estimated market value. This tool allows you to select any location and examine information (like housing prices and other details) that refers to this location.

Expansion Questions

- Use Zillow to check out your own local area. What does the housing market look like for your own area (in terms of how many houses are for sale and, more importantly, where are there any patterns or clusters of houses for sale)?

- How do the asking prices of the houses for sale compare with their estimated values?

HANDS-ON APPLICATION 1.6

Geospatial Technology and Emergency Preparedness

Geospatial technologies are used in many applications related to disaster management and emergency preparedness. A fun (but still useful) example of this is available in the Map of the Dead online at **www.mapofthedead.com/map**, which includes a free mapping tool intended to be used to help survive the "zombie apocalypse." When the map opens, zoom in on your own local area—you'll see the delineation of urban areas, as well as locations helpful in a disaster or emergency, such as gas stations, police stations, hospitals, and grocery stores. By clicking on the Menu button, you can access a Map Key (to show what each symbol represents) and a set of Place Names (for more information about the locations). All of this location information is taken from users of the Foursquare app (see page 18) and translated to places on the map.

While surviving fictional outbreaks of the walking dead makes for good entertainment, this kind of Web mapping tool highlights how geospatial technology can be used to enable individuals to quickly find the locations of necessary resources in the event of an actual emergency, such as a flood, tornado, or hurricane. Along these lines, Google has set up an online emergency preparedness tool for the state of Florida at **www.google.org/crisismap/florida_emergency_preparedness**, which shows current weather radar and traffic conditions, shelters, and evacuation routes in order to aid in emergency preparation and disaster management for the state.

Expansion Questions

- From examining both maps, what types of resources would be best to highlight on an emergency preparedness mapping tool and why?

- How can these Web resources be used or adapted for other types of disaster mitigation or emergency preparation (such as floods, fires, or earthquakes)?

further into words of one syllable, geography is not just "where things are" but the study of "why things are where they are." Geography deals with concepts of the spatial characteristics of our planet and the spatial relationships and interactions of the people and features that occupy it. The notion of "Geographic Information Science" has been around for years as the study of the multiple concepts that surround the handling of spatial data, and today the use of geospatial technologies is tightly interwoven with the discipline of geography. See *Hands-On Application 1.7: The Geospatial Revolution Will Be Televised* for several examples of geospatial technologies being used for solving geographic problems.

At the higher education level, courses in GIS, remote sensing, GPS, and various applications of these technologies are key components of a geography curriculum. Certificate programs in Geographic Information Science (GISci) or Geospatial Technologies are also being increasingly offered by colleges and universities. Geospatial degree programs at the bachelor's, master's, or doctoral level have been developed at numerous universities (see Chapter 15 for more information about geospatial technology in education).

HANDS-ON APPLICATION 1.7

The Geospatial Revolution Will Be Televised

Penn State University's Geospatial Revolution Project showcases a series of excellent videos that detail how geospatial technology affects our everyday lives, especially when it comes to concerns such as monitoring climate change and tracking disease. Visit **http://geospatialrevolution.psu.edu** and watch the available videos. There are additional videos aimed at a K–12 audience (but also present relevant information for all education levels) in the For Educators section as well.

Expansion Questions

- How can geospatial technologies be used for aiding in monitoring climate change conditions?
- How have geospatial technologies helped aid workers in dealing with food shortages or in mitigating the effects of disease outbreaks?

Even in basic activities like collecting data in the field for a class project or producing a map of results, geospatial technologies have become an integral part of the geography discipline.

In a 2004 article, Jack Dangermond, CEO of the Environmental Systems Research Institute, Inc. (Esri; see Chapter 5), heralded GIS as a way of "Speaking the Language of Geography," and in a 2009 article noted that GIS was seen as "Geography in Action." Both of these are great descriptions—for example, when you use an in-vehicle navigation system to navigate your way through unfamiliar territory or use an online tool to examine housing values for your neighborhood, you're using basic geographic concepts of space and place through technology. Geospatial technology gives us the tools to more easily apply location-based principles to real-world situations. For example, how is the new freeway bypass going to affect the local community regarding traffic flows, new commercial development, and individual properties? Or, what are the potential impacts of the site chosen for a new casino development in relation to land use, economic and social effects, and the draw of gamers from far outside the local area? Throughout this book, you'll be dealing with geospatial technology concepts and applications, using some sort of geospatial data, measuring spatial characteristics, or examining spatial relations.

One of the most welcome recent developments is the opportunity for people without geospatial training to get involved in this field by applying their geographic knowledge. With geospatial tools becoming more widely utilized, people now have the chance to create their own geospatial content. Whether they're creating new maps, updating existing maps, tying digital pictures to locations on a map, or adding to geocaching databases, more and more people are using geospatial technologies to add to or enhance geospatial data and knowledge. Dr. Michael Goodchild coined the term volunteered geographic information (**VGI**) to give a name to data generated in this way.

A **wiki** is a Web utility that allows users to freely create, manage, update, and change the content of a Website. Possibly the best known one is

VGI volunteered geographic information, a term used to describe user-generated geospatial content and data

wiki a database available for everyone to utilize and edit

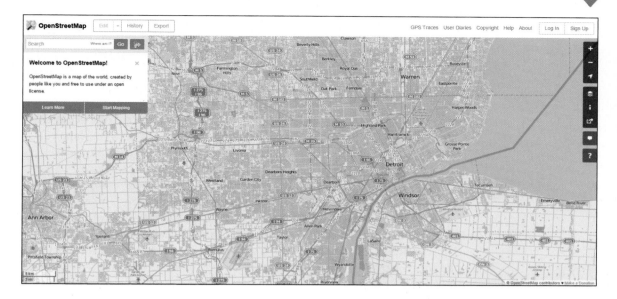

Wikipedia, an online encyclopedia that anyone can edit. By extending this wiki concept to geospatial data, individuals can contribute their own maps, images, and updates to geospatial content available on the Internet, thereby generating VGI. Websites such as Google Map Maker or OpenStreetMap (see **Figure 1.9** and *Hands-On Application 1.8: User-Generated Geospatial Content Online* for more information) allow users to generate or contribute geospatial content as VGI.

FIGURE 1.9 Detroit, Michigan, and Windsor, Canada, as shown in OpenStreetMap. *[Source: www.openstreetmap.org]*

HANDS-ON APPLICATION 1.8

User-Generated Geospatial Content Online

Volunteered geographic information (VGI) represents user-generated geospatial content online. Use your Web browser to investigate some of these sites (and see how your own local area is represented with each one):

- Google Map Maker: **www.google.com /mapmaker**
- MapQuest Open: **http://open.mapquest.com**
- OpenStreetMap: **www.openstreetmap.org**
- The National Map Corps: **https://my.usgs. gov/confluence/display/nationalmapcorps /Home**
- Wikimapia: **http://wikimapia.org**

Expansion Questions

- What types of data and information are available via these sites, and how can you add your own geospatial content?

- From examining your local area on all five sites, what kinds of things could you potentially update or add using your own local geographic knowledge (and have any of these additions or updates been already made to one site but not another)?

For example, Google Map Maker allows users to contribute to the content of Google Maps by adding places, locations, and hiking trails, or by helping to improve on directions given by the system. Similarly, the USGS has established The National Map Corps to allow users to make edits and updates to the data of The National Map. After all, who would know more than you about new businesses that have opened, new roads that have been put in, or local hiking trails in your area? By contributing your geographic knowledge to sources like Google Map Maker or OpenStreetMap, you're helping to build an extensive database of geospatial content as well as improving available online resources used by others.

It's becoming easier to apply geographic principles using geospatial technologies. For instance, on May 4, 2012, *The Saugerties Times* published a call for volunteers to take part in the launch of a free smartphone app for mapping the spread of invasive species, and explained that the collected data would be utilized in a New York invasive species database. Activities like these are referred to as **crowdsourcing,** in which the contributions of untrained volunteers collect data, content, or observations as part of a larger project (often referred to as **citizen science**). In geospatial terms, crowdsourcing projects allow huge numbers of people to contribute VGI to ongoing projects. Geospatial technologies are enabling everyone to become involved with collecting and analyzing geographic data. In particular, as mobile technologies and smartphones become more and more widespread, geospatial technology applications are ending up, quite literally, in the palm of your hand. There are numerous apps available that use geospatial technologies, and a technique called geolocation lies at the heart of many of them.

crowdsourcing the activities of untrained volunteers to create content and resources that can be utilized by others

citizen science the activities of untrained volunteers conducting science

 ## What Is Geolocation?

geolocation the technique of determining where something is in the real world

Geolocation is the technique of determining where something is in the real world. By using geolocation, your smartphone, tablet, laptop, or desktop computer can determine what its current location is, give you the coordinates (such as latitude and longitude; see Chapter 2), and then plot that information on a map. This position can be found in several different ways—by using the IP address of your computer, your wi-fi connection, the cell towers your phone is connected to, or by the GPS receiver in your mobile device. Of course, once your phone, tablet, or computer knows where it is, it can share that location information with others via the Internet.

For instance, Websites often ask you if they can use your location. When you use a Website such as fandango.com to look up movie times or chilis.com to find information about a Chili's restaurant, these Websites will ask if they can use your computer's location. If you agree to share your location, you will receive information about movie theatres nearby or address information about the Chili's closest to you. Websites like these can leverage knowledge about where you are (i.e. geolocation) to provide services to you based on your location.

This is done through the **W3C Geolocation API**, a set of commands and techniques used by programmers to obtain the location information of your phone, tablet, or computer (see Chapter 15 for more about APIs). Basically, with a few lines of computer code, a Web or app developer can use geolocation to determine where your computer or mobile device is and in turn use that location in its services. For an example of geolocation in action, see *Hands-On Application 1.9: Where Am I Right Now?*

Often an app or device will give you the ability to **geotag** your media (such as a photo) by assigning your location to it through geolocation. For instance, when you take a photo with your phone, you may have the ability to include the location information of where you took the photo. Thus, you can connect the photos to their locations on a map. This use of geolocation is becoming a large component of social media as well. Whenever you use the Facebook app on your smartphone to make a post or send a tweet with the Twitter app, you have the option of adding your current location to that post or tweet. Geotagged tweets can then be mapped to see the locations of where people are sending tweets from.

For instance, in the aftermath of the January 7, 2015, terrorist attack on the offices of the *Charlie Hebdo* newspaper in Paris, France, people worldwide tweeted "Je suis Charlie" (which is French for "I am Charlie") to show support for the murdered journalists. When the people sending the tweets included their location information with the tweet, those locations can be

> **W3C Geolocation API** the commands and techniques used in programming to obtain the geolocation of a computer or mobile device

> **geotag** connecting real-world location information to an item

HANDS-ON APPLICATION 1.9

Where Am I Right Now?

To see the W3C Geolocation API in action, go to **http://whereamirightnow .com**. When prompted by your browser, indicate that you will allow your location to be shared with the Website. Give it a few seconds and the Website will display a map of your current location (as determined using the W3C Geolocation API). This feature will also work on a smartphone or tablet using its Web browser as well.

Expansion Questions

- What kind of information did the Website display about your current location?

- How accurate was the Website as to identifying where you were using the W3C Geolocation API? What methods did Geolocation use to find you?

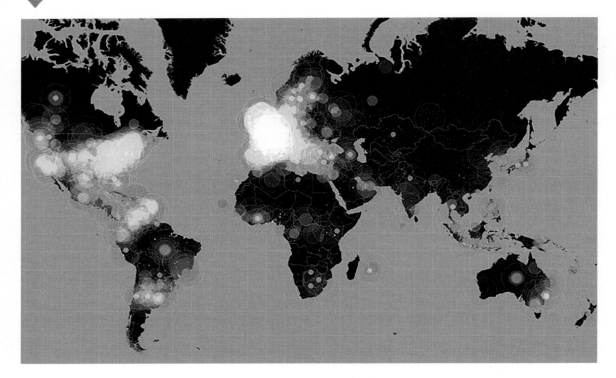

FIGURE 1.10 Geotagged tweets of "Je suis Charlie" sent in the aftermath of the terrorist attacks on the *Charlie Hebdo* newspaper on January 7, 2015. *[Source: Twitter]*

mapped. **Figure 1.10** shows an example of these worldwide geotagged "Je suis Charlie" tweets.

Through geolocation, a mobile device can know where it is. The "Find My iPhone" service uses geolocation to track your iPhone or iPad if it should go missing and displays its location on a map. Geolocation fuels many other common applications used on mobile devices today. For instance, the popular Yelp app allows users to see nearby restaurants' locations and read (or write) reviews for them. The Foursquare app works in a similar way—it collects crowdsourced information about restaurants, shopping, or nightlife from users and then can provide that information to you for places near your location. The Swarm app allows users to share their location with others (referred to as a "check in") at different venues, such as shops or clubs. Even online dating uses geolocation; the popular Tinder app will show you how far away other users of the dating site are from your location. Beyond geolocation, there are many different free geospatial technology apps available today, and a widely used example is Google Earth.

Google Earth a freely available virtual globe program first released in 2005 by Google

virtual globe a software program that provides an interactive three-dimensional map of Earth

 ## What Is Google Earth?

An example of a widespread geospatial technology application is the popular **Google Earth** software program. Google Earth is a **virtual globe**—a program

that provides an interactive three-dimensional (3D) representation of Earth. Google Earth brings together several aspects of geospatial technology into one easy-to-use program. You can examine satellite imagery or aerial photography of the entire globe, zoom from Earth's orbit to a single street address, navigate through 3D cities or landscapes, measure global distances, and examine multiple layers of spatial data together, all with an intuitive interface. In many ways, the coolness and simplicity of Google Earth make it feel like a video game of geospatial technology, but it's far more than that. Google Earth is able to handle vast amounts of geospatial data swiftly and easily, and it enables users to specify and create their own location-based spatial data (see **Figure 1.11** for a look at Google Earth and some of its layers). For instance, Google Earth works with Google's Picasa program to give you the capability to geotag photos—so by clicking on a geotagged location in Google Earth, you can view a photo taken at that spot.

Before Google Earth, there was Keyhole's Earth Viewer, a program that enabled you to fly across a virtual Earth, zooming in and out of locations and imagery. Google purchased the Keyhole company, and the first version of Google Earth was released in 2005. Google Earth is freely available for download off the Web, and relies on a high-speed Internet connection to send the data from Google servers to your computer. For instance, when you use Google Earth to fly from space to your house, the imagery you see on the screen is being streamed across the Internet to you.

The imagery on Google Earth is "canned"—that's to say, it's fixed from one moment in time, rather than being a live view. For instance, if the imagery of your house used by Google was obtained in 2005, then that's the snapshot in time you'll be viewing, rather than a current image of your house. However, Google Earth has a feature that enables you to examine past images. For instance, if imagery of your town was available in 2010, 2012, and today, you can choose which of these images you want to view. Many

FIGURE 1.11 A view of downtown Boston, Massachusetts, from Google Earth. *[Source: Google Earth]*

FIGURE 1.12 The Google Earth app running on a tablet.
[Source: © Iain Masterton/Alamy]

other data layers have been updated and added, such as 3D buildings and other location-based application layers. Also, like many other geospatial technologies, Google Earth can be accessed as an app on smartphones and tablets (see **Figure 1.12**).

Geospatial Lab Application 1.1: Introduction to Geospatial Concepts and Google Earth at the end of this chapter walks you through several different ways of utilizing Google Earth. It assumes no prior experience with Google

THINKING CRITICALLY WITH GEOSPATIAL TECHNOLOGY 1.1

What Happens to Privacy in a Geospatial World?

Think about this scenario. If you own property (house or business), then that information is part of the public record with the county auditor's office and likely available online. Someone can type your address into an online GIS (or mapping Website such as MapQuest or Google Maps) and get a map showing how to travel to your property. In addition, the dimensions, boundaries, and a high-resolution image of your property can be easily obtained. It is possible that other information can be acquired, such as a close-up photo of your house taken from the street. All of this is free, publicly available information that can be had with just a few clicks of a mouse. Anonymity about where you live or work is suddenly becoming a thing of the past. What does the widespread availability (and access to) this type of spatial information mean with regard to privacy? If anyone with an Internet connection can obtain a detailed image of the house you live in, has your personal privacy been invaded?

Earth, but even veteran users may find some new tricks or twists in how geospatial data is handled. You'll use a variety of Google Earth tools to investigate several of the avenues this book will explore, from aerial image interpretation and 3D representations of terrain and objects to ways of determining the shortest path between two destinations.

Chapter Wrapup

So what are you waiting for? Move on to this chapter's lab (see *Geospatial Lab Application 1.1: Introduction to Geospatial Concepts and Google Earth*) and start getting to work. This chapter's lab will have you doing all sorts of things with Google Earth while touching on ideas that will be explored in every chapter of this book. There are also questions throughout the lab for you to answer based on the tasks you'll be doing. Also, check out the *Geospatial Apps* box for a sample of downloadable apps for your mobile device (including a Google Earth app). For further ways to get connected with geospatial technology applications through blogs, Facebook, Twitter, and YouTube, check out the options in *Geospatial Technologies in Social Media*.

Important note: The references for this chapter are part of the online companion for this book and can be found at **macmillanhighered.com /shellito/catalog.**

Geospatial Apps

Here's a sampling of available representative geospatial technology apps for your phone or tablet. Note that some apps are for Android, some are for Apple iOS, and some may be available for both.

- **3D Geo Globe:** An app that will determine your coordinates and plot them on a map of Earth
- **Field Photo:** An app that allows you to take photos with your smartphone camera, geotag them, then upload them to a crowdsourced database for use in environmental studies
- **Foursquare:** An app that provides information on restaurants, shopping, and nightlife near your location
- **Google Earth:** An app that allows you to use Google Earth on your mobile device
- **Open Maps:** An app that allow you to view OpenStreetMap maps
- **Real Estate by Zillow:** An app that allows you to view maps and information about homes for sale

- **Swarm:** An app that allows you to "check in" your location with your smartphone and share that information with others
- **ViewRanger:** An app that allows you to view your location (or other locations) on a variety of different maps
- **Where Am I At?:** An app that determines where you are and shows your location on a map
- **Yelp:** An app for locating restaurants near your location and showing reviews of them
- **You Need A Map:** An app that displays several different types of maps and map layers and uses your own location for detailed geospatial mapping

Geospatial Technologies in Social Media

Here's a sampling of some representative Facebook and Twitter accounts, along with some YouTube videos and blogs related to this chapter's concepts.

On Facebook, check out the following:

- **Google Earth Outreach:**
  www.facebook.com/earthoutreach
- **OpenStreetMap:**
  www.facebook.com/OpenStreetMap

Become a Twitter follower of:

- **Directions Magazine** (geospatial news magazine): **@directionsmag**
- **Geofeedia: @Geofeedia**
- **Geospatial News: @geospatialnews**
- **Geospatial Newsfeed: @GeoMurmur**
- **GIS and Science: @GISandScience**
- **GIS GeoTech News: @gisuser**
- **Google Earth: @googleearth**
- **Google Earth Blog: @googleearthblog**
- **Openstreetmap: @Openstreetmap**

You Tube **On YouTube, watch the following videos:**

- **Geofeedia** (a YouTube channel demonstrating the location-based content of social media): **www.youtube.com/user/GeofeediaTV**
- **Introduction to The National Map** (a USGS video about The National Map): **www.youtube.com/watch?v=ISzUlINbB4o**
- **Jobs in Geography** (a video sponsored by the Association of American Geographers): **www.youtube.com/watch?v=mJz0YTHvCY8**
- **Weird Earth: Overview** (a video of some weird and wonderful places around the planet as seen by geospatial technologies): **www.youtube.com/watch?v=DDfSxMCof2w**

For further up-to-date info, read up on these blogs and news sites:

- **Directions Magazine** (online geospatial news source): **www.directionsmag.com**
- **Geofeedia Blog** (a blog concerning the location-based content of social media): **http://blog.geofeedia.com**
- **The Google Earth Blog** (Google's official blog): **www.gearthblog.com**
- **Mapperz:** The Mapping News Blog (for all sorts of geospatial technology applications): **http://mapperz.blogspot.com**

Key Terms

aerial photography (p. 2)
citizen science (p. 16)
crowdsourcing (p. 16)
GEOINT (p. 8)
geospatial data (p. 8)
geospatial technology (p. 2)
Geographic Information System (GIS) (p. 2)
geolocation (p. 16)
geotag (p. 17)
Google Earth (p. 18)

Global Positioning System (GPS) (p. 2)
The National Map (p. 10)
non-spatial data (p. 9)
remote sensing (p. 2)
satellite imagery (p. 2)
VGI (p. 14)
virtual globe (p. 18)
wiki (p. 14)
W3C Geolocation API (p. 17)

Introduction to Geospatial Concepts and Google Earth

This chapter's Geospatial Lab Application will introduce you to some basic concepts of geospatial technology through the use of the Google Earth software program. This lab also provides an introduction on how to use Google Earth and will help familiarize you with many of its features. Although the Geospatial Lab Applications in later chapters will use a variety of software packages, you'll also be using Google Earth in many of them. The investigations in this introductory lab may seem pretty basic, but labs in later chapters will be more in-depth and will build on concepts learned here.

Objectives

The goals for you to take away from this lab are:

▶ To familiarize yourself with the Google Earth (GE) environment, basic functionality, and navigation using the software

▶ To use different GE layers and features

Using Geospatial Technologies

The concepts you'll be working with in this lab are used in a variety of real-world applications, including the following:

▶ Cable and network news, which frequently incorporate Google Earth imagery or flyovers into news segments to show areas of interest, to visually demonstrate the spatial relationships or distances between cities, or to make visual links between locations of news items

▶ K–12 teachers, which utilize Google Earth imagery (overhead views, 3D models, flyovers, or directions) to help illustrate a variety of concepts in class, including geography, history, and science

Obtaining Software

The current version of Google Earth Pro (7.1) is available for free download at **www.google.com/earth/explore/products/desktop.html**. You used to have to pay for Google Earth Pro; however, as of early 2015, Google has released Google Earth Pro for free. As such, you'll be using Google Earth Pro instead of the regular Google Earth in this lab (and others in the book) to take advantage of its many additional features.

Note that when downloading Google Earth Pro you'll require a username and license key when installing it. The download page will give you these same instructions, but use your e-mail address for your user name and GEPFREE as the license key (you'll only have to input this information once when installing the software and then not bother with it again when you use it).

Important note: Software and online resources sometimes change fast. This lab was designed with the most recently available version of the software at the time of writing. However, if the software or Websites have significantly changed between then and now, an updated version of this lab (using the newest versions) is available online at **macmillanhighered.com/shellito/catalog**.

Lab Data

There is no data to copy in this lab. All data comes as part of the GE data that is installed with the software or is streamed across the Internet when using GE.

Localizing This Lab

Although this lab visits popular locations in a tour around South Dakota, the techniques it uses can be easily adapted to any locale. Rather than using GE imagery and features of South Dakota, find nearby landmarks or popular spots and use GE to tour around your local area.

1.1 Starting to Use Google Earth Pro (GE)

1. Start GE (the default install folder is called Google Earth Pro). GE will usually open with a startup tip box, which you can close. Also, make sure the GE window is maximized, or else some features may not show up properly.

[Source: Google]

A popular use of GE is to view the ground (whether physical landscapes or structures) from above by using the aerial photography or satellite imagery streamed in from Google's servers. The Search box is used for this.

2. In the Search box, type in Mitchell, SD. Next, click on the Search button.

3. GE will rotate and zoom the view down to eastern South Dakota. One of the popular attractions in Mitchell is the Corn Palace, a historic building of fanciful design that now serves as a civic arena, presenting cultural and sports events and attracting the attention of millions each August when the murals that cover the building's façade (which are completely constructed from corn) are reconfigured to illustrate the theme of that year's Corn Palace Festival. More information about the Corn Palace is available at **www.cornpalace.com/index.php**.

4. In the Layers box, place a checkmark next to Photos. This enables linking of locations on the ground to photos that users have taken. The locations that have linked photos have small blue squares over the imagery. You'll see numerous photo locations clustered together in one place—that would be the Corn Palace. Double-click on some of them to see some photos (taken from the ground) of the Corn Palace.

5. In GE, you can rotate and zoom your view to get a closer look at the imagery on the ground. Move the mouse to the upper right-hand side of the screen and a set of controls will appear. These fade out when not in use and reappear when the mouse rolls over them.

6. There are five controls. The first one (the ring with the N at the top) allows you to change the direction you're facing. Grab the N with the mouse (by placing the pointer on N and holding down the left mouse button) and drag the N (that is, slowly rotate it) around the ring (this is commonly called "grab and drag"). You'll see the view change (north will still be the direction in which the N is pointed). Clicking on the N will return the imagery so that the north direction is facing the top of the screen.

7. The second control is the Look control (the arrows surrounding the eye). By selecting one of the directional arrows and pressing it with the mouse, you can tilt your view around the terrain as if you were standing still and looking about. You can also grab the control and move it (by holding down the left button on your mouse), to simulate looking around in various directions.

8. The third control is the Move control (the arrows surrounding the hand). By grabbing this control, you can pan around the imagery in whichever direction you choose.

9. The fourth control is the Street View icon (the yellow figure of a human—referred to as the "pegman"—standing on a green circle). This icon will appear when Google Street View imagery is available to see on the ground. We'll spend more time on Google Street View in Chapter 8, as it's a utility that allows you to see what places look like as if you were standing on the street in front of them. To use Street View, you would grab the icon from the controls with the mouse and drag it to a street that's visible in the view to enter the Street View mode. This control will only be visible if there are streets in the view through which you can enter Street View mode (which is why you won't initially see it when you start GE and are looking at the entire planet).

[Source: Google]

10. The last control is the Zoom Slider. By pushing forward (moving the bar toward the plus sign), you zoom in, and by pulling backwards (moving the bar toward the minus sign), you zoom out. By default, when GE zooms in, it tilts the terrain to give you a perspective view, rather than looking from the top directly down. To always use the top-down option rather than tilting the ground when zooming, select the Tools pull-down menu and choose Options. In the dialog box that appears, click on the Navigation tab, and then select the radio button that says "Do not automatically tilt when zooming." To have GE tilt while zooming, select one of the other radio buttons.

11. You can also use the mouse for navigating around GE. Grab the imagery and drag it to sweep across the map. To zoom in and out, roll the mouse wheel back and forth. Pan and zoom over to the location of the Corn Palace. To confirm that you're in the right place, type the address of the Corn Palace into the Search box as follows: 604 North Main Street, Mitchell, SD 57301 (GE will automatically adjust to that location).

12. For now, turn off the Photos options by removing the checkmark from the box.

13. Zoom in on the Corn Palace so that it fills most of the view and you can make out some details about it.

14. Grab the Street View icon from the controls and drag it to the street right in front of the Corn Palace. You'll see the street turn blue (this indicates that Street View imagery is available for this street). Release the icon on the street and the view will shift from an aerial view to imagery that looks like you're actually standing on Main Street in front of the Corn Palace.

[Source: Google]

15. In this Street View, use the mouse to pan around the image (you can see 360 degrees around your position) until you can see the entrance of the Corn Palace. You can pan and zoom with the mouse as necessary (and also click on the road—or elsewhere in the image—to advance your view).

16. Examine the Corn Palace and its surrounding area. Then answer Question 1.1. To return to the aerial view, click on the Exit Street View text in the upper right corner of the Street View. We'll do more with Street View (and street networks) in Chapter 8.

> **Question 1.1** From viewing the Corn Palace from above and in front, what details from the aerial view can help identify just what the building is? (You may need to switch back and forth between the aerial view and the street view to answer this question.)

17. When dealing with remotely sensed imagery, it's important to consider that you're viewing the ground from above and that many features you can view from a street level are difficult to perceive from an aerial view. Chapter 9 delves further into interpreting aerial images and offers some strategies and tips for doing so, while Chapter 10 deals with how this type of remotely sensed imagery is acquired. Chapters 11 and 12 explore the acquisition of imagery from various satellites.

18. Here's another way of looking at the Corn Palace—a 3D representation. From the GE Layers options, select 3D Buildings. A 3D representation of the Corn Palace will appear on top of the imagery of the building.

Use the zoom and tilt controls and navigate around the building to get a better look at the 3D model of the Corn Palace, including its façade and turrets. Note that if the 3D model of the Corn Palace doesn't appear, select the Tools pull-down menu, select Options, then select the 3D View tab and make sure a checkmark is not in the Use 3D imagery (disable to use legacy 3D buildings) box.

19. After viewing the 3D model of the Corn Palace, turn off the option for 3D Buildings when you're done. There are numerous buildings and objects that have 3D representations designed for use in GE. In Chapter 14, you'll investigate more about working with 3D versions of geospatial data, as well as constructing your own 3D buildings and viewing them in GE.

1.2 Finding Your Way Around with Google Earth

1. From the Layers box, turn on the Roads layer. You'll see major roads (interstates, state highways) identifiable as their labels appear, and local roads will also be identifiable when you zoom in close.

2. We're going to leave the Corn Palace behind and continue west through South Dakota to Wall Drug, a famous tourist location. More information on Wall Drug can be found at **www.walldrug.com**.

3. In the Search box, click on the Get Directions option.

4. In the A option (this is where you're traveling from), type in the Corn Palace's address as follows: 604 N Main St, Mitchell, SD 57301.

5. In the B option (this is where you're traveling to), type in Wall Drug's address as follows: 510 Main Street, Wall, SD 57790.

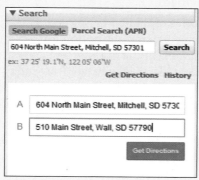

[Source: Google]

6. Click the Get Directions button. GE will zoom out to show you the path (in purple) it has calculated for driving distance between the Corn Palace and Wall Drug, while the Search Box will fill with directions featuring stops and turns along the way.

Question 1.2 By viewing the expanse of South Dakota between Mitchell and Wall, you'll realize there are many possible routes between the two. Why do you suppose GE chose this particular route to go from the Corn Palace to Wall Drug?

Question 1.3 Based on the route that GE calculated, what is the driving distance (and approximate time equivalent) to get from the Corn Palace to Wall Drug?

7. This capability to take a bunch of letters and numbers and turn them into a mapped location and to calculate the shortest driving distance between points are further developed in Chapter 8. Matching addresses and calculating shortest routes are key reasons for utilizing geospatial technologies, and the Chapter 8 lab will go into more depth on how these types of operations are performed.

8. In the Search box, scroll down through the list of directions and click on the Wall Drug address next to B. Google Earth will zoom to Wall Drug's location. Press the X at the bottom of the Search Box to clear the directions and remove the route.

1.3 Using Google Earth to Examine Landscapes

1. There's a lot more to the South Dakota landscape than tourist attractions. Wall Drug is located in the town of Wall, which is named for the line of magnificent land formations nearby and lies directly adjacent to Badlands National Park. The Badlands are located to the south and southwest of Wall. For more information about the Badlands visit **www.nps.gov/badl**.

2. To see the boundaries of Badlands National Park in Google Earth, go to the Layers box and expand the option for More (click on the triangle to the left of the name). In the options that appear under the expanded heading, put a checkmark in the Parks/Recreation Areas option. Zoom out from the boundaries of Wall, and to the south you'll see the northernmost boundary of Badlands National Park highlighted in green.

3. Zoom out so that you can see the entire park in the view. New icons for the locations of Visitors Centers and other features will also appear.

4. Pan over to the eastern edge of the park and you'll see a large Question Mark icon indicating a park entrance as well as a green arrow indicating an overlook.

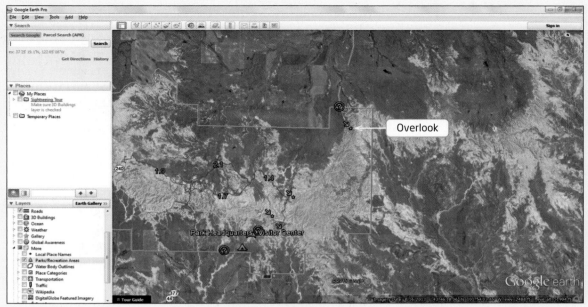

[Source: Google]

5. Zoom in very closely to the point representing the overlook. At the
    bottom of the view you'll see numbers representing the latitude and
    longitude of the point, as well as the real-world elevation of that spot.

    Question 1.4 What is the elevation for this particular overlook in
    Badlands National Park?

6. The imagery in GE is placed on top of a model of Earth's vast array of
    terrains and landscapes (hills, valleys, ridges, and so on). It's very
    difficult to make out the pseudo three-dimensional (3D) nature of the
    terrain model from above, so use the Zoom Slider to tilt the view down
    (you can also hold down the Ctrl key and move the mouse to help
    change your perspective) so that you can look around as if you were
    standing on the overlook point (in a perspective view of the planet).
    Once you've tilted all the way down, use the Look controls to examine
    the landscape. From here, use the Move controls to fly over the
    Badlands from this overlook point. When you're done cruising around
    the Badlands, answer Question 1.5.

    Question 1.5 How does the terrain modeling (with the tilt
    function) aid in the visualization of the Badlands?

7. This ability to model the peaks and valleys of the landscape with aerial
    imagery "draped" or "stretched" over the terrain for a more realistic
    appearance is often used with many aspects of geospatial technology.
    Chapter 13 greatly expands on this by getting into how this is actually
    performed as well as doing some hands-on terrain analysis.

1.4 Using Google Earth to Save Imagery

It's time to continue to the next leg of our South Dakota journey by heading to Mount Rushmore. Carved out of the side of a mountain in the Black Hills, Mount Rushmore is a monument featuring the faces of presidents George Washington, Thomas Jefferson, Theodore Roosevelt, and Abraham Lincoln. For more information about Mount Rushmore visit **www.nps.gov/moru**.

1. In GE's Search box, type Mount Rushmore. GE will zoom around to an overhead view of the monument. Like the Badlands, Mount Rushmore is a national park—zoom out a little bit and you'll see the extent of the park's boundaries (still outlined in green).

2. You can also save an image of what's being shown in the view. GE will take a "snapshot" and save it as a JPEG graphic file (.jpg), which is like a digital camera picture. Position the view to see the full outlined extent of Mount Rushmore, select the File pull-down menu, then select Save, and finally select Save Image. Name the image moru (GE will automatically add the .jpg file extension), and save it to your computer's hard drive.

3. Minimize GE for now and go to the location on your computer where you saved the image and then open it to examine it (using a simple program like Microsoft Office Picture Manager).

> **Question 1.6** Note that even though the graphic contains the locations of Mount Rushmore, the outline of the park, and information concerning location (including latitude, longitude, and elevation) at the bottom, it doesn't have any spatial reference for measurements. Why is this?

4. Close the image and return to Google Earth. Even though the saved image doesn't have any spatial reference, Chapter 3 describes how to take unreferenced imagery (and other unreferenced data) and transform it to match referenced data. In Chapter 3's lab, you do this in a hands-on fashion. You can also turn off the Parks/Recreation layer for now.

1.5 Using Google Earth to Make Placemarks and Measurements

While you're examining Mount Rushmore imagery, you can set up some points of reference that you can return to. GE allows you to create points of reference as Placemarks. There are three points to mark on the map—the top of the mountain, the amphitheater, and the parking area (see the following graphic for their locations).

[Source: Google]

1. From the GE toolbar, select the Add Placemark button:

[Source: Google]

2. A yellow pushpin (labeled "Untitled Placemark") will appear on the screen. Using your mouse, click on the Pushpin and drag it to the rear of the amphitheater so that the pin of the placemark is where the path meets the amphitheater.

3. In the Google Earth New Placemark dialog box, type Mount Rushmore Amphitheater.

4. By pressing the Placemark icon button next to where you typed the name, you can select a different icon rather than the yellow pushpin. Choose something more distinctive.

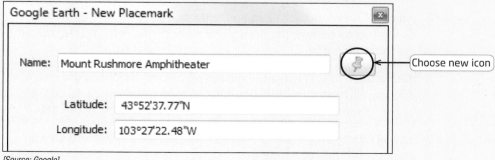

[Source: Google]

5. Click OK to close the dialog box.

6. Repeat the process by putting a new placemark at the top of the mountain where the monument is. Name this new placemark Top of the Monument.

7. Zoom GE's view in tightly, so that you can see the two placemarks clearly in the view (with the two placemarks in diagonally opposite corners of the view).

8. Next, select the Show Ruler button from the GE toolbar:

[Source: Google]

9. In the Ruler dialog box that appears, select Feet from the Length pull-down menu.

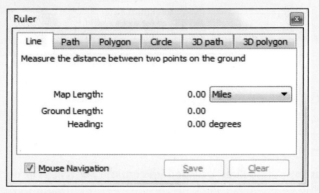

[Source: Google]

10. Use the options under the Line tab. These will let you compute the distance between two points. The Path option will enable you to find out the total accumulated distance between more than two points.

11. Click the mouse on the point of the placemark in the amphitheater and drag it to the placemark on top of the mountain (your heading should be about 310 degrees or so). Click the mouse at the top of the mountain placemark.

12. Answer Question 1.7. When you're done, click on Clear in the Ruler dialog box to remove the drawn line from the screen, and then close the Ruler dialog box.

> **Question 1.7** What is the measured distance between the rear of the amphitheater and the top of the monument (keep in mind this is the ground distance, not a straight line between the two points)?

13. This ability to create points of reference (as well as lines and area shapes) and then compute the distance between them are functions of

a very basic GIS tool, which forms the basis for performing numerous tasks that will be discussed in upcoming chapters. Chapter 5 introduces GIS concepts, Chapter 6 expands on how spatial analysis can be performed with GIS, and Chapter 7 demonstrates how to take your data and create a professional quality map of the results.

14. Use the Tilt functions of GE to get a perspective view on Mount Rushmore (as you did with the Badlands), and take a look at the mountain from a pseudo-3D view. Make sure the 3D Buildings option is turned off. Although you can see the height and dimensions of the mountain, the famous four presidents' faces can't be seen, however you zoom or rotate the model.

> **Question 1.8** Even with the terrain turned on and the view switching over to a perspective, why can't the presidents' faces on the side of the monument be seen?

15. Turn the 3D Buildings option on. If you cannot see the separate Mount Rushmore 3D model (which lets you see the faces), select the Tools pull-down menu, select Options, then select the 3D View tab, and make sure a checkmark is not in the Use 3D imagery (disable to use legacy 3D buildings) box.

1.6 Using Google Earth to Examine Coordinates

1. You'll notice next to the elevation value at the bottom of the GE screen there's a set of coordinates for lat (Latitude) and lon (Longitude). Move the mouse around the screen and you'll see the coordinates change to reflect the latitude and longitude of whatever the mouse's current location is.

2. Zoom in closely on the road that enters the parking structure area of Mount Rushmore.

> **Question 1.9** What are the latitude and longitude coordinates of the entrance to the Mount Rushmore parking area?

3. You can also reference specific locations on Earth's surface by their coordinates instead of by name. In the Search box type the following coordinates: 43.836584, −103.623403. GE will rotate and zoom to this new location. Turn on Photos to obtain more information on what you've just flown to. You can also turn on the 3D Buildings layer (if necessary, again using the legacy 3D Buildings option, as you did with the Corn Palace and Mount Rushmore), to look at the location with a 3D version of it there. Answer Questions 1.10 and 1.11, and then turn off the Photos (and the 3D Buildings option) when you're done.

> **Question 1.10** What is located at the following geographic coordinates: latitude 43.836584, longitude −103.623403?

> **Question 1.11** What is located at the following geographic coordinates: latitude 43.845709, longitude −103.563499?

Chapter 2 deals with numerous concepts related to coordinate systems, reference systems, and projections of a three-dimensional Earth onto a two-dimensional surface.

1.7 Other Features of Google Earth

As noted before, you will be using GE for several other labs in this book and exploring some of these applications of geospatial technology. GE has a multitude of other features beyond those covered in this first lab. For instance, if you had a GPS receiver (say, one from Garmin or Magellan) you could import the data you've collected directly into GE.

1. From the Tools pull-down menu, select GPS. You'll see several options for offloading information you collected in the field (such as waypoints) from the receiver into GE.

   Chapter 4 will introduce you to several concepts related to GPS, as well as collecting data in the field and some pre-fieldwork-planning options.

2. Close the GPS Import dialog box now (unless you have a receiver you want to plug into a computer on which you can view your data in GE).

   There are many other GE layers and features to explore, especially in relation to how geospatial data is visualized. To look at one of these features, return to the Badlands once more and do the following:

3. In the Layers box, expand the options for Gallery, and then place a checkmark in the box next to 360 Cities.

4. Fly back to the Badlands and scroll to the western edge of the park and you'll see another symbol on the map—a red circle marked "360." Hover your cursor over it, and the text should read "Mako Sica aka The Badlands."

5. Double-click on the 360 circle. You'll zoom down and see a sphere on the surface, and then the view should dive inside the sphere. The view will change to a new view of the Badlands (use the mouse to move about and look around a full 360 degrees). In essence, this tool simulates what the Badlands would look like from that position, all

around you. You can look in any direction, as well as zooming in or out of places. Answer Question 1.12. When you're done looking around, click on the Exit Photo button in the view to return to regular GE. Chapter 14 will present information about some more cool visualization techniques for geospatial data.

> **Question 1.12** How does the 360-degree image of the Badlands aid in visualizing the scenery (keep in mind this image has been tied to a particular location)?

Geospatial technology is such a rapidly changing field that new advances come quickly. Chapter 15 also explores some of these current developments on the frontiers of geospatial technology.

Google Earth is always changing, and new versions and upgrades are constantly being made available. Some of the newer options you may want to explore are Google Mars, Google Sky, and Flight Simulator.

Google Mars

From the View pull-down menu, select Explore, and then select Mars. The view will shift to the familiar-looking globe, but this time covered with imagery (from NASA and the USGS) from the surface of Mars. The usual controls work the same way as they do with GE, and there's a lot of Martian territory to be explored. When you're ready to return to GE, from the View pull-down menu, select Explore, and then select Earth.

Google Sky

From the View pull-down menu, select Explore, and then select Sky. GE's view will change: instead of looking down on Earth, you're looking up to the stars—and instead of remotely sensed aerial or satellite imagery, you're looking at space-telescope (including the Hubble) imagery. There's plenty of imagery to be explored in this new mode. When you want to get back "down to Earth" again, select the View pull-down menu, select Explore, and then select Earth.

Flight Simulator

From the Tools pull-down menu, select Enter Flight Simulator. A new dialog box will appear asking if you want to pilot an F-16 fighter jet or an SR-22 propeller airplane. Select whichever plane you want to fly; GE will switch to the view as seen out the cockpit of your chosen plane, and you can fly across South Dakota (and the rest of GE). Try not to crash (although you can easily reset if you do).

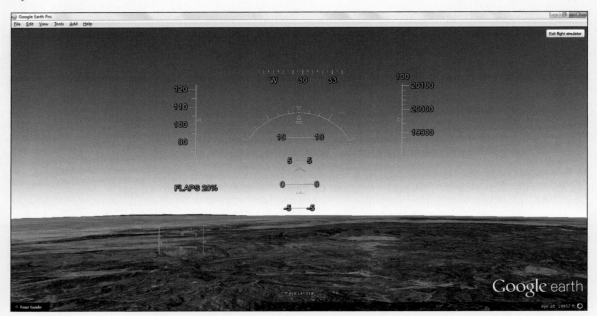

[Source: Google]

Closing Time

This introductory lab was pretty straightforward, but it served to introduce you to the basic functions of Google Earth Pro, as well as providing examples of many geospatial concepts you'll be working with over the course of this book. Chapter 2 starts looking at location concepts (like various coordinate systems) and returns to Google Earth Pro to start applying some of the topics you've worked with in this chapter. You're all set, and now you can exit GE by selecting **Exit** from the **File** pull-down menu. There's no need to save any data (or Temporary Places) in this lab.

2

Where in the Geospatial World Are You?

Locations in a Digital World, Position Measurements, Datums, Coordinate Systems, GCS, Map Projections, UTM, and SPCS

Think about this. You're at the mall and your cell phone rings. It's your friend Sara calling you with a basic question: "Where are you?" For such a simple question, there's really a wide variety of answers. You could say "I'm at the mall," which gives Sara some basic location information—but only if she knows where the mall is. If she's from out of town and unfamiliar with the location of the mall, you haven't given her the information she needs to know where you are. If you tell her, "I'm in Columbus," then she'll have another type of information—but Columbus is a big city, and she still won't know where in the city you are. You could give her the name of the cross streets by the mall, the address of the mall, or the name of the store you're in at the mall, and even though all of the these contain different levels of information, your friend still won't have the information she needs to know precisely where you are.

"I'm at the mall" effectively conveys your location only if your friend has some sort of knowledge of the mall's location. If you give her a reference like "The mall's on the north side of town," or "The mall is just off exit 5 on the freeway," then she'll have some idea of your location. For the location information to make sense, she needs to have something (perhaps a map of the city you're in) to reference the data to. If you're asked to provide specific information about your exact location, it'll have to be much more detailed than "I'm at the mall." In a medical emergency, emergency medical technicians (EMTs) need to know your exact location (they can't afford to waste time searching the entire mall to find you). You need to provide precise location information so that the EMTs will be able to find you, no matter where you are.

Geospatial technology works the same way. First, since it deals with geospatial data (which is linked to non-spatial data), there has to be a way of assigning location information to the data being handled. There has to be some way of referencing every location on Earth. In other words, every location has to be clearly identifiable and measurable. Second, these measurements require some sort of reference system, so that a location at one point on Earth can be compared with locations at other points. Third, since we're dealing with points on a three-dimensional (3D) Earth but only have two-dimensional (2D) surfaces to examine (i.e., maps of various kinds), there has to be a way of translating real-world data to an environment we can use. This chapter examines how these concepts are treated in all forms of geospatial technology. If you're going to be able to precisely pin down the coordinates of a location, you must first have a datum and a coordinate system.

 ## What Is a Datum?

A **datum** is a reference surface, or model of Earth that is used for plotting locations anywhere on the actual surface of Earth. The datum represents the size and shape of Earth—which, contrary to popular belief, isn't perfectly round. Earth is actually an **ellipsoid** (or spheroid), meaning that it's larger at its center than it is at its poles. Think of taking a basketball and squeezing the top and bottom of the ball to compress it slightly and you'll have a basic spheroid shape. However, Earth isn't a smooth ellipsoid—gravitational forces affect different parts of Earth in different ways, causing some areas to be out of sync with the ellipsoid. Another model of Earth, called a **geoid**, places Earth's surface at mean sea level to try to account for these differences (see **Figure 2.1** to examine how the geoid and the ellipsoid stack up against each other and against the variable topography of Earth). The science of measuring Earth's shape to develop these kinds of models and reference surfaces is referred to as **geodesy**.

Developing a datum means creating a mathematical model against which to reference locations and coordinates. Thus, when you're plotting a

datum a reference surface of Earth

ellipsoid a model of the rounded shape of Earth

geoid a model of Earth using mean sea level as a base

geodesy the science of measuring Earth's shape

FIGURE 2.1 How the real-world Earth, the ellipsoid, and the geoid match up.

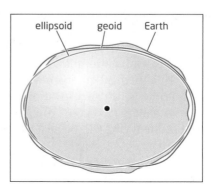

point, you have a system for referencing this location. The difficulty in mapping with datums arises because there isn't just one datum to use for all measurements of Earth's locations. In fact, there are hundreds of datums in use. Some datums are global and measure the entire world, whereas some are localized to particular continents or regions. These are some of the more common datums you'll likely encounter with geospatial data:

▶ **NAD27** (The North American Datum of 1927): This datum was developed for measurements of the United States and North America. It has its center point positioned at Meades Ranch in Kansas.

> NAD27 the North American Datum of 1927

▶ **NAD83** (The North American Datum of 1983): This datum was developed by the National Geodetic Survey, together with Canadian agencies, and is used as the datum for much data for the United States and the North American continent as a whole. While NAD83 is very commonly used, the U.S. National Geodetic Survey is currently preparing a new datum to take its place, but this won't be ready until 2022.

> NAD83 the North American Datum of 1983

▶ **WGS84** (The World Geodetic System of 1984): This datum was developed by the U.S. Department of Defense, and is used by the Global Positioning System (GPS) for locating points worldwide on Earth's surface (see Chapter 4 for more information about GPS).

> WGS84 the World Geodetic System of 1984 datum (used by the Global Positioning System)

Measurements made with one datum don't necessarily line up with measurements made from another datum. Problems arise if you take one set of data measured in one datum and your friend takes a second set of data measured from a different datum and you put both sets of data together. Both datasets refer to the same place, but they won't match up because you and your friend are using different forms of reference for measurement. For instance, data measured from the NAD27 datum could be off by a couple hundred meters from data measured from the NAD83 datum. This problem is compounded by the availability of several local datums, or reference surfaces, designed for smaller geographic areas. With the measurement differences that may arise, the best strategy when you're dealing with geospatial data is to keep everything in the same datum. A **datum transformation** is required to alter the measurements from one datum to another (for instance, changing the measurements made in NAD27 to NAD83). Datum transformation is a computational process, a standard one in many geospatial software packages. Datums can be used to set up geographic coordinate systems (GCS) for measuring coordinates.

> **datum transformation**
> changing measurements from one datum to measurements in another datum

> **geographic coordinate system (GCS)** a set of global latitude and longitude measurements used as a reference system for finding locations

What Is a Geographic Coordinate System?

A **geographic coordinate system (GCS)** is a global reference system for determining the exact position of a point on Earth (see **Figure 2.2** on page 42 for an example of GCS). GCS consists of lines of latitude and longitude that cover the entire planet and are used to find a position. Lines of **latitude** (also

> **latitude** imaginary lines on a globe north and south of the Equator that serve as a basis of measurement in GCS

FIGURE 2.2 The geographic coordinate system (GCS) covers the globe, and locations are measured using latitude and longitude. *[Source: 2012 Google. Data SIO, NOAA, U.S. Navy, NGA, GEBCO. Image 2012 TerraMetrics, Image IBCAO, 2012 Cnes/Spot Image]*

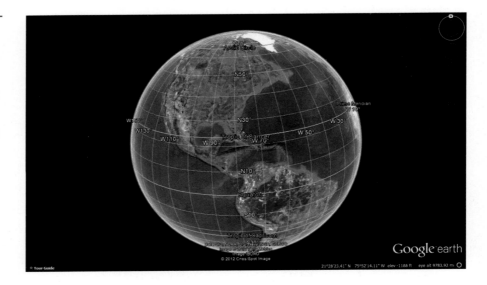

Equator the line of latitude that runs around the center of Earth and serves as the 0 degree line from which to make latitude measurements

longitude imaginary lines on a globe east and west of the Prime Meridian that serve as a basis of measurement in GCS

Prime Meridian the line of longitude that runs through Greenwich, England, and serves as the 0 degree line of longitude from which to base measurements

degrees, minutes, and seconds (DMS) the measurement system used in GCS

referred to as parallels) run in an east-to-west direction around the globe. The **Equator** serves as the starting point of zero, with measurements north of the Equator being numbers read as north latitude and measurements south of the Equator being numbers read as south latitude. Lines of **longitude** (also called meridians) run north-to-south from the North Pole to the South Pole. The **Prime Meridian** (the line of longitude that runs through the Royal Observatory in Greenwich, England) serves as the starting point of zero. Measurements made east of the Prime Meridian are numbers read as east longitude, while measurements made west of the Prime Meridian are numbers read as west longitude. The breaking point between east and west longitude is the 180th meridian.

Using this system (where measurements are made north or south of the Equator and east or west of the Prime Meridian), any point on the globe can be located. GCS measurements are not made in feet or meters or miles, but in **degrees, minutes, and seconds (DMS)**. A degree is a unit of measurement that can be broken into 60 subsections (each one referred to as a minute of distance). A minute can be subsequently broken down into 60 subsections (each one referred to as a second of distance). A single minute of latitude is equivalent to roughly one nautical mile (or about 1.15 miles). Lines of latitude are measured in angular units from the equator while lines of longitude are measured in angular units from the Prime Meridian. It should be noted that the distance between degrees of longitude varies at different places in the world—lines of longitude get closer together as they approach the poles and are furthest apart at the Equator.

When GCS coordinates are being written for a location, they are expressed in terms of the location's distance (that is, in terms of degrees, minutes, and seconds) from the Equator and the Prime Meridian. Latitude values run from 0 to 90 degrees north and south of the Equator, while longitude values run from 0 to 180 east and west of the Prime Meridian. For example, the GCS coordinates of a location at the Mount Rushmore National Monument are 43 degrees, 52 minutes, 53.38 seconds north of the Equator (43°52'53.38" N latitude) and 103 degrees, 27 minutes, 56.46 seconds west of the Prime Meridian (103°27'56.46" W longitude).

Negative values can also be used when expressing coordinates using latitude and longitude. When distance measurements are made west of the Prime Meridian, they are often listed with a negative sign to note the direction. For instance, the location at Mount Rushmore, being west longitude, could also be written as −103°27'56.46" longitude. Negative values are used similarly in measurements of south latitude (below the Equator).

GCS coordinates can also be expressed in their decimal equivalent, referred to as **decimal degrees (DD)**. In the decimal degrees method, values for minutes and seconds are converted to their fractional number of degrees. For instance, 1 degree and 30 minutes of latitude (in DMS) equates to 1.5 degrees in DD (since 30 minutes is equal to 0.5 degrees). Using the Mount Rushmore coordinates as an example, the latitude measurement in decimal degrees would be 43.881494 north latitude (since 52 minutes and 53.38 seconds is equal to 0.881494 of one degree) and 103.465683 west longitude.

Using this system, any point on Earth's surface can be precisely measured and identified. Keep in mind that these locations are being made on a three-dimensional surface (in reference to a datum), so measurements between these locations have to take this into account. When making measurements on a sphere, the shortest distance between two points is referred to as the **great circle distance**. Thus, if you want to find the shortest distance between the cities of New York and Paris, you have to calculate the great circle distance between two sets of coordinates (see *Hands-On Application 2.1: Great Circle Distance Calculations* on page 46 for more information about using the great circle distance).

The latitude and longitude system of GCS also serves as the basis for breaking Earth up into multiple **time zones** to account for difference in time around the world. Time zones set up the concept of a 24-hour day, with each zone having a different time as it relates to the current time in Greenwich, England (also called Greenwich Mean Time or GMT). Earth is spanned by 360 degrees of longitude, and there are 24 hours in a day. 360 divided by 24 equals 15: there are 15 degrees of longitude for each hour. Thus, every 15 degrees away from the time zone containing the Prime Meridian marks the start of a new time zone (see **Figure 2.3** on page 44).

From 7.5 degrees west longitude to 7.5 degrees east longitude is one time zone, 7.5 to 22.5 degrees longitude is a second time zone, and so on. Following these sections of longitude around the world, the east coast of the United States is five time zones to the west of the Prime Meridian, which makes the time in New York and Boston five hours behind the time in

decimal degrees (DD) the fractional decimal equivalent to coordinates found using degrees, minutes, and seconds

great circle distance the shortest distance between two points on a spherical surface

time zones a method of measuring time around the world, created by dividing the world into subdivisions of longitude and relating the time in that subdivision to the time in Greenwich, England

FIGURE 2.3 The world's time zones are separated by lines of longitude (and geographic boundaries).

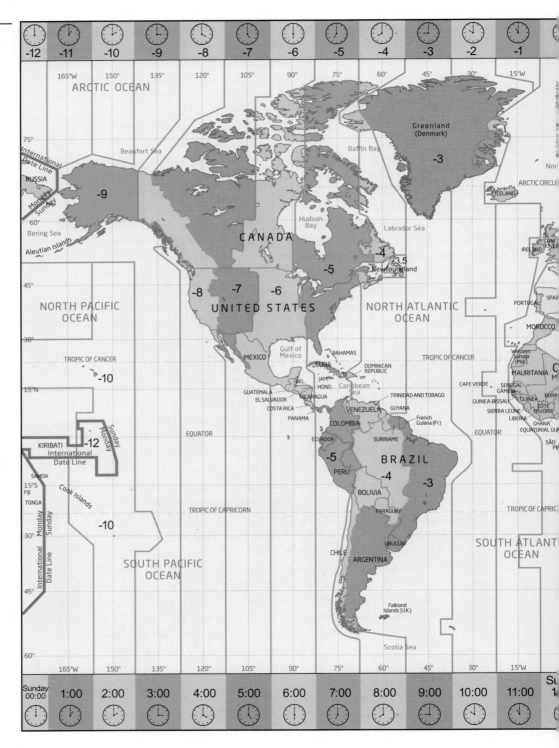

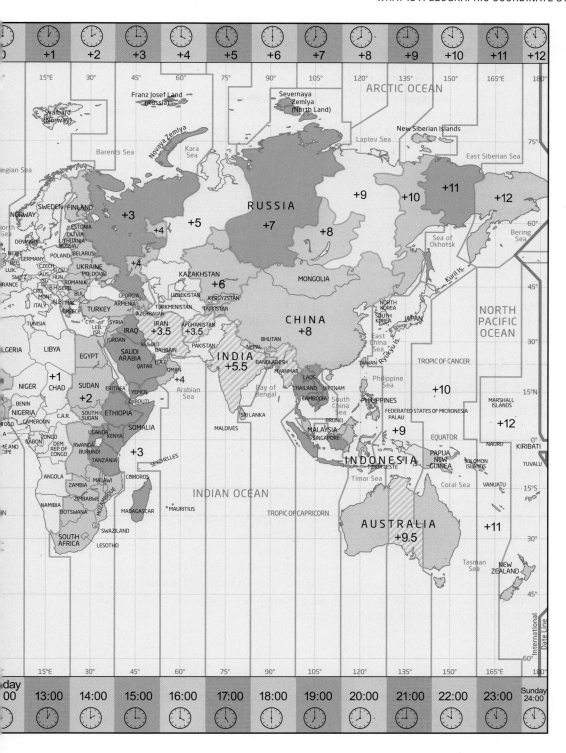

HANDS-ON APPLICATION 2.1

Great Circle Distance Calculations

To find the real-world shortest distance around the planet between two cities—New York and Paris—you'll need to calculate the great circle distance between them. To calculate great circle distances quickly, use the Surface Distance application online at **www.chemical-ecology .net/java/lat-long.htm**. Use the option for Great Circle Distance Between Cities to find the real-world shortest distance between New York and Paris. Use it to find the city closest to where you are, and calculate the great circle distances between various pairs of cities. Also, use a program like Google Earth to obtain the GCS coordinates (latitude and longitude) for a location close to you and find the surface distance from there to a nearby city.

Expansion Questions

• What is the great circle distance between New York and Paris?

• What is the great circle distance between the city closest to you and London, Paris, Rome, and Shanghai?

• Use Google Earth to obtain the GCS (latitude and longitude) coordinates (in degrees, minutes, and seconds) for a location near you and GCS coordinates for a location at the White House in Washington, D.C., and calculate the surface distance between the two points. How many miles are you from the White House?

THINKING CRITICALLY WITH GEOSPATIAL TECHNOLOGY 2.1

Do You Really Need Printed Maps in a Digital World?

Let's face it, these days we're living in a digital world—road maps, directions, and property maps are all available in digital form with a few clicks of a mouse. With everything available digitally (and easily portable and accessible, thanks to laptops, tablets, and phones), is there still a need for printed maps? Do you really need to carry around a paper map of the road network in Los Angeles, California, or will a mapping app on your smartphone suffice to find your location and get you where you want to go? Are paper map collections at universities and libraries relics of the past, or are they still useful resources? When might it be necessary to have access to such a collection?

When you're driving a long distance and have GPS, satellite maps, and Internet mapping capabilities, is it necessary to have a printed map of the area you're traveling through? If your answer is yes, describe a situation in which it would be necessary. If your answer is no, why are printed maps still available?

Greenwich. The west coast of the United States is eight time zones to the west of the Prime Meridian (and three to the west of the United States east coast), which makes the time in San Francisco eight hours behind Greenwich (and three hours behind New York).

Take a closer look at Figure 2.3, and you'll see that time zones don't exactly follow the lines of longitude. Political or geographic boundaries are often the reason why a particular area is entirely in one time zone even though that area may straddle two zones, and why lines of longitude shouldn't always be read as dividing lines between one time zone and the next. China, for instance, adopts one time measurement for the entire country, even though its geography crosses five time zones. The **International Date Line** marks the division between 24-hour periods and divides one day from another, but it also bends to accommodate considerations of geography, political boundaries, and convenience, and makes detours to include islands in one time zone or another, rather than strictly following the line of the 180th meridian.

> **International Date Line** a line of longitude that uses the 180th meridian as a basis (but changes away from a straight line to accommodate geography)

 ## How Can Real-World Data Be Translated onto a Two-Dimensional Surface?

These measurements belong to a three-dimensional world, but are all translated, for ease of use, into two dimensions (flat surfaces like pieces of paper and computer screens). A **map projection** is a translation of coordinates and locations from their places in the real world to a flat surface, and this type of translation is necessary if we want to be able to use this geospatial data. The biggest problem with a map projection is that not everything is going to translate perfectly from a three-dimensional object to a two-dimensional surface. In fact, the only representation of the world that can accurately capture every feature would be a globe. Since it's hard to keep a globe in the glove compartment of the car, we're going to have to make do with flat surface representations of the world (like paper maps and electronic screens), and realize that such representations will always have some sort of distortions built into them, simply because they're two-dimensional versions of a three-dimensional reality. One or more of the following will be distorted in a map projection: the shape of objects, the size of objects, the distance between objects, or the direction of objects. Some map projections may retain the shape of continents but misrepresent their land area. Other projections may present the land area accurately but throw the shapes out of place.

> **map projection** the translation of locations on the three-dimensional (3D) Earth to a two-dimensional (2D) surface

An example of this translation of the world into two dimensions is the Mercator projection (see **Figure 2.4** on page 48), a map developed in 1569 by a Flemish cartographer known as Gerardus Mercator for use as a navigation aid—it was designed so that every straight line on the map was a line of constant direction. The Mercator projection encompasses the whole globe, and although it keeps the shapes of areas intact (that is, the continents look the way they should), the sizes (that is, the areas of land masses) are very distorted, and become more distorted the further you go from the Equator. For example, take a look at Greenland and Africa—they look to be about the same size on the map, even though Africa is really more than 13 times larger in area than Greenland. Similar problems can be seen with Antarctica and

FIGURE 2.4 The Mercator projection retains the shapes of countries and continents, but greatly warps their sizes.

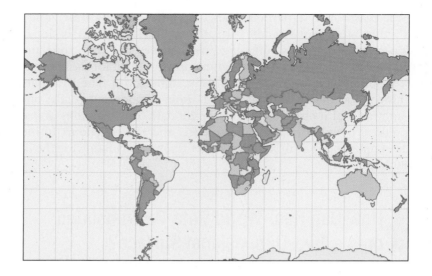

Russia, which (although they have the right shape) look overwhelming in land area compared with their actual size.

Think of a projection like this. Imagine a see-through globe of Earth with the various landforms painted on it, and lines of latitude and longitude clearly marked. If you could place a light bulb inside the globe and turn it on, it would cast shadows of the continents and the lines of latitude and longitude on the walls of the room. In essence, you've "projected" the three-dimensional world on the surfaces of the walls. However, to make a map projection, you'd have to wrap a piece of paper around the globe and preserve or copy (i.e. project) the lines on to that paper. Then, when you unrolled the paper, you'd have the globe translated onto the map.

The real process of mapmaking is a much more complicated mathematical translation of coordinates, but the basic concept is similar. A map projection is created through the use of a developable surface, or a flat area on which the locations from Earth can be placed. In this way, the locations on Earth's surfce get translated onto the flat surface. The three main developable surfaces (**Figure 2.5**) that are used in this fashion are a cylinder (which creates a cylindrical projection), a cone (which creates a conical projection), and a flat plane (which creates an azimuthal surface). The place where Earth "touches" the surface is called the point of tangency, which is where the map's distortion is at its least.

Like datums, there are numerous map projections available, each with its own specific uses and its own built-in distortions (see *Hands-On Application 2.2: Examining the Effects of Different Map Projections* for more information). Two projections that you'll often encounter with geospatial data are the Lambert Conformal Conic and the Transverse Mercator.

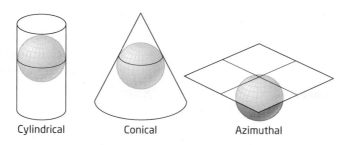

FIGURE 2.5 The three types of developable surfaces in map projections.

Cylindrical   Conical   Azimuthal

Lambert Conformal Conic

This is a conic projection, in which a cone intersects Earth at two parallels. The choice of these tangency lines is important, as the distortion on the map is reduced as you get closer to them and increases the farther you move away from them. The Lambert Conformal Conic projection is commonly used for geospatial data about the United States (and other east–west trending areas).

HANDS-ON APPLICATION 2.2

Examining the Effects of Different Map Projections

Different map projections misrepresent Earth in different ways, each one distorting certain features and representing other features accurately. To interactively examine the characteristics of different map projections, go to **http://indiemapper.com/app**. Indiemapper is a free online application that enables you to quickly make a variety of different maps and also examine how your map will appear in different projections. When you first start using Indiemapper, you'll have several choices as to the source of the data you want to map. For now, just choose "data library" and then select a theme to map (such as Water Consumption and Resources). Next, you'll have to choose an attribute to map (such as People), a map type (select Choropleth), and for now choose "no" when asked about data standardization. By clicking on Okay the data will be mapped. In order to see how the data looks in different projections, click on the Projections button in the upper-right of the screen. You'll

have 14 different choices of map projections that you can select from—choose a projection from the available options and then click Apply to see your data with a different map projection. Switch the projections to view the world map as Mercator, Lambert Conformal Conic, Sinusoidal, and Transverse Mercator, and then examine each one. Note that the Indiemapper controls allow you to zoom in and out from the map.

Expansion Questions

- How does each of these map projections alter the map as drawn? What kinds of distortions (shape, area, distance, or direction) appear to be built into each projection?

- Try some of the other map projection options (such as Orthographic, Bonne, and Robinson). What types of distortions and changes enter the map with each projection?

Transverse Mercator

This is a cylindrical projection, with the tangency being the intersection between the cylinder and Earth. Measurements are most accurate at that point, and the distortions are worst in the areas of the map farthest away from that point, to the east or west. It's considered "transverse" because the cylinder is wrapped around the poles, rather than around the Equator. Transverse Mercator is used most for north–south trending areas.

GCS coordinates can find locations across the entire globe, down to precise measurements of degrees, minutes, and seconds. However, latitude and longitude is not the only coordinate system used today—there are several coordinate systems that use a map projection to translate locations on Earth's surface to a Cartesian grid system using x and y measurements. Two grid systems widely utilized with geospatial data are UTM and SPCS.

 ## What Is UTM?

Universal Transverse Mercator (UTM) the grid system of locating coordinates across the globe

The **Universal Transverse Mercator (UTM)** grid system works by dividing the world into a series of zones, and then determining the x and y coordinates for a location in that zone. The UTM grid is set up by translating real-world locations to where their corresponding places would be on a two-dimensional surface (using the Transverse Mercator projection). However, UTM only covers Earth from between 84° N latitude and 80° S latitude (and thus can't be used for mapping the polar regions—a different system, the Universal Polar Stereographic Grid System, is used to find coordinates in the Arctic and Antarctic).

UTM zone one of the 60 divisions of the world set up by the UTM system, each zone being 6 degrees of longitude wide

The first step in determining UTM coordinates is to find which **UTM zone** the location is in. UTM (**Figure 2.6**) divides the world into 60 zones, each one 6 degrees of longitude wide and set up as its own cylindrical projection. These zones are numbered 1 through 60, beginning at the 180th meridian and moving east. For instance, Mount Rushmore, outside of Keystone, South Dakota, is located in Zone 13. Beijing, China, is located in Zone 50. Sydney, Australia, is in Zone 56.

easting a measurement of so many units east (or west) of some principal meridian

northing a measurement of so many units north (or south) of a baseline

The next step is to determine the x and y grid coordinates of the point you want to locate. Unlike GCS, which uses degrees, minutes, and seconds, UTM uses meters as its unit of measurement. UTM's x and y coordinates are referred to as the **easting** and **northing** coordinates. The UTM northing is found by counting the number of meters the point is north of the Equator. For instance, a particular location at Mount Rushmore is located 4,859,580 meters to the north of the Equator.

Locations in the southern hemisphere handle northing a little differently, since UTM doesn't use negative values. For southern hemisphere locations, measure the distance in meters (as a negative value) from the

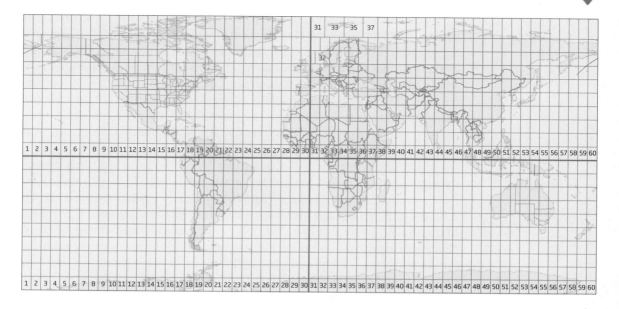

Equator to the particular location, then add 10,000,000 meters to the number. For example, a location at the Opera House in Sydney, Australia (in the southern hemisphere), has a northing of 6,252,268 meters. This indicates it would be approximately 3,747,732 meters south of the Equator ($-3,747,732 + 10,000,000$). Another way of conceptualizing this is to assume that (for southern hemisphere locations) rather than measuring from the Equator, you're measuring north of an imaginary line drawn 10,000,000 meters south of the Equator. This extra 10 million meters is referred to as a **false northing** since it's a value set up by the system's designers to avoid negative values.

The next step is to measure the point's easting. Unlike GCS, there is no one Prime Meridian to make measurements from. Instead, each UTM zone has its own central meridian which runs through the zone's center. This central meridian is given a value of 500,000 meters, and measurements are made to the east (adding meters) or west (subtracting meters) from this value. For example, a specific spot at Mount Rushmore is located 123,733.22 meters to the east of Zone 13's central meridian—so the easting for this location at Mount Rushmore is 623,733.22 meters ($500,000 + 123,733.22$). A location in the town of Buffalo, Wyoming, is also in Zone 13, and has an easting value of 365,568 meters. This indicates that this location is approximately 134,432 meters to the west of Zone 13's central meridian ($500,000 - 134,432$). No zone is more than 1,000,000 meters wide, so no negative values will be assigned to eastings in a zone. This value for easting is referred to as a **false easting**, since the easting calculation is based on an imaginary central meridian in each zone.

false northing a measurement made north (or south) of an imaginary line such as is used in measuring UTM northings in the southern hemisphere

false easting a measurement made east (or west) of an imaginary meridian set up for a particular zone

When UTM coordinates are written, the false easting, northing, and zone must all be specified. If a false northing is used (that is, the coordinates are found in a location south of the Equator), then a notation is included, explaining that the coordinates are from the southern hemisphere. For example, the UTM coordinates of a location at Mount Rushmore would be written: 623733.22 E, 4859580.26 N, Zone 13. (See **Figure 2.7** for a simplified graphical example of finding the location at Mount Rushmore's coordinates with UTM. Also, see *Hands-On Application 2.3: Converting from Latitude/Longitude to UTM*).

UTM values are also utilized as the basis for a related projected coordinate system, the **United States National Grid (USNG),** also known as the military grid reference system (MGRS), which is used extensively by the military and by disaster relief organizations like the Federal Emergency Management Agency (FEMA). USNG is particularly well-suited to emergency management and military applications since it uses a single identifier for the northing, easting, and zone information for every location—ensuring that the same coordinates are labeled and used in the same way by everyone. USNG measures coordinates in meters, using values derived from UTM northing and easting. It also uses UTM zone numbers in conjunction with smaller 100,000 meter gridded zones to derive location information. Depending on the number of digits computed for the position measurement, USNG accuracy ranges between 1000 meters and 1 meter.

For example, the UTM coordinates of our previously computed Mount Rushmore location were 623733.22 E, 4859580.26 N, Zone 13. In the USNG system, using 1 meter precision, the coordinates of the same location would

United States National Grid (USNG) a grid system of identifying locations in the United States

FIGURE 2.7 Determining the UTM coordinates (easting, northing, and zone) of a location at Mount Rushmore (a simplified representation).

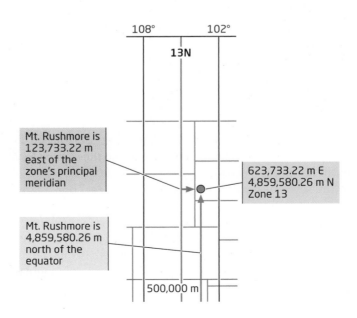

108°   102°

13N

Mt. Rushmore is 123,733.22 m east of the zone's principal meridian

623,733.22 m E
4,859,580.26 m N
Zone 13

Mt. Rushmore is 4,859,580.26 m north of the equator

500,000 m

HANDS-ON APPLICATION 2.3

Converting from Latitude/Longitude to UTM

GCS and UTM are completely different systems, but there are utilities to make conversion from one system to another simple and easy. One of these is available at **www.earthpoint.us/Convert.aspx**. This Web resource can be used to convert from GCS to UTM. Input a coordinate for latitude and one for longitude, and click the Calc button. The corresponding UTM easting, northing, and zone will be computed for you and returned in a box below the inputs (along with other coordinate system conversions as well). You can also click View on Google Earth and a marker will be placed in Google Earth at the coordinates you specify.

First, you're going to need GCS coordinates. A good source of GCS coordinates can be found at **http://lat-long.com**. This Website allows you to input the name of a location (try the Empire State Building) and the state (New York) or county, and it will return the latitute and longitude (in both DD and

DMS). Copy those coordinates into the converter, and click the Calc button to determine the UTM coordinates of the Empire State Building.

Expansion Questions

- What are the UTM coordinates for the latitude and longitude coordinate given for a location at the Empire State Building?

- What are the UTM coordinates for the latitude and longitude coordinate given for the Jefferson Memorial in the District of Columbia, the Space Needle in Washington, and the Transamerica Pyramid in California?

- Use the two Websites in conjunction with each other to determine the UTM coordinates of prominent locations around your local area. What locations did you identify, and what are their UTM coordinates?

be written as a single string: 13TFJ2373359580—"13T" represents UTM Zone 13, "FJ" identifies which one of smaller 100,000-meter grid zones the location is in, "23733" represents the value for easting (a shortened version of a UTM easting value of 623733), and "59580" represents the value for northing (a shortened version of a UTM northing value of 4859580).

 ## What Is SPCS?

The **State Plane Coordinate System (SPCS)** is another grid-based system for measuring and determining coordinates. It was designed back in the 1930s, before computers. Today, SPCS is used as a coordinate system for United States data, especially city and county data and measurements. SPCS is a projected coordinate system that uses a translation method similar to UTM to set up a grid coordinate system to cover the United States. SPCS data was originally measured using the NAD27 datum, but now there are datasets using the NAD83 datum.

Like UTM, SPCS requires you to specify a zone, an easting, and a northing in addition to the state in which the measurements are being made. SPCS

> **State Plane Coordinate System (SPCS)** a grid-based system for determining coordinates of locations within the United States

SPCS zone one of the divisions of the United States set up by the SPCS

divides the United States into a series of zones. Unlike UTM zones, which follow lines of longitude, **SPCS zones** are formed by following state or county boundaries. A state is often divided into multiple zones—for example, Ohio has two zones: Ohio-North and Ohio-South. Nevada has three zones: Nevada-West, Nevada-Central, and Nevada-East. Texas has five zones, and California has six, while all of Montana is represented by a single zone (**Figure 2.8**). To determine the coordinates for a point in SPCS, you must first know which state the point is in, and then you must identify the zone within that state. For example, to map data from Miami, the state is Florida and the zone is Florida-East, while Cleveland is in the state of Ohio and in the Ohio-North zone.

The next step is to determine the northing and easting. As with UTM, both of these are measured north and east from a predetermined baseline and meridian, but in SPCS, measurements are made in feet (using the NAD27 datum), or in feet or meters (using the NAD83 datum). Each of the SPCS zones has its own baseline and its own principal meridian. Two projected versions of SPCS are used—the Transverse Mercator version, which is usually used for states that have a north–south orientation (like Indiana), and the Lambert Conformal Conic version, which is usually used for states that have a more east–west orientation (like Ohio). Baselines and principal meridians will be positioned differently, depending on which version is being used.

FIGURE 2.8 SPCS zones are formed by following state or county boundaries.

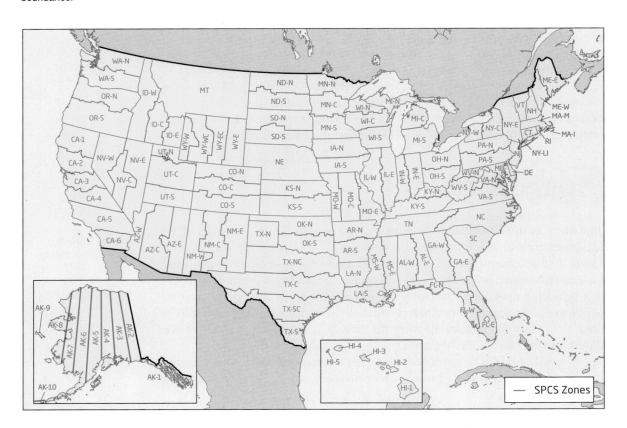

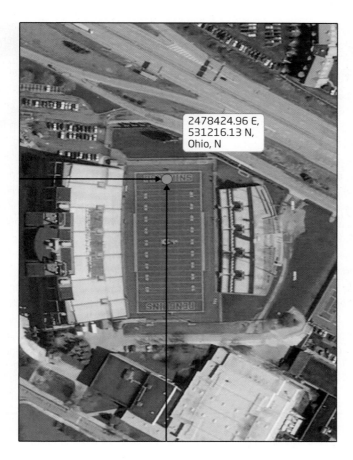

2478424.96 E,
531216.13 N,
Ohio, N

FIGURE 2.9 A simplified representation of SPCS measurements and coordinates for a location on the YSU football field in Youngstown, Ohio.
[Source: Ohio Geographically Referenced Information Program (OGRIP), Ohio Statewide Imagery Program (OSIP), April 2006]

Baselines are placed a certain distance under the zone's southern border and northings are measured from there (to ensure all positive values). In the NAD27 version, states using the Transverse Mercator format have their zone's origin 500,000 feet to the west of the principal meridian, while states using the Lambert Conformal Conic projection have their zone's origin 2,000,000 feet to the west of the meridian. Coordinates are determined by measuring distance using these false origin points, creating a false northing and a false easting. Thus, as each of the zones uses its own coordinate measurements, each zone has a different false northing and false easting from all other zones. Note that in the NAD83 version of SPCS, different values may be used for each zone's origin.

SPCS coordinates are referenced by listing the false easting, false northing, state, and zone. For example, the SPCS (NAD83 feet datum) coordinates of a location on the football field of Stambaugh Stadium in Youngstown, Ohio, are approximately: 2478424.96 E, 531216.31N, Ohio, N (North Zone; see **Figure 2.9** for a simplified graphical representation of SPCS measurements—and see also *Hands-On Application 2.4: Using the State Plane Coordinate System* for a utility that will convert latitude and longitude coordinates into SPCS values).

HANDS-ON APPLICATION 2.4

Using the State Plane Coordinate System

Like the conversion tool in *Hands-On Application 2.3: Converting from Latitude/ Longitude to UTM*, there are Web-based utilities to convert GCS coordinates to their SPCS counterparts. A good one is available free online at **www .earthpoint.us/StatePlane.aspx,** a Web resource similar to the one used in *Hands-On Application 2.3.* Enter the latitude and longitude coordinates of a location, click the Calc button, and it will calculate the SPCS coordinate for that location (it will also convert the other way, going from SPCS to GCS). Again, use the GCS coordinates you received from **www.lat -long.com** to compute the SPCS equivalents for them. Note that you also have to specify the SPCS zone when entering the latitude and longitude of the points. When calculating the SPCS coordinates, you may want to use the View option on Google Earth to verify your results.

Expansion Questions

- What are the SPCS coordinates for the latitude and longitude coordinate given for a location at the Empire State Building (New York Long Island Zone)?

- Working from their given latitude and longitude coordinates, what are the SPCS coordinates for the Jefferson Memorial in the District of Columbia (which uses Maryland's Zone), the Space Needle in Washington (Washington North Zone), and the Transamerica Pyramid in California (California Zone 4)?

- Use the two Websites in conjunction to determine the SPCS coordinates of prominent locations around your local area (check Figure 2.8 to identify which SPCS zone you are located in). What locations did you choose, and what are their SPCS coordinates?

Chapter Wrapup

This chapter provided a look at what goes into setting up coordinates and references for data used with geospatial technology. Whether you're reading coordinates from a GPS receiver or making measurements from a map, knowing how that geospatial information is put together and referenced has an impact on how you're able to make use of that data, especially if you're trying to fit several types of geospatial data together. Check out *Projection and Coordinate System Apps* for some downloadable apps and *Coordinate Systems and Projections in Social Media* for some YouTube videos that deal with this chapter's concepts.

In the next chapter, we're going to switch gears and look at how you can take your data (whether it's a remotely sensed image, some information created using GIS, or a paper map of the local fairgrounds) and get it to match up with all the other geospatial data you have. Before that, *Geospatial Lab Application 2.1: Coordinates and Position Measurements* will have you start applying the coordinate and measurement concepts described in this chapter with some more work with Google Earth.

Important note: The references for this chapter are part of the online companion for this book and can be found at **www.macmillanhighered.com /shellito/catalog.**

Projection and Coordinate System Apps

Here's a sampling of available representative coordinate systems apps for your phone or tablet. Note that some apps are for Android, some are for Apple iOS, and some may be available for both.

- **Coordinates—Calculate and Convert a Position:** An app that calculates many types of coordinates (latitude and longitude, UTM, MGRS, etc.) for a location
- **Coordinate Tools Free:** An app that converts coordinates and plots them on a map
- **Globe Earth 3D Pro:** An app that shows a time zone map of the world on a virtual globe and also allows you to view flags and hear national anthems of countries from around the world
- **Latitude/Longitude Convert:** An app that converts from DMS to decimal degrees and also plots the location using Google Maps
- **Map Projections—Ultimate Guide for Cartography:** An app that allows you to view a world map in many different map projections and also provides information about each projection
- **MyLocations:** An app that computes the latitude and longitude of your current location
- **World Clock—Time Around the World Free:** An app that allows to you to select locations on a map and see its time zone as well as its current time

Coordinate Systems and Projections in Social Media

Here's a sampling of some representative YouTube videos related to this chapter's concepts.

You Tube On YouTube, watch the following videos:

- **Latitude and Longitude** (a video demonstrating concepts of latitude and longitude using Google Earth):
  www.youtube.com/watch?v=swKBi6hHHMA
- **Map Projections** (an Esri video showing map projection concepts):
  www.youtube.com/watch?v=2LcyMemJ3dE
- **Maps That Prove You Don't Really Know Earth** (a video about the distortions in the Mercator projection):
  www.youtube.com/watch?v=KUF_Ckv8HbE

Key Terms

datum (p. 40)
datum transformation (p. 41)
decimal degrees (DD) (p. 43)
degrees, minutes, and seconds
  (DMS) (p. 42)
easting (p. 50)
ellipsoid (p. 40)
Equator (p. 42)
false easting (p. 51)
false northing (p. 51)
geodesy (p. 40)
geographic coordinate system
  (GCS) (p. 41)
geoid (p. 40)
great circle distance (p. 43)
International Date Line (p. 47)
latitude (p. 41)

longitude (p. 42)
map projection (p. 47)
NAD27 (p. 41)
NAD83 (p. 41)
northing (p. 50)
Prime Meridian (p. 42)
SPCS zone (p. 54)
State Plane Coordinate System
  (SPCS) (p. 53)
time zones (p. 43)
United States National Grid
  (USNG) (p. 52)
Universal Transverse Mercator
  (UTM) (p. 50)
UTM zone (p. 50)
WGS84 (p. 41)

2.1 Geospatial Lab Application

Coordinates and Position Measurements

This chapter's lab will continue using Google Earth Pro (GE), but only to examine coordinate systems and the relationships between various sets of coordinates and the objects they represent in the real world. In addition, you'll be making some measurements using GE and using some Web resources for comparing measurements made by using different coordinate systems.

Note that this lab makes reference to things like "coordinates for the White House" or "coordinates for Buckingham Palace"—these represent measuring a set of coordinates at one specific location at these places and are used as simplifications of things for lab purposes.

Objectives

The goals for you to take away from this lab are:

▶ To set up a graticule of lines in GE

▶ To locate places and objects strictly by their coordinates

▶ To make measurements across long and short distances and then compare measurements with surface distance calculations

▶ To translate latitude/longitude coordinates into UTM

Using Geospatial Technologies

The concepts you'll be working with in this lab are used in a variety of real-world applications, including:

▶ Emergency rescue and relief operations, which need exact coordinates in order to locate incidents (like stranded civilians and downed power lines) and send the closest available emergency assistance

▶ Outdoor event planners, which need to know the coordinates of every entrance and exit to all buildings, temporary structures, and enclosures, as well as the locations of essential resources and amenities, like first aid rooms and other emergency services (on- and off-site), handicapped-accessible ramps, and restrooms (for both the disabled and the able-bodied)

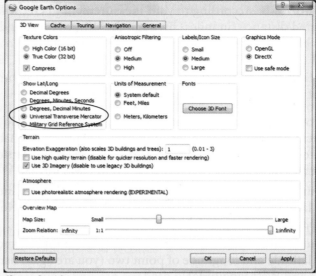

[Source: Google]

2. Zoom out to see the whole world—you'll also see that the graticule has changed from a web of latitude/longitude lines to the UTM grid draped over Earth. You should see the UTM zones laid out across the world, numbered from 1 through 60.

> **Question 2.8** What UTM zone is Buckingham Palace located in? What UTM zone is the White House located in?

3. Double-click on the **White House** placemark (in the **Places** box) and GE will rotate and zoom back to the White House.

4. Scroll the mouse around the White House area. You'll see the new coordinates appear at the bottom of the view, but this time they'll be the zone, easting, and northing measurements. Open the **Properties** of the **White House** placemark by right-clicking on it. The UTM easting, northing, and zone will appear.

> **Question 2.9** What are the UTM coordinates of the White House?

5. In the **Places** box, double-click on the **Lincoln Memorial** placemark, and GE will zoom and rotate to it.

> **Question 2.10** What are the UTM coordinates of the Lincoln Memorial?

6. UTM coordinates are measured in meters rather than degrees of latitude or longitude —this enables an easier method of determining coordinates. If you move the mouse east, your easting will increase,

and moving it west will decrease the easting. The same holds true for northing—moving the mouse north will increase the value of northing, while moving it south will decrease the northing value.

7. Change the view so that you can see both the White House and the Lincoln Memorial. Using your answers for Questions 2.9 and 2.10 (and perhaps the Ruler tool), answer Question 2.11.

> **Question 2.11** How many meters away to the north and east is the White House from the Lincoln Memorial?

8. UTM can also be used to determine the locations of other objects. Return to the White House placemark again. Answer Questions 2.12 and 2.13.

> **Question 2.12** What object is located approximately 912 meters south and 94.85 meters east of the White House?

> **Question 2.13** What are the full UTM coordinates of this object?

Closing Time

This lab served as an introduction for using and referencing coordinates and measurements, using Google Earth. Chapter 3's lab will use a new software program, but we'll return to Google Earth in Chapter 8, and we'll continue to use its functions in other chapters. You can now exit GE by selecting Exit from the File pull-down menu. There's no need to save any data (or Temporary Places) in this lab.

Getting Your Data
to Match the Map

Reprojecting, Georeferencing, Control Points, and Transformations

Because there are so many datums and map projections out there, it's important to know which one you're using when you're making measurements. Say, for instance, you have two sets of geospatial data about your town that you need to bring together for analysis on one map (that is, a street map with building footprints). Each of these pieces of data are in a different projection and coordinate system, and each use a different datum—the street map is in State Plane NAD27 feet and the buildings are in UTM NAD83 meters. When you put these things together, nothing is going to match up—the streets and buildings are going to be in different places in relation to each other because each one uses a different basis for its coordinates.

It's important to note when you're dealing with geospatial data that it can be constructed using a variety of different datums, coordinate systems, and projections. With all the different variations involved with these items, it's best to try and work with all geospatial data on the same basis (that is, to use the same datum, coordinate system, and projection). If you have multiple data layers in different map projections and you try to overlay them, those pieces of data won't properly align with each other (**Figure 3.1**).

How Can You Align Different Geospatial Datasets to Work Together?

Back in Chapter 2, we discussed the concept of having to transform data measured from different datums to a common datum so that everything would align correctly. It's best to **reproject** your data (or transform all of the datasets to conform to a common datum, coordinate system, and projection) before moving forward. Reprojecting data will translate the coordinates, map projections, and measurements from one system (for

reproject changing a dataset from one map projection (or measurement system) to another

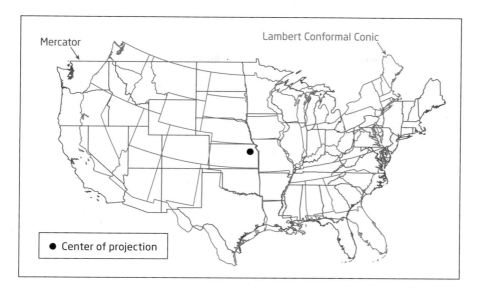

FIGURE 3.1 Measuring the same data with different projections produces maps that don't match.

instance, SPCS) to another (such as UTM). Many geospatial software packages give you the capability to take your initial dataset (like your street map in State Plane NAD27) and reproject it to create a new dataset with new coordinates (so now you have a new street map measured in UTM NAD83). This can also be done to change data from one map projection into another. In some cases, the software can "project on the fly"—calculate the necessary transformations without actually altering the base dataset. Whatever method you choose, you'll save yourself a lot of headaches later on by using a common basis for the various geospatial datasets you use. This is especially important when you're trying to get several different types of data to match up.

THINKING CRITICALLY
WITH GEOSPATIAL TECHNOLOGY 3.1

What Happens When Measurements Don't Match Up?

It's not uncommon to have multiple data sources as in the previous scenario, with different projections, coordinate systems, and datums that all need to be used together in a project. If everything doesn't align properly, you can end up with points being plotted in the wrong location with respect to the base map they're being located on, or street data that doesn't line up with a satellite image of the same area. What kind of effects could this type of simple data mismatch have on real-world companies or government agencies? What would happen if a road renovation crew is using sewer data and street data that don't align properly because they use different datums or projections? What could happen if the misalignment of data caused one dataset to be significantly off from its proper place on a map? Similarly, what effects could such misalignments have on projects like shoreline construction, zoning maps, or species habitat mapping?

Problems are going to arise if one (or more) of your pieces of data doesn't have coordinates assigned to it, or if you don't have any information as to what projection or datum was used to construct it. Suppose you scan an old map of your town into your computer, and you want to add other layers to it (like a road network or parcel layer) in a GIS to see how things have changed since that old map was made. You add the layers into the GIS, but the scanned historic map won't match up with any of your data. This is because that image doesn't contain any **spatial reference** (like a real-world coordinate system). Although it might show features like old roads or the dimensions of particular buildings, the only coordinate system the old map possesses is a grid that begins at a zero point at the lower left-hand corner and extends to the dimensions of the scanned image. For instance, coordinates would not be referenced by feet or degrees, but by whatever units you're measuring (such as 200 millimeters over from the left-hand corner and 40 millimeters up from the left-hand corner).

> **spatial reference** the use of a real-world coordinate system for identifying locations

Items in the image can't be referenced by latitude and longitude because those types of coordinates haven't been assigned to anything in the image. Similarly, when you add this scanned image as a layer in a GIS, the software won't know how to match it up with other data layers that do have referenced coordinate systems. Without knowing "where" your scanned image matches up with the rest of the data, you won't be able to use this image as geospatial data or reproject it to match your other data. The same concept holds for other types of data used with geospatial technology: If all types of data automatically had the correct spatial reference when they were created, things would be a lot easier. So what you'll have to do is match the data to the map, or assign a real-world set of coordinates to your data so it can sync up with other geospatial data. In other words, you're going to have to georeference your data.

 ## What Is Georeferencing?

> **georeferencing** a process of aligning an unreferenced dataset with one that has spatial reference information

Georeferencing is the process of aligning an unreferenced dataset with one that has spatial reference information. Other terms used for similar processes include "image to map rectification," "geometric transformation," and "registration." Suppose you're hosting a meeting at your school and you want the invitation to include a map to direct people not only to the school, but also to your specific building on the campus. You go to your school's Website and download a map that shows the campus boundaries and specific building locations within those boundaries. So far so good—but you want to place that downloaded map in the larger context of a map of the city, so that people will be able to see where your campus is in relation to the surrounding roads and neighborhoods. The problem is that while the data you have of the surrounding area (whether aerial photos, satellite imagery, or GIS

HANDS-ON APPLICATION 3.1

David Rumsey Historical Map Collection

Information about this ongoing project for the collection and preservation of historic maps can be found at **www.davidrumsey .com**. The maps on the Website are part of a collection that stretches from the seventeenth century to the present. Although there are many ways of viewing these maps, some have been georeferenced to make them match up with current geospatial imagery. By looking at georeferenced maps in relation to current data, you can see how areas have changed over time.

On the Website, select the link for Google Maps under the Quicklinks section, then click on the map that appears in your Web browser. A new window will open (also available at **http://rumsey.geogarage.com**) showing you a map of the world, along with several symbols that indicate the presence of georeferenced maps. Clicking on a symbol will open a dialog box about its corresponding map and selecting the Open Image link in the box will show you the georeferenced map overlaid on the world map. A slider bar in the upper right-hand corner will allow you to adjust the transparency to see both the historic map and the current imagery together. Check out some of the map icons for a variety of places around the globe and see how this historical information syncs up with modern boundaries and images.

The Rumsey Historical Map collection is also available as a layer in Google Earth itself. To use the maps as another Google Earth layer (like the roads or buildings), open the Gallery options, open the Rumsey Historical Maps option, and click on the Map Finder icon. You'll see a new icon representing the maps appear in Google Earth that you can use to view the georeferenced maps. There's also an "About the Maps" pop-up where you can read more information about the map collection and how these maps were scanned and georeferenced for use in Google Earth.

Expansion Questions

- From examining some of the historic maps in both the Web interface and on Google Earth, how do the areas represented match up with their modern-day counterparts?

- Examine the "Chicago 1857" map. How does this historic map match up with modern-day Chicago? You can use Google Earth to adjust the transparency of the map by using the tools in the Places Box (select the slider bar and move it about halfway over)—this will enable you to see current imagery of Chicago beneath the georeferenced layer. Are there any areas that have changed significantly between then and now?

- Examine the "United States 1833" map (the "Eagle" map with its icon located in Ohio), and make the layer semi-transparent. How does this map of the United States match up with current boundaries and coastlines?

data layers) contains real-world spatial references, your downloaded map is just a graphic or PDF without any spatial reference at all. The campus map won't match up with the spatially referenced data, and your plan of a snappy geospatial-style invitation will be ruined. Or will it? If you could just georeference that campus map, everything would fit together. Well, the good news is, you can. The georeferencing procedure requires a few steps of work to proceed, but you'll be able to make your map (see *Hands-On Application 3.1: David Rumsey Historical Map Collection* and **Figure 3.2** on page 70 for an example of an online georeferenced map collection).

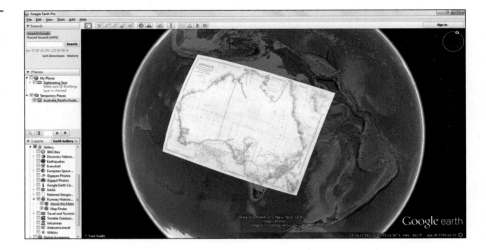

How Can Data Be Georeferenced?

When data (such as an old scanned map or other unreferenced data) is georeferenced, each location in that data is aligned with real-world coordinates. So
to get started, you're going to need a source of real-world coordinates to align
your unreferenced data against. The choice of a source to use for the matching
is important, because you have to be able to find the same locations on both
the unreferenced image and the source. This is crucial to georeferencing,
since the process needs some common locations to "tie" the unreferenced
data to the source data. In the georeferencing process, these spots (where you
can find the coordinates on the source data) are referred to as **control points**
(sometimes called Ground Control Points or GCPs). Field work, such as surveying or making GPS measurements (see Chapter 4), can provide sources
for these control points, but you can avoid having to go into the field by finding common locations on the unreferenced image and the source.

control points point
locations where the
coordinates are known,
used to align the
unreferenced image to the
source

   Clearly, the source should be something on which you can find the same
locations as the ones you can see in your unreferenced data. For instance, a
low-resolution satellite image isn't going to be a good match with the detailed
map of your campus buildings. Trying to match individual locations like
buildings to an image like this isn't going to work simply because the resolution is too low. A high-resolution image (where you can determine features of
buildings) would be a more useful choice for this specific task. Datasets such
as road networks (see Chapter 8) or orthophotos (see Chapter 9) are also good
sources to use in finding these common points. In addition, your source data
should be a similar projection to your unreferenced data. If the projections are
different, some features will match up well and others will end up distorted.
Whatever the source, the georeferencing process will align an unreferenced
image against the projection and coordinates of the source map (**Figure 3.3**).

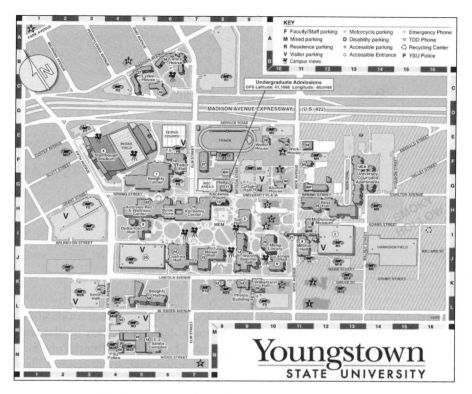

FIGURE 3.3 The map of Youngstown State University is unreferenced. The fully-referenced aerial image of the same area is a potential source for the references the map requires. *[Source: Mahoning County Enterprise GIS, April 2004]*

Where to select the control points is the key to the whole georeferencing procedure. Because control points are the objects that are going to connect the unreferenced data to the source data, they have to be placed in the same location in both sets of data; real-world coordinates for these points will be required. By finding these same locations on the source (that has real-world coordinates) and on the unreferenced data (that doesn't have real-world coordinates), we can start to match the two datasets together. Features that you can clearly find in the same spot in the unreferenced data and the source data will be good choices for the placement of control points. For instance, the intersections of roads and the corners of buildings make good control point locations, because they're well-defined and can be found on both sets of data. These are also locations that aren't likely to change over time, so even if the unreferenced data and the source aren't from the same time period, they're usually valid as control point selections. Poor locations for control point selection spots would be those that can't be accurately determined between the two sets of data, either because they're poorly defined or because they might be in different places at different times—for example, the top of a building, the center of a field, the shoreline of a beach, or a shrub on a landscaping display. See **Figure 3.4** for some examples of good and poor control point location selections on the campus map.

Another consideration to keep in mind when you're selecting control point locations is to not bunch them all together in one area of the data and ignore the rest of the map. Placing all of the control points in one-quarter of the image is going to ensure that one section of the map is well referenced, but the rest of the map may not fit very well. Make sure to spread out the control points to representative locations throughout the data so that all areas fit well in relation to their newly referenced spots. You'll want to use several control points in the process—some georeferencing processes may only need a few, whereas others may require many more.

FIGURE 3.4 Good (in red) and poor (in blue) locations for control point selection. *[Source: Mahoning County Enterprise GIS/Esri]*

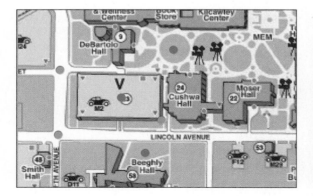

The minimum number of control points required to fit the unreferenced image to the source is three. Three control points are often enough to enable you to make a rough fit of the image to the source. With this initial fit, some programs (for instance, MapCruncher, the program used in this chapter's Geospatial Lab Application) will attempt to match up the next control point location with a spot you select on the source. As you add more control points, the fit between the unreferenced data and the source should (hopefully) improve. Proper selection of control points is critical for georeferencing, since the procedure needs to have some sort of real-world coordinates to match to, and it's these points that will be used to calculate coordinates for the other locations in the image. Often, you'll find that deleting a control point, adding another one, and tinkering with the placement of others is necessary to get the best fit of the data (see *Hands-On Application 3.2: Online Georeferencing Resources* for several online examples of placing control points in relation to old maps). Once your data is properly aligned by using control points, you can then transform it to have the same coordinate system information as the source.

HANDS-ON APPLICATION 3.2

Online Georeferencing Resources

Georeferencing of maps has become a popular online activity. Several free georeferencing tools are available for anyone who wants to create spatially referenced versions of maps, especially old ones. You can select control points and transform the old maps to match up with current ones. For instance, the British Library is encouraging people to use online georeferencing tools to align and save old maps so that they can become part of the collection. Check out the following online resources to become part of the georeferencing effort (some sites will ask you to register for free before you can do any georeferencing, but you can usually view completed maps without registering):

1. The British Library Online Gallery: **www.bl.uk /maps**

2. David Rumsey Map Collection Georeferencer: **www.davidrumsey.com/view/georeferencer**

3. Georeferencer: **www.georeferencer.com**

4. The National Library of Scotland Map Georeferencer: **http://maps.nls.uk/projects /georeferencer**

5. The New York Public Library Map Warper: **http://dev.maps.nypl.org/warper**

6. WorldMap Warp at Harvard University: **http:// warp.worldmap.harvard.edu**

Expansion Questions

*   What kinds of historic georeferenced maps are available on these sites? How do the already georeferenced maps match up with modern maps?

*   What kinds of difficulties did you find in trying to match unreferenced historic maps to modern maps (especially when it comes to selecting control points)?

How Is Data Transformed to a Georeferenced Format?

rotated when the unreferenced image is turned during the transformation

skewed when the unreferenced image is distorted or slanted during the transformation

scaled when the unreferenced image is altered during the transformation

translated when the unreferenced image is shifted during the transformation

affine transformation a linear mathematical process by which data can be altered to align with another data source

Take another look at Figure 3.3, which shows both the unreferenced campus map and the referenced source image. In order to make the unreferenced map align properly with the source image, there are a number of actions that have to be taken with the map. First, check out the football stadium (Stambaugh Stadium), which you'll find in the upper left quarter of both the map and the referenced image. In order to match the basic position and alignment of the stadium as it appears on the map with the stadium in the image, it must be rotated about 45 degrees clockwise. As you look at the other features, it becomes clear that the whole map must be rotated by some amount, since the image is oriented to the north and the map is oriented to the northeast. It also becomes apparent that the map might have to be stretched or skewed to match up with the image.

When unreferenced data is transformed or "rectified," it's warped to make it match up with the source. It can be **rotated** (turned, like changing the orientation of the map), **skewed** (by pulling the image, like at a slant), **scaled** (by altering the size and/or dimensions of the map to match the image), and/or **translated** (by altering the location and placement of the map). See **Figure 3.5** for examples of the effects of each of these transformations. This transformation of the data changes it to align with the new coordinate system. There are a number of transformation methods that are used, but a common one is the **affine transformation**, which will properly rotate, skew, scale, and translate the data.

The affine transformation calculates real-world X and Y coordinates for each unreferenced x and y coordinate. The formulae used for this are:

$$X = Ax + By + C$$

and

$$Y = Dx + Ey + F$$

FIGURE 3.5 Four ways that unreferenced data can be warped to align it with the source.

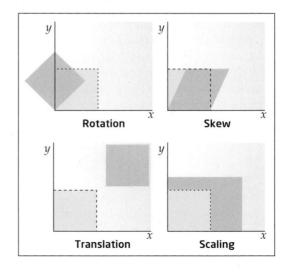

where A, B, C, D, E, and F are the values calculated internally by the procedure and applied to the mathematical formulae as coefficients (and also control the properties like skewing and rotation). The affine transformation is a first-order transformation, which is the type of transformation that's usually utilized in this procedure. The end result is that unreferenced data will be turned into spatially referenced data.

Once you select some control points, the software runs the transformation and the data is georeferenced. Everything should work just fine. However, there are two questions that you should be asking yourself: "How do I know it worked?" and, more importantly, "How do I know the results are any good?" After all, just because the unreferenced data was transformed doesn't necessarily mean that it's a good match with the source. Since you know the coordinates of each place where you've put a control point, you should be able to match up those locations and see how closely your selected control points match the source. **Figure 3.6** shows a report from a software program that measures error values for each control point.

A transformation procedure uses a value called **root mean square error (RMSE)** as an indicator of how well the now-referenced data matches with the source. After the initial set of control points has been chosen and the unreferenced data is aligned with the source, coordinates are computed for locations of the selected control points. The known coordinate location of each control point (in the source) is compared with the newly computed value for its corresponding location in the unreferenced data. If the control points are in exactly the same location in both the unreferenced data and the source, then the difference in their coordinates will be zero. However, the chances are that when placing coordinates—even if you select good locations like road intersections—the two points aren't going to match up exactly, though they might be very close. The software will examine all error values for all control points and report back a value for the RMSE.

> **root mean square error (RMSE)** an error measure used in determining the accuracy of the overall transformation of the unreferenced data

The smaller the differences between where you put the control point on the unreferenced data (the estimated location) and its true location on the source data, the better the match between them will be. Thus, the lower the overall RMSE is, the better the transformation has worked. Usually, the georeferencing program will allow you to examine the difference between each

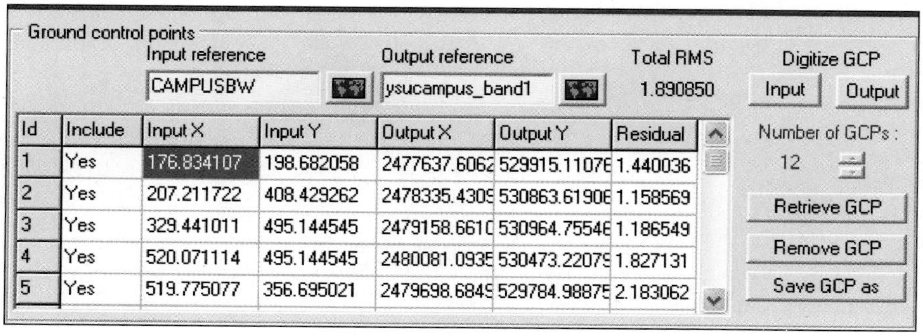

Id	Include	Input X	Input Y	Output X	Output Y	Residual
1	Yes	176.834107	198.682058	2477637.6062	529915.1107€	1.440036
2	Yes	207.211722	408.429262	2478335.4309	530863.6190€	1.158569
3	Yes	329.441011	495.144545	2479158.6610	530964.7554€	1.186549
4	Yes	520.071114	495.144545	2480081.0935	530473.22079	1.827131
5	Yes	519.775077	356.695021	2479698.6849	529784.98875	2.183062

Ground control points. Input reference CAMPUSBW. Output reference ysucampus_band1. Total RMS 1.890850. Digitize GCP: Input Output. Number of GCPs: 12. Retrieve GCP. Remove GCP. Save GCP as.

FIGURE 3.6 Some control point coordinate locations in the unreferenced map and the referenced image, and the computed error term for each one. [Source: Clark Labs, Clark University, IDRISI Taiga]

point, so that you can see how closely your control point selections matched each other. If the RMSE is too high, it's a good idea to check the difference in coordinates for each point—if a few of them are significantly off, they can be adjusted, changed, or deleted, and new points can be selected to try to get a better fit. If some parts of the data match well and others don't, perhaps the control points need to be moved to other places to balance out the alignment better, or perhaps other control points need to be added or removed.

When you're georeferencing an image, new locations for the image's pixels will need to be calculated, and in some cases new values for the pixels will need to be generated as well. This process is referred to as *resampling*. The end result of a good georeferencing process will have your unreferenced data properly aligned with the source data (**Figure 3.7**). Maps that were not referenced can now match up with current maps and datasets (see *Hands-On Application 3.3: Georeferenced Historic Maps and the Spyglass* and *Hands-On Application 3.4: An Overview of the Georeferencing Process in ArcGIS* for examples). Georeferencing can also be performed on other data created without a spatial reference, such as unreferenced drawings and plans. A related procedure can also be used for remotely sensed images that need to be corrected.

FIGURE 3.7
Georeferencing the campus map: A detail from the end result.
[Source: Clark Labs, Clark University, IDRISI Taiga]

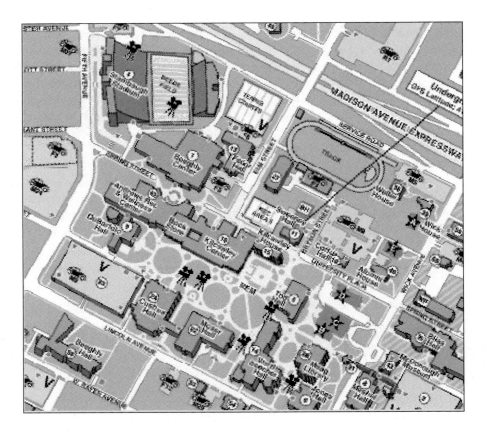

THINKING CRITICALLY WITH GEOSPATIAL TECHNOLOGY 3.2

What Happens When the Georeferencing Is Wrong?

Think about this scenario. Surveying is being done for a new beachfront development, and a base map of the area will be created (using GIS) from current high-resolution remotely sensed imagery. Because the base map (and all layers created from it, such as property boundaries, utilities, and local land cover) will be derived from it, this initial layer is critical. However, something goes wrong in the georeferencing process of the image used to create the initial base map, and it has incorrect coordinates—and then each layer created from it has incorrect coordinates. When on-site GPS coordinates and measurements are reprojected to match the new base map, they'll also be inaccurate, since they use an incorrectly referenced base map. What's going to be the end result for the land owners, developers, construction crews, and others involved in the project? What kind of long-term or widespread problems could be caused by this type of "cascading error" when the initial referencing is incorrect?

HANDS-ON APPLICATION 3.3

Georeferenced Historic Maps and the Spyglass

Esri Story Maps are great Web applications that have a variety of different layouts and tools (see Chaper 15 for more about Story Maps). One particularly useful tool is the "Spyglass," which allows you to view a portion of one map on top of another. In the case of historic maps, the Spyglass app allows you to view an old geoferenced map on top of current satellite imagery, as viewed through a movable lens (the "Spyglass"). Wherever you move the Spyglass you'll see the historic map in the georeferenced location.

For example, check out the Spyglass Story Map of 1851 Washington, D.C., at **http://bit.ly/1ch2gb5.** When the Story Map opens, you'll see the spyglass placed over current satellite imagery of Washington, D.C. Within the spyglass you'll see a David Rumsey historical map of Washington, D.C., from 1851. By dragging the spyglass around the imagery, you'll see the historic georeferenced map underneath. You can also zoom in and out and the amount of area shown under the spyglass will change as well.

The Spyglass Story Map is a very neat way to look into the past by using a georeferenced historic map and current imagery. There are several other historic map examples that use the Spyglass available here: **http://storymaps.arcgis.com/en/app-list/swipe /gallery-spyglass/#s=0&md=storymaps-community: history** (such as historic San Francisco, Chicago, Los Angeles, and New York City).

Expansion Questions

- From examining the 1851 Washington, D.C., map using the spyglass, what prominent items were to the immediate east and west of the U.S. Capitol in 1851 that are not there now?

- From examining the 1851 Washington, D.C., map using the spyglass, how was the White House referred to?

- From examining the 1851 Washington, D.C., map using the spyglass, were any of the following three things built in 1851: the Washington Monument, the Lincoln Memorial, or the Smithsonian Institute?

HANDS-ON APPLICATION 3.4

An Overview of the Georeferencing Process in ArcGIS

Although you'll use the free Map-Cruncher to align unreferenced data in the lab, georeferencing tools are common in other GIS software such as ArcGIS (which we'll discuss further in Chapter 5). In order to demonstrate georeferencing in ArcGIS, the company has provided a video showing the whole process, step by step, at **http://video.arcgis.com/watch/376/georeferencing -rasters-in-arcgis**. There's also a second video that demonstrates how georeferencing can be performed automatically, without having to set up control points manually: **http://video.arcgis.com/watch/1553 /automatic-georeferencing-in-arcgis-101**. Check out both videos.

Expansion Questions

- What other kinds of data besides imagery can be used as a source for georeferencing?
- What are the advantages and disadvantages of the automatic georeferencing option?

Chapter Wrapup

Proper georeferencing of data is a key element of working with geospatial technology and allows for all types of maps and imagery to work side by side with other forms of geospatial data. This chapter's lab will have you using the Microsoft MapCruncher program. You'll start with an unreferenced campus map, and by the end it should be properly aligned with a real-world dataset. Also, check out *Georeferencing Apps* for a mobile app using georeferenced historic maps as well as *Georeferencing in Social Media* for more about historic maps online, the David Rumsey Historic Maps on Facebook and Twitter, and a YouTube video about online georeferencing tools.

In the next chapter, we'll be looking at the Global Positioning System, an extremely useful (and freely available) system for determining your position on Earth's surface and providing you with a set of coordinates so that you can find where you are.

Important note: The references for this chapter are part of the online companion for this book and can be found at **www.macmillanhighered .com/shellito/catalog**.

Georeferencing Apps

Here's a sampling of available representative georeferencing-related apps for your phone or tablet. Note that some apps are for Android, some are for Apple iOS, and some may be available for both.

- **Mapmatcher:** An app that allows you to compare current satellite imagery with georeferenced maps (including old historic ones) of different cities (and view the same position on up to four maps at one time simultaneously). There are separate downloadable apps for each city and each app works the same way, for the following cities:

- **Mapmatcher Boston**
- **Mapmatcher Chicago**
- **Mapmatcher London**
- **Mapmatcher New York**
- **Mapmatcher San Francisco**

Georeferencing in Social Media

Here's a sampling of some representative Facebook and Twitter accounts, along with a YouTube video and a blog related to this chapter's concepts.

On Facebook, check out the following:

- **David Rumsey Map Collection: www.facebook.com/pages /David-Rumsey-Map-Collection/216302804001**

Become a Twitter follower of:

- **David Rumsey Map Collection:** @DavidRumseyMaps

On YouTube, watch the following video:

- **Georeferencer: David Rumsey Introduction** (a guide to using the online Georeferencer tools): **www.youtube.com/watch?v=0OrSVn8k8Mc**

For further up-to-date info, read up on this blog:

- **Old Maps Online Blog** (a blog from the makers of Oldmapsonline): **http://blog.oldmapsonline.org**

Key Terms

affine transformation (p. 74)

control points (p. 70)

georeferencing (p. 68)

reproject (p. 66)

root mean square error (RMSE) (p. 75)

rotated (p. 74)

scaled (p. 74)

skewed (p. 74)

spatial reference (p. 68)

translated (p. 74)

Georeferencing an Image

This chapter's lab will introduce you to some of the basics of working with georeferencing concepts by matching an unreferenced map image against a referenced imagery source. You'll be using the free Microsoft MapCruncher utility for this exercise.

Important note: MapCruncher will take a map (in an image or PDF format) and align it to its Virtual Earth (and transform it into a Mercator projection). The end result of the process will be a series of "tiles" used by Virtual Earth to examine your map and match it up with their data. While you'll be using geo-referencing concepts from the chapter, the end result will be that your map will be turned into a "mashup" that could later be used on a Web page.

Objectives

The goals for you to take away from this lab are:

▶ To familiarize yourself with the MapCruncher operating environment

▶ To analyze both the unreferenced image and the source for appropriate control point locations

▶ To perform a transformation on the unreferenced data

▶ To evaluate the results of the procedure

Using Geospatial Technologies

The concepts you'll be working with in this lab are used in a variety of real-world applications, including:

▶ The historical preservation of old maps, which utilizes georeferencing techniques with scanned versions of the maps, so that they may be matched up with current geography for comparison and analysis

▶ Engineering and real estate analysis, which utilizes scanned and georef-erenced versions of old railroad property valuation maps (often hand-drawn), to compare land parcel plans and values from the past with current maps and imagery

Obtaining Software

The current version of MapCruncher (3.2) is available for free download at **www.microsoft.com/en-us/download/details.aspx?id=22420**.

Important note: Software and online resources can change fast. This lab was designed with the most recently available version of the software at the time of writing. However, if the software or Websites have changed significantly between then and now, an updated version of this lab (using the newest versions) will be available online at **www.macmillanhighered.com /shellito/catalog**.

[Source: British Library Board]

Lab Data

Copy the Chapter 3 folder containing both a PDF version (parkmap2010. pdf) and a BMP version (parkmap2010.bmp) of the same file, which is a map of the Youngstown State University (YSU) campus with no spatial reference attached to it. Maps of YSU can also be downloaded from the campus map Web page at **http://web.ysu.edu/gen/ysu/Directions_and_Virtual _Tour_m162.html**.

Important note: If you are working with MapCruncher on a 32-bit machine, use the PDF version of parkmap2010, and if you're working with MapCruncher on a 64-bit machine, use the BMP version of parkmap2010 instead (MapCruncher will not work properly with PDF files on a 64-bit machine). Information on how to work with PDFs in MapCruncher on a 64-bit machine is available at the bottom of this Web page: **http://research.microsoft.com/en-us/um/redmond/projects /mapcruncher**.

Localizing This Lab

MapCruncher is capable of georeferencing a graphic or PDF map of an area. If you have a graphic (such as a BMP) or PDF of a map of your school, it may be substituted for the YSU campus map and you can match your own local data instead. The same steps will still apply; you'll just be finding appropriate local control points for your school rather than the ones for Youngstown, Ohio.

Alternatively, you could use a graphic or PDF of a map of a local park, hike or bike trail, fairgrounds, or local shopping district, and georeference that instead.

3.1 Getting Started With MapCruncher

1. Start MapCruncher. It will open with the default screen referred to as "Untitled Mashup." Maximize the MapCruncher window to give yourself plenty of room to work with.

2. The left side of the screen will contain the unreferenced image you want to match, while the right side of the screen will contain the referenced imagery from Microsoft's Bing Maps "Virtual Earth" (this is what you'll be matching your image to).

3. As the text box notes, select the File pull-down menu, and choose Add Source Map.

4. Navigate to the Chapter 3 folder and select the parkmap2010.pdf file (or the parkmap2010.bmp file if you're working on a 64-bit machine).

5. In the upper right corner of the MapCruncher screen you'll see the Virtual Earth Position box.

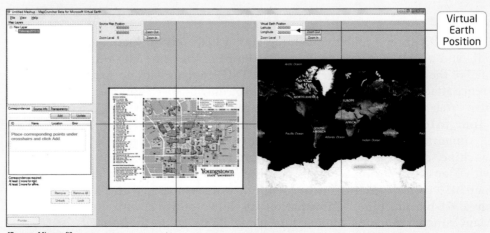

Virtual
Earth
Position

[Source: Microsoft]

6. Type in the following coordinates and then click on Zoom In to move the view to YSU:

   - **Latitude:** 41.10687400
   - **Longitude:** −80.64670900

7. The crosshairs in the image will center on a point in northeast Ohio. Press the Virtual Earth Position's Zoom In button several times to zoom in to YSU's campus (Zoom Level 17 is probably a good starting point).

8. Both of the views will now show the campus—the Source Map (left-hand view) will show the PDF map image (or graphic) while the Virtual Earth (right-hand view) will show the aerial imagery from Virtual Earth.

3.2 Selecting Correspondences

1. The next step is to choose the same positions on both images to use as control points (MapCruncher refers to them as "correspondences"). A good starting point will be the intersection of Wick Avenue and Lincoln Avenue (a major road intersection that borders the southeastern portion of the campus).

2. Click on Zoom In on the Source Map until you reach Zoom Level 7.

3. The cursor will change to a hand—use this to pan around the map until the intersection of Wick and Lincoln Avenues is centered under the Source Map's crosshairs.

4. In the Virtual Earth view, pan around the map until that same intersection is in the Virtual Earth's crosshairs.

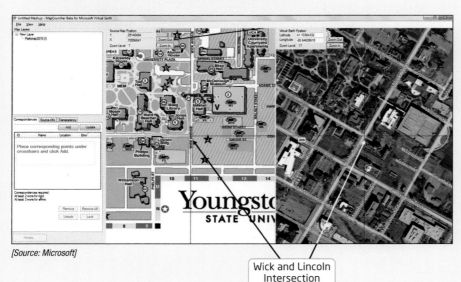

[Source: Microsoft]

Wick and Lincoln
Intersection

5. When you have both crosshairs accurately positioned, click Add in the Correspondences box.

[Source: Microsoft]

6. The point will appear as ID 0. This will be the first correspondence (or control point) to be selected.

> **Question 3.1** Why is this a good location for a correspondence?

> **Question 3.2** Just by visual examination of the two maps, what types of transformative operations will have to be performed to the Source Map to make it match Virtual Earth?

7. Use the combination of the Source Map and Virtual Earth to select another two correspondence points at clearly defined locations on both sources. Be sure to spread out the points geographically across the Source Map image, then click Lock. MapCruncher will perform an initial fit of the data, but if you pan around you'll see that things don't match up just yet. Press Unlock and select a total of 10 correspondences (you can use the Lock option to see how things are fitting together as you go, then Unlock to adjust the points).

> **Question 3.3** Which locations did you choose for the next nine correspondence points, and why?

> **Question 3.4** Why would you want to spread the correspondence point selection across the image instead of clustering the points together?

8. Once you have about 10 good locations chosen, click the Lock button in the Correspondences box.

Correspondences	Source Info	Transparency	
ID	Name	Location	Error
0	Pin0	41.1038352...	
1	Pin1	41.1063898...	
2	Pin2	41.1097850...	
3	Pin3	41.1092111...	
4	Pin4	41.1077964...	
5	Pin5	41.1059533...	
6	Pin6	41.1049751...	
7	Pin7	41.1019758...	
8	Pin8	41.1052499...	

Ready to lock; add more points to increase accuracy.

Remove   Remove All

Unlock   Lock

Render...

[Source: Microsoft]

9. A value for error will be computed for each of the points, and they will be sorted from highest error value to lowest error value (rather than by ID number).

Question 3.5 What are the error values associated with your chosen control points?

10. Pan around the source image to the ID number with the highest error value. With the maps locked you'll see both the Source Map and Virtual Earth move together.

Important note: To zoom directly to one of the correspondence points, double-click on its ID in the box, and MapCruncher will zoom in and center on that control point.

If any of your error values are very high, you can change them by doing the following:

a. Zoom in to the selected point. You'll see a green dot indicating where MapCruncher thinks the point should be and a line showing the distance from where you placed the correspondence point to where MapCruncher thinks it should be.

b. Click Unlock.

c. Move the crosshairs of the Source Map to the new spot.

d. Click Update.

e. Click Lock again to view the updated error terms.

11. Pan around the Source Map, viewing how well points on the interior and on the fringes match up between the Source Map and the Virtual Earth representation. Add a few more correspondence points if you feel it's necessary, to adjust for areas that don't match up well.

3.3 Matching the Source Map to Virtual Earth

When you have the Source Map set up the way you want it, it's time to complete the process.

1. You may first want to save your work by selecting the **File** pull-down menu and choosing **Save Mashup**. Give your work a file name and save it on your machine (it will save with the file extension ".yum").

2. Back in MapCruncher, click the **Render** button in the lower left-hand corner. The Render dialog box will open.

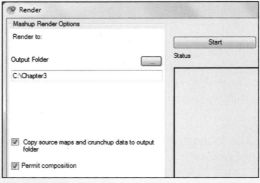

[*Source: Microsoft*]

3. Press the **Pick** button (the one with the ". . ." symbol on it) next to the Output Folder, and then navigate to a folder on your machine to save the aligned file to.

4. Put checkmarks in the **Copy source maps and crunchup data to output folder** and the **Permit composition** boxes.

5. Click the **Start** button to begin the process. (*Important note:* The Render process may take several minutes and could fill up about 200 MB of space on your machine.)

6. You'll see the new layer being "built" in the Status window of the Render dialog box.

7. After a couple of minutes, the **Render Complete** box will appear. Click OK in this box.

8. Click on the **Preview rendered results** text at the bottom of the Render dialog box. The Mashup Viewer will open to show the map lined up with Virtual Earth.

9. From the Mashup Viewer's **VE Background** pull-down menu, select the **Hybrid** option.

10. Zoom in on the image. You'll see that the map has been transformed and aligned, but it'll be tough to judge how good the match actually is, with the solid colors of the map overlaid on the roads and imagery. It can be made easier to judge if some of those solid colors (including the white border of the map) are made transparent. Other colors can also be made transparent, making it easier to match up the new map with the aerial imagery.

11. Close the **Mashup Viewer**. Back in MapCruncher, select the **Transparency** tab.

12. Put a checkmark in the **Use document transparency** box.

13. Choose the **Make selected colors transparent** option.

14. In the Source Map, pan across the map so that the green color (for instance, the circular greenery in the campus center) is under the crosshairs, and then click the **Add Color Under Crosshairs** button. The green color will appear in the box, while the green on the Source Map should change to pink. If necessary, change the **Fuzziness** option to 3 if the colors don't change to pink.

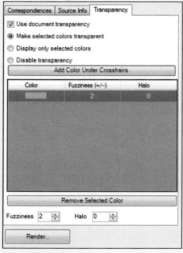

[Source: Microsoft]

15. Now pan across the Source Map so that the color white (you could choose one of the roads going through campus, such as Lincoln Avenue) is under the crosshairs, and then click the **Add Color Under Crosshairs** button. The white color will appear in the box, while the white on the Source Map should also change to pink.

16. Click the **Render** button again.

17. In the Render dialog box, save the results to a new folder. Click Start and the image will re-render, this time with the colors green and white made transparent. Preview the rendered results again.

18. In the new Mashup Viewer, the map will be rendered, but it will be different this time.

19. From the Mashup Viewer's VE Background pull-down menu, select the Roads option.

> **Question 3.6** How does the map match up with the roads leading in and out of campus? Are there any locations where the roads don't match?

20. From the VE Background pull-down menu, select the Aerial Photos option.

> **Question 3.7** How does the map match up with the aerial imagery around campus? Are there any locations where the images don't really match (that you can see with the greenery and roads made transparent)?

Closing Time

MapCruncher is a highly versatile tool that is used for making data (such as unreferenced images) match the map. Any type of map that is drawn to scale (such as a park trail map, a fairgrounds map, a theme park diagram, or an unreferenced image) can be aligned using MapCruncher. You can now exit MapCruncher by selecting Exit from the File pull-down menu.

4

Finding Your Location with the Global Positioning System

GPS Origins, Position Measurement, Errors, Accuracy, GNSS around the World, Applications, and Geocaching

Here's something for you to think about: You're out hiking in a forest. A storm comes up, and you quickly take shelter well off the trail. When things are all clear, you've gotten completely disoriented, and you can't find your location: you're well and truly lost. Or consider this: You're driving in unfamiliar territory. It's late at night, you're miles from anywhere, you missed an important turn a few miles back, and none of the names or route numbers on the signs mean anything to you. In both cases, you're out of cell phone range, and even though you have a printed map, you really have no idea where you are. How do you start navigating yourself out of this mess?

A few years ago, situations like these could have been potentially disastrous, but today there's a geospatial technology tool that's ready and available to help get you out of these situations. Using a small, hand-held, relatively inexpensive device, you can find the coordinates of your position on Earth's surface to within several feet of accuracy. Simply stand outside, turn the device on, and shortly your position will be found and perhaps even plotted on a map stored on the device. Start walking around, and it'll keep reporting where you are and even take into account some other factors like your elevation, speed, and the compass direction in which you're moving. Even though it all sounds like magic, it's far from it—there's an extensive network of equipment across Earth and in space continuously working in conjunction with your hand-held receiver to locate your position. This, in a nutshell, is the operability of the Global Positioning System (**GPS**).

GPS receivers have become so widespread that they seem to be everywhere. In-car navigation systems have become standard items on the shelves of electronics stores. GPS receivers are also on sale at sporting goods stores, because they're becoming standard equipment for campers, hikers, fishing enthusiasts, and golfers. Runners can purchase a watch containing a GPS

GPS the Global Positioning System, a technology using signals broadcast from satellites for navigation and position determination on Earth

receiver that will tell them their location, the distance they run, and their average speed. Smartphones will often have GPS capability as a standard function. Use of the GPS is free for everyone and available worldwide, no matter how remote the area you find yourself lost in. So the first questions to ask are—who built this type of geospatial technology and why?

 ## Who Made GPS?

The acronym "GPS" is used in common vocabulary to describe any number of devices that use Global Positioning System technology to tell us where we are, but the term is often used inaccurately. The term "GPS" originated with the initial designers and developers of the setup, the U.S. Department of Defense, and is part of the official name of the system called **NAVSTAR GPS**. Other countries besides the United States have developed or are currently engaged in developing systems like the NAVSTAR GPS, so strictly speaking we should only use the term "GPS" when we're referring to the United States system. Since the United States isn't the only nation with this technology, it would be more accurate to describe GPS as one type of global navigation satellite system (**GNSS**).

GPS isn't the first satellite navigation system. During the 1960s, the U.S. Navy used a system called Transit, which used satellites to determine the location of sea-going vessels. The drawback of Transit was that it didn't provide continuous location information—you'd have to wait a long time to get a fix on your position rather than always knowing where you were. Another early satellite navigation program of the 1960s, the Naval Research Laboratory's Timation program, used accurate clock timings to determine locations from orbit. The first GPS satellite was launched in 1978, and the 24th GPS satellite was launched in 1993, completing an initial full operational capability of the satellite system for use. Today, GPS is overseen and maintained by the 50th Space Wing, a division of the U.S. Air Force headquartered at Schriever Air Force Base in Colorado.

 ## What Does the Global Positioning System Consist Of?

Finding your position on the ground with GPS relies on three separate components, all operating together: a space segment, a control segment, and a user segment.

Space Segment

The **space segment** is made up of numerous GPS satellites (also referred to as "SVs" or "space vehicles") orbiting Earth in fixed paths. GPS satellites make two orbits around Earth every day (their orbit time is actually about 11 hours and 58 minutes) at an altitude of 20,200 kilometers (12,552 miles).

NAVSTAR GPS the U.S. Global Positioning System

GNSS the global navigation satellite system, an overall term for the technologies that use signals from satellites to find locations on Earth's surface

space segment one of the three segments of GPS, consisting of the satellites and the signals they broadcast from space

FIGURE 4.1 The constellation of GPS satellites in orbit around Earth.

The satellites are set in a pattern of specific orbits called a **constellation**, in which the satellites are specifically arranged for maximum coverage over Earth (**Figure 4.1**). The way the GPS constellation is designed allows for a person to be able to receive enough signals to find their location wherever they are on the planet. Twenty-four satellites is the minimum for a full constellation, and there are currently several additional operational GPS satellites in orbit to improve global coverage.

The job of GPS satellites is to broadcast a set of signals down to Earth from orbit. These signals (discussed later) contain information about the

constellation the full complement of satellites comprising a GNSS

position of the satellite and the precise time at which the signal was transmitted from the satellite (GPS satellites measure time using extremely accurate atomic clocks carried onboard). The signals are sent on carrier frequencies L1 (broadcast at 1575.42 MHz) and L2 (broadcast at 1227.60 MHz). Additional carrier frequencies are planned for future GPS upgrades, including an L5 frequency to be used for safety-of-life functions. You need to have a direct line of sight to the satellites to receive these frequencies (which means that you have to be in the open air, not indoors).

Control Segment

control segment one of the three segments of GPS, consisting of the control stations that monitor the signals from the GPS satellites

The **control segment** of GPS represents a series of worldwide ground stations that track and monitor the signals being transmitted by the satellites. These ground stations, also known as control stations, are spread out to enable continuous monitoring of the satellites. The control stations collect the satellite data and transmit it to the master control station at Schriever Air Force Base. In the control segment, corrections and new data are uploaded to the satellites so that the satellites will be broadcasting correct data. The control stations also monitor the satellites' positions and relay updated orbit information to the satellites (**Figure 4.2**).

FIGURE 4.2 A control station monitors the signals from the satellites and sends correction data back to the satellites.

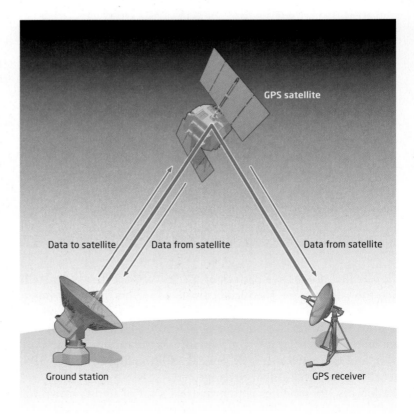

GPS satellite

Data to satellite　Data from satellite　　Data from satellite

Ground station　　　　　　　　　　GPS receiver

User Segment

The **user segment** represents a GPS unit somewhere on the Earth that is receiving satellite signals. The receiver can then use the information in these signals to compute its position on Earth. A key component of a GPS receiver is how many **channels** the receiver has—the number of channels reflects the number of satellites the receiver can obtain signals from at one time. Thus, a 12-channel receiver can potentially pick up signals from up to 12 satellites. The type of unit you're using is also important—a **single frequency** receiver can only pick up the L1 frequency, but a **dual frequency** receiver can pick up both L1 and L2. One thing to keep in mind about the receiver is that no matter what size, capabilities, or number of channels it has, a receiver (like the name implies) can only receive satellite data—it can't transmit data back up to the satellites. This is why GPS is referred to as a one-way ranging system: satellites broadcast information, and receivers receive it.

How Does GPS Find Your Position?

At this point, the three segments of GPS have satellites broadcasting signals, ground stations monitoring and correcting those signals, and users receiving those signals—and all three of these segments work together to determine your position on Earth's surface. The signals being sent from the satellites are the key to the whole process—the signals contain information about the satellite's status, the orbit, and the location (referred to as the **almanac**) that's sending them, as well as more precise data about satellite location (this information is referred to as the **ephemeris**).

The information is sent in one of two digital pseudo-random codes—the **C/A code** (coarse acquisition code) and the **P code** (precise code). The C/A code is broadcast on the L1 carrier frequency and is the information that civilian receivers can pick up. The P code is broadcast on the L1 and L2 carrier frequencies and contains more precise information, but a military receiver is required to pick up this signal. The **Y code** is an encrypted version of the P code and is used to prevent false P code information from being sent to a receiver by hostile forces. This "anti-spoofing" technique is commonly used to make sure that only the correct data is being received. Basically, the satellites are transmitting information in the codes about the location of the satellite, and they can also be used to determine the time when the signal was sent. By using this data, the receiver can find its position relative to that one satellite.

The signals are transmitted from space using high-frequency radio waves (the L1 and L2 carrier frequencies). These radio waves are forms of electromagnetic energy (discussed in Chapter 10) and will therefore be moving at the speed of light. Your receiver can compute the time it took for the signal to arrive from the satellite. If you know these high-frequency

user segment one of the three segments of GPS, consisting of the GPS receivers on the ground that pick up the signals from the satellites

channels the number of satellite signals a GPS unit can receive

single frequency a GPS receiver that can pick up only the L1 frequency

dual frequency a GPS receiver that can pick up both the L1 and L2 frequency

almanac data concerning the status of a GPS satellite, which is included in the information being transmitted by the satellite

ephemeris data referring to the GPS satellite's position in orbit

C/A code the digital code broadcast on the L1 frequency, which is accessible by all GPS receivers

P code the digital code broadcast on the L1 and L2 frequencies, which is accessible by the military

Y code an encrypted version of the P code

pseudorange the calculated distance between a GPS satellite and a GPS receiver

radio waves move at the speed of light (c), and you also know how long the transmission time (t) was, then you can calculate the distance between you and that one satellite by multiplying t by c. This result will give you the **pseudorange** (or distance) between your receiver and the satellite transmitting the signal.

Unfortunately, this still doesn't tell us much—you know where you are on the ground in relation to the position of one satellite over 12,000 miles away, but that by itself gives you very little information about where you are on Earth's surface. It's like waking up in an unknown location and being told by a passerby that you're 500 miles from Boston—it gives you very little to go on in determining your real-world location.

trilateration finding a location in relation to three points of reference

The process of determining your position is referred to as **trilateration**, which means using three points of reference to find where you are. Trilateration in two dimensions is commonly used when plotting a location on a map. Let's say, for example, that you're on a road trip to Michigan—you've been driving all night, and you've arrived in an unknown location with no identifying information to tell you where you are. The first person you bump into tells you (somewhat unhelpfully) that you're 50 miles away from Ann Arbor, Michigan (see **Figure 4.3**). That puts you somewhere on a circle with Ann Arbor at its center 50 miles away, and gives you a choice of far too many possibilities to be of any use to you.

The second person you run into tells you (again, not being overly helpful) that you're located 150 miles away from Traverse City, Michigan. This again puts your location somewhere on a circle 150 miles from Traverse City—but when you combine this with the information that places you 50 miles away from Ann Arbor, you can limit your options down considerably. There are only two cities in Michigan that are 50 miles from Ann Arbor and 150 miles away from Traverse City—you're either in Lansing or Flint (see **Figure 4.4** on page 95).

Luckily, the third person you see tells you that you're 60 miles from Kalamazoo. Your first thought should be "Who are all these geographically-minded people I keep running into?" but your second thought is that you now know exactly where you are. Lansing is the only option that fits all three

FIGURE 4.3 Measuring distance from one point.

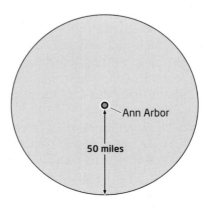

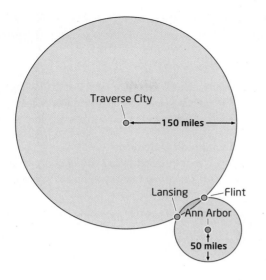

FIGURE 4.4 Using distances from two points is the second step in the process of trilateration.

of the distances from your reference points, so you can disregard Flint and be satisfied that you've found your location (see **Figure 4.5** and *Hands-On Application 4.1: Trilateration Concepts*).

The same concept applies to finding your location using GPS, but rather than locating yourself relative to three points on a map, your GPS receiver is finding its position on Earth's surface relative to three satellites. Also, because a position on the three-dimensional (3D) Earth is being found with reference to positions surrounding it, a spherical distance is calculated rather than a flat circular distance. This process is referred to as trilateration in three dimensions (or **3D trilateration**). The concept is similar, though

3D trilateration
finding a location on Earth's surface in relation to the positions of three satellites

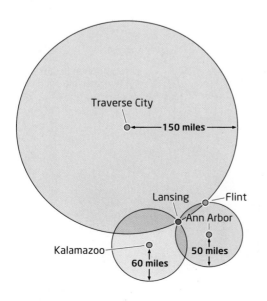

FIGURE 4.5 The measurement of distances from the third point completes the process of trilateration by finding a location.

HANDS-ON APPLICATION 4.1

Trilateration Concepts

Trilateration can be used to determine your location on a map. Go to **http://electronics.howstuffworks.com/gadgets/travel/gps1.htm** to view the article by Marshall Brain and Tom Harris at *How Stuff Works*. Check out how three measurements are made to determine a single location (like in the previous Michigan example), then start Google Earth and zoom to your location. Use Google Earth and its measuring tools to set up a similar scenario—find your location relative to three other cities and figure out the measurements to do a similar trilateration. Then zoom in closer to identify three local locations and calculate the measurements from each one that you need to trilaterate your position.

Expansion Questions

- What three cities did you choose, and what measurements did you use to trilaterate your city?

- What three local locations did you choose, and what measurements did you use to trilaterate your current position?

finding the receiver's position relative to one satellite means it's located somewhere on one sphere (similar to only having the information about being 50 miles from Ann Arbor). By finding its location in relation to two satellites, the receiver is finding a location on a common boundary between two spheres. Finally, by finding the location relative to three satellites, there are only two places where those three spheres intersect, and the way the geometry works out, one of them is in outer space and can be disregarded. That leaves only one location where all three spheres intersect, and thus one location on Earth's surface relative to all three satellites. That position will be the location of the GPS receiver (**Figure 4.6**).

FIGURE 4.6 Finding a location using GPS: the results of 3D trilateration.

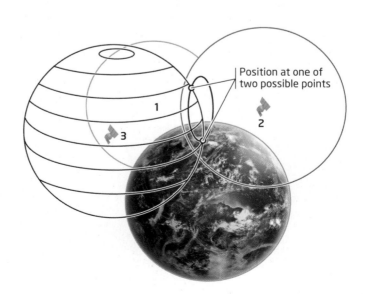

There's just one problem with all this—and it's serious. Remember, what you're calculating is a pseudorange, based on the speed of light (which is a constant) and the time it takes for the signal to transmit from space to Earth. If that time is just slightly off, a different value for distance will be calculated, and your receiver will be located somewhere different from where it really is. This becomes an even bigger issue because the satellite has a super-precise and very expensive atomic clock (somewhere in the tens of thousands of dollars range), while the off-the-shelf GPS receiver has a less precise (and very much less expensive) quartz clock. Obviously, GPS receivers can't have atomic clocks in them, because they'd be far too expensive to buy. The receiver's clock isn't as accurate as the atomic clock, and timing errors can cause inaccuracies in calculating distances when it comes to finding a position. Fortunately, there's an easy solution to this mess: your receiver picks up additional information from a fourth satellite. By using a fourth measurement, clock errors can be corrected and your position can be accurately determined. The position you'll receive will be measured using the WGS84 datum.

 ## Why Isn't GPS Perfectly Accurate?

If GPS can be used to find a location on Earth's surface, the next question to ask is: "Just how accurate is GPS?" Through the use of pseudorange measurements from four satellites, you might reasonably expect the system to give the user one accurately measured location. However, in normal everyday use with an ordinary off-the-shelf civilian receiver, accuracy is usually somewhere around 10–15 meters horizontally. This doesn't mean that the location is always off by 15 meters—it could be off by 5, 10, or 15 meters. This is due to a number of factors that can interfere with or delay the transmission or reception of the signal—and these delays in time can cause errors in position.

One of the biggest sources of error was actually worked into the system intentionally to make GPS less accurate. The United States introduced **selective availability** (SA) into GPS with the goal of intentionally making the C/A signal less accurate, presumably so that enemy forces couldn't use GPS as a tool against the U.S. military. Selective availability introduced two errors into the signals being transmitted by the satellites—a delta error (which contained incorrect clock timing information) and an epsilon error (which contained incorrect satellite ephemeris information). The net effect of this was to make the accuracy of the C/A about 100 meters. As you can imagine, this level of accuracy certainly limited GPS in the civilian sector—after all, who wants to try to land an airplane with GPS when the accuracy of the runway location could be hundreds of feet off? In the year 2000, the U.S. government turned selective availability off and has no announced plans to

selective availability
the intentional degradation of the timing and position information transmitted by a GPS satellite

turn it back on (in fact, the U.S. Department of Defense has announced that new generations of GPS satellites will not have the capability for selective availability).

Although selective availability was an intentional error introduced into the signals broadcast by the satellites, other errors can occur naturally at the satellite level. Ephemeris errors can occur when a satellite broadcasts incorrect information about orbit position. These types of errors indicate that the satellite is out of position, and the ephemeris data needs to be corrected or updated by a control station. Typically, these errors can introduce 2 meters or so of error.

Even the arrangement of satellites in space (that is, how they are distributed) can have an effect on the determination of position. The further the satellites are spread out from one another and the wider the angle between them, the more accurate a measurement obtained by the receiver will be. The error introduced by a poor geometric arrangement of the satellites is referred to as the position dilution of precision or **PDOP**. Some receivers will calculate a value for PDOP for you to indicate what relative error could be introduced because of the arrangement of the satellites being used (basically, the lower the value the better).

You should also keep in mind that the signals are being broadcast from space and have to pass through Earth's atmosphere before reaching the receiver. Though some of these atmospheric effects can be modeled, and attempts can be made to compensate for them, atmospheric problems can still cause errors in accuracy by making changes to the speed of the signals (**Figure 4.7**). Atmospheric interference, especially in the ionosphere and troposphere, can cause delays in the signal on its way to the ground. The ionosphere can alter the propagation speed of a signal, which can cause an inaccurate timing measurement, while the amount of water vapor in the troposphere can also interfere with the signals passing through it and also cause delays. It's

PDOP the position dilution of precision; describes the amount of error due to the geometric position of the GPS satellites

FIGURE 4.7 The effects of the ionosphere and troposphere on GPS signals.

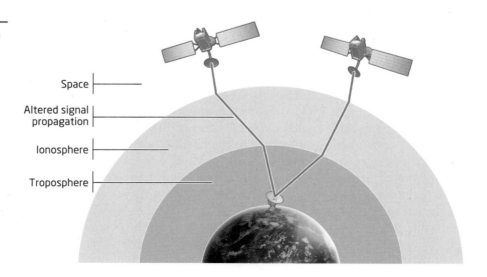

estimated that the problems caused by conditions in the ionosphere can cause about 5 meters (or more) of position inaccuracy, while conditions in the troposphere can add about another 0.5 meters of measurement error.

The atmosphere is not the only thing that can interfere with the signal reaching the receiver. The receiver needs a good view of the sky to get an initial fix, so things that would block this view (like, say, a heavy tree canopy) may interfere with the signal reception. In addition, the signals may be reflected off objects (such as tall buildings or bodies of water) before reaching the receiver rather than making a straight path from the satellite (**Figure 4.8**). This **multipath** effect can introduce additional delays into the reception of the signals, since the signal is reaching the receiver later than it should (and will add further error to your position determination).

> **multipath** an error caused by a delay in the signal due to reflecting from surfaces before reaching the receiver

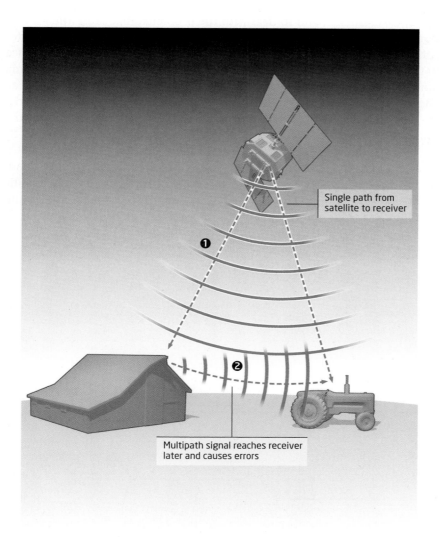

FIGURE 4.8 The effect of multipathing on the GPS signal.

Single path from satellite to receiver

❶

❷

Multipath signal reaches receiver later and causes errors

It's quite possible for any or all of these various types of error measurements to combine together to contribute several meters worth of errors into your position determination. When the ephemeris introduces a few meters of error, and the ionosphere and troposphere add a few more, and then multipathing and PDOP add a few more on top, before you know it they're starting to add up to a pretty significant difference between your position as plotted and where you actually are. So, there should be some manner of solutions to improve the accuracy of GPS and get a better fix on your position.

How Can You Get Better Accuracy from GPS?

DGPS differential GPS; a method of using a ground-based correction in addition to the satellite signals to determine position

A few methods have been developed to improve the accuracy of GPS position measurements, and to deal with the problems that cause the kinds of errors described above. The first of these is a process referred to as differential GPS (or **DGPS**). This method uses one or more base stations at locations on the ground to provide a correction for GPS position determination. The base station is positioned at a place where the coordinates are known, and it receives signals from the satellites. Because the base station knows its own coordinates, it can calculate a correction to any error that causes the position measurement from the satellites to locate it in the wrong place. The base station can then broadcast this correction out to receivers equipped to collect this type of differential correction. Thus, when you're using DGPS, your receiver is picking up the usual four satellite signals plus an additional correction from a nearby location on the ground (see **Figure 4.9** for an example of the operation of DGPS). As a result, the

FIGURE 4.9 The operation of DGPS.

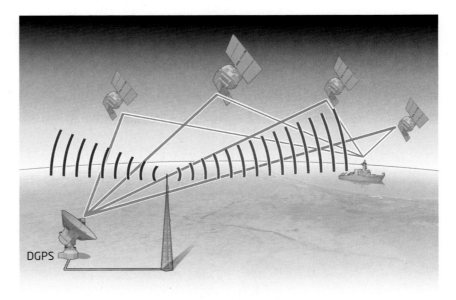

position measurement can be reduced to 5 meters or less (and often less than 1 meter).

Today, there are DGPS locations all over the world. In the United States, the U.S. Coast Guard has set up several reference stations along the coasts and waterways to enable ships to find their exact location and to improve navigation. The U.S. Department of Transportation operates the Nationwide Differential GPS (**NDGPS**) system and aims to provide drivers and travelers with the most accurate position information that's available. NDGPS reduces position error to 1 to 2 meters.

A related program, **CORS** (Continuously Operating Reference Stations), is operated by the National Geodetic Survey and consists of numerous base stations located across the United States and throughout the world. CORS collects GPS information and makes it available to users of GPS data (like surveyors and engineers) to make more accurate position measurements. CORS data provides a powerful tool for location information for a variety of applications (such as the U.S. Army establishing CORS in Iraq as an aid in the rebuilding efforts).

Satellite-based augmentation systems (**SBAS**) are additional resources for improving the accuracy of position measurements. SBAS works in a similar way to DGPS, in that a correction is calculated and picked up by your receiver in addition to signals from the regular four satellites. However, in this case, the correction is sent from an additional new satellite, not from an Earth-bound station. A good example of an SBAS is the Wide Area Augmentation System (**WAAS**), which was developed by the Federal Aviation Administration (FAA) to obtain more accurate position information for aircraft. WAAS has developed widespread uses beyond the world of aviation as a quick and easy method of improving overall GPS position accuracy.

WAAS operates through a series of base stations spread throughout the United States that collect and monitor the signals sent by GPS satellites. These base stations calculate position correction information (similar to the operation of DGPS) and transmit this correction to geostationary WAAS satellites. These WAAS satellites then broadcast this correction signal back to Earth, and if your receiver can pick up this signal, it can use the information to improve its position calculation (see **Figure 4.10** on page 102). WAAS reduces the position error to 3 meters or less. One drawback to WAAS is that it will only function within the United States (including Alaska and Hawaii) and nearby portions of North America (although WAAS base stations have been added to Mexico and Canada, greatly expanding the availability of WAAS advantages). There are other systems that improve the accuracy of position calculation, but they all operate on the same principle as WAAS and DGPS (some sort of correction is calculated and transmitted to a receiver, which uses it to determine its position).

NDGPS National Differential GPS, which consists of ground-based DGPS locations around the United States

CORS Continuously Operating Reference Stations; a system operated by the National Geodetic Survey to provide a ground-based method of obtaining more accurate GPS positioning

SBAS Satellite-based augmentation system; a method of using correction information sent from an additional satellite to improve GPS position determination

WAAS Wide Area Augmentation System; a satellite-based augmentation system that covers the United States and other portions of North America

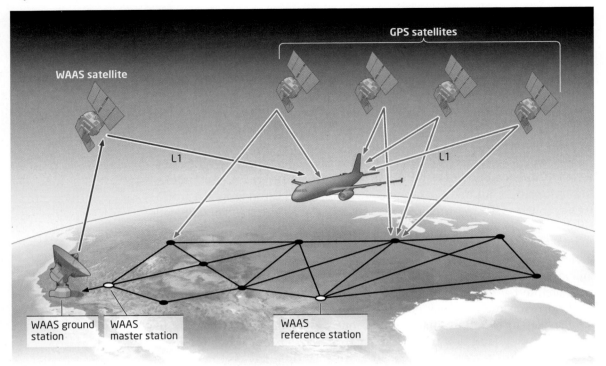

FIGURE 4.10 The setup and operation of the WAAS system.

 # What Other GNSS Are There Beyond GPS?

This chapter has been devoted to GPS: the way it operates, the types of errors it's liable to make, and how its accuracy issues can be resolved. However, as we touched on earlier, "GPS" properly refers to the U.S. NAVSTAR GPS. Other countries have developed (or are in the process of developing) their own versions of GPS or GPS augmentation systems, presumably so as not to be completely dependent on U.S. technology. There are other full constellations of GPS-style satellites being developed and put into orbit, and numerous other enhancement systems are either in operation or in the planning stage all around the globe.

GLONASS was the name of the USSR's counterpart to GPS and operated in a similar fashion. The GLONASS equivalent to the C/A code operated at horizontal accuracies of about 100 meters (or better) with an equivalent of the P code for more accurate use. GLONASS consisted of a full constellation of satellites with an orbital setup similar to GPS. By 2001, however, there were only a few GLONASS satellites in operation. In recent years Russia has renewed its GLONASS program—by the end of 2011 a fully operational constellation was back in place, and at the time of writing there are plans to launch additional satellites in the near future. In 2007, Russia announced that the civilian signal

GLONASS the former USSR's (now Russia's) GNSS

from GLONASS would be made freely available for use, and at the time of writing GLONASS accuracy was reported to be approximately 5 meters (or less).

Galileo is the European Union's version of GPS. When it's completed, it will have a constellation of 30 satellites and operate in a similar fashion to GPS. The first of the GIOVE (Galileo In Orbit Validation Element) satellites was launched in 2005, and since then the program has continued to develop, with several satellites now orbiting as part of its constellation (at the time of writing it was projected to be completed by 2019). Galileo promises four different types of signals, including information available to civilians as well as restricted signals. Also in development is China's **Compass** (also referred to as Beidou-2), which is also projected to have a full constellation of satellites. The first Compass satellite was launched in 2007, and at the time of writing the system was servicing the Asia/Pacific region, and planned to be in service globally by 2020.

Other satellite-based augmentation systems have been developed across the world to operate like WAAS but to function in regions outside of North America. **EGNOS** (European Geostationary Navigation Overlay System) is sponsored by the European Union and provides improved accuracy of GPS throughout most of Europe. EGNOS functions the same way WAAS does—a series of base stations throughout Europe monitor GPS satellite data, calculate corrections, then transmit these corrections to geostationary EGNOS satellites over Europe, which in turn broadcast this data to Earth. An EGNOS-enabled receiver is necessary to utilize this correction data, and the system is capable of accuracies of about 1.5 meters. Japan also operates its own SBAS, known as **MSAS** (Multifunctional Satellite Augmentation System), which covers Japan and portions of the Pacific Rim. MSAS operates in a similar way to WAAS and EGNOS, but it covers a different region of the world and uses just two satellites to provide coverage.

Many other GPS enhancement systems exist or are in development throughout the world, including the following:

▶ Beidou: an SBAS operating in China

▶ GAGAN: an SBAS in development for India

▶ IRNSS: an Indian satellite navigation system, consisting of seven satellites, currently in development and at the time of writing expected to be operational in 2016

▶ QZSS: an expanded SBAS based in Japan, currently in development

Galileo the European Union's GNSS, currently in development

Compass China's GNSS, currently in development

EGNOS an SBAS that covers Europe

MSAS an SBAS that covers Japan and nearby regions

What Are Some Applications of GPS?

GPS is used in a wide variety of settings. Any application that requires location-based data to be collected firsthand can apply GPS to the task. For instance, geologists use GPS to mark the location of water wells when testing

GPX a standard file format for working with GPS data collected by GPS receivers

groundwater, while civil engineers use GPS to determine the coordinates of items such as culverts, road signs, and manhole covers. Coordinates measured with a GPS receiver are commonly used with other geospatial technologies in **GPX** format, a standard for storing and mapping collected GPS data. In the following sections you'll see just a handful of the ways GPS is being used in businesses, jobs, and the public sector.

Emergency Response

When you make a 911 call, the call center is able to fix your location and route emergency services (police, firefighters, or ambulances) to where you are. When you make a 911 call from a cell phone, the Enhanced 911 (E-911) system can determine your position—either using the cell towers to determine where you are, or using a GPS receiver in your phone to find your location.

Farming

Precision farming methods rely on getting accurate location data about factors such as watering levels, crops, and soils. Having the ability to map this data gives farmers information about where crops are thriving and what areas need attention. GPS serves as a data-collection tool to quickly (and accurately) obtain location information for farmers, which can then be combined with other geospatial technologies, such as using Geographic Information Systems (GIS) to place the GPS data on a map.

Forensics

Recovery of human remains has been greatly assisted through the use of GPS. Rather than having to rely solely on surveying equipment, GPS can provide quick data and measurements in addition to other measurements made in the field. The same concepts have been applied to identifying the locations of artifacts and remains at archeological sites.

Public Utilities

City or county data for items that need to be measured manually (such as culverts, road signs, or manhole locations) can be done quickly and easily with GPS. When field-based data is necessary to create or update utility data, workers equipped with GPS can collect this type of data quickly and accurately.

Transportation

GPS provides a means of tracking travel continuously, whether on the road, in the water, or in the air. Since GPS receivers can receive signals constantly from satellites, your travel progress can be monitored to make sure that you stay on track. When your GPS-determined position is plotted in real time on a map, you can see your exact point in your journey and measure your progress. Delivery and shipping companies can use this type of data to keep tabs on their fleets of vehicles and to check if they deviate from their routes.

Wildlife Management

In much the same way you track vehicles, wildlife and endangered species can be tagged with a GPS receiver to monitor their location and transmit it to a source capable of tracking those movements. In this way, migration patterns can be determined and other large-scale movements of wildlife can be observed.

Beyond its uses in the private and public sectors, GPS plays a big part in personal recreation. If you're a runner, GPS can track your position and measure your speed. If you're a hiker, GPS can keep you from getting lost when you're deep into unfamiliar territory. If you're into fishing, GPS can help locate places reputed to be well stocked with fish. If you're a golfer, you can use GPS to measure your ball's location and the distance to the pin. In fact, there's a whole new form of recreation that's grown in popularity along with GPS, called **geocaching**.

In geocaching, certain small objects (referred to as "geocaches") are hidden in an area and the coordinates of the object are listed on the Web (a log is maintained). From there, you can use a GPS receiver to track down those coordinates (which usually involve hiking or walking to a location) and locate the cache. In essence, geocaching is a high-tech outdoor treasure hunt (see *Hands-On Application 4.2: Things to Do Before You Go Geocaching*).

geocaching using GPS and a set of coordinates to find locations of hidden objects

---

HANDS-ON APPLICATION 4.2

Things to Do Before You Go Geocaching

Geocaching is a widespread and popular recreational activity today. As long as you have a GPS receiver and some coordinates to look for, you can get started with geocaching. This chapter's lab will have you investigating some specific geocaching resources on the Web, but before you start looking for caches, there are some online utilities to explore first. Open your Web browser and go to **www.groundspeak.com**. Groundspeak operates a main geocaching Website as well as provides some tools to use. On the main page there are links to several GPS-related sites:

1. Geocaching.com: **www.geocaching.com**. Check around for locations of geocaches near you (we'll use this during the lab).

2. Waymarking.com: **www.waymarking.com**. This Website provides a way to find the coordinates of interesting locations ("waymarks") that have been uploaded by

users. Type in your zip code to see what's been labeled as a waymark near you and collect its coordinates.

3. Wherigo.com: **www.wherigo.com**. This Website allows you to download a player and a toolset for constructing games (referred to as "cartridges") and other applications for portable GPS devices.

4. Cache In Trash Out: **www.geocaching.com /cito**. This environmental initiative is dedicated to cleaning up areas where geocaches are hidden.

Expansion Questions

- What geocaches are located around your own local area or zip code?

- What locations have been labeled as waymarks near your zip code, and what are the coordinates of these waymarks?

FIGURE 4.11 A VirginiaView geocoin and a map showing how it can be tracked as it moves from location to location.
[Source: VirginiaView/Virginia Geospatial Extension Program]

This type of activity (using a GPS receiver to track down pre-recorded coordinates to find objects or locations) is also used in educational activities, in and out of the classroom, to teach students the relationship between a set of map coordinates and a real-world location, as well as give them practice in basic land navigation and GPS usage.

An example of the kind of item that's used in geocaching is a **geocoin**, a small object about the size of a casino chip. A geocoin is "trackable" in that the unique ID number on the coin can be logged into geocaching.com, and the location of where the coin is taken from and the location it's placed in can then be mapped (see **Figure 4.11**). Geocoins are primarily for recreational geocaching but can also be used in several educational activities involving

geocoin a small coin used in geocaching with a unique ID number that allows its changing location to be tracked and mapped

THINKING CRITICALLY
WITH GEOSPATIAL TECHNOLOGY 4.1

What Happens if GPS Stops Working?

What would happen if the Global Positioning System stopped working—or at least working at 100% capability? If the satellites stopped operating and weren't replaced, GPS would essentially stop functioning. If the number of satellites dropped below full operational capacity, GPS would only work some of the time and its accuracy would likely be greatly reduced. An operational failure of GPS forms the scenario of this article from GPS World: **www.gpsworld.com /gps-risk-doomsday-2010**.

If something like this came to pass, how would the lack of GPS affect life on Earth? What industries would be impacted by sudden deprivation of ac-

cess to GPS? What would be the effect on navigation at sea and in the air, and how would other kinds of travel be disrupted? What would be the military implications if the world were suddenly deprived of GPS? How would all of these factors affect your own life?

Is society too reliant on GPS? Since selective availability was turned off in 2000, there's been an overwhelming increase in GPS usage in many aspects of everyday life. Is the availability of GPS taken for granted? If the system could fail or be eradicated, what impacts would this have on people who believe that this technology would always be available to them?

geographic studies. For example, VirginiaView (one of the StateView programs of the AmericaView consortium—see Chapter 15 for more information) has put together a set of lesson plans that involve students tracking geocoin movements and relating these activities to larger projects like the tracking of invasive species and watershed characteristics. See VirginiaView's Website, **http://virginiaview.cnre.vt.edu**, for further information on combining trackable geocoins with geographic education.

Chapter Wrapup

This chapter provided an introduction to how the Global Positioning System operates as well as how GPS is able to find where a receiver is located on the ground. An increasingly popular GPS application is the kind of in-car navigation system manufactured by companies like Garmin, Magellan, or TomTom (many of these also operate on smartphones). These types of systems plot not only your location on a map using GPS, but can also show surrounding roads and nearby locations (such as restaurants or gas stations), locate an address, and show you the shortest route from your current location to any destination. The only one of these features that's actually using GPS is the one that finds your current location—the rest are all functions that are covered in Chapter 8. Until then, starting in the next chapter, we'll begin to focus on the actual computerized mapping and analysis that underlies those types of in-car systems, all of which are concepts of GIS.

This chapter's lab will start putting some of these GPS concepts to work, even if you don't have a GPS receiver. Before that, check out *GPS Apps* for information on some available apps that use the GPS capabilities of your mobile device, as well as *GPS and GNSS in Social Media* for Facebook, Twitter, and Instagram accounts, along with a blog related to GPS.

Important note: The references for this chapter are part of the online companion for this book and can be found at **www.macmillanhighered .com/shellito/catalog**.

GPS Apps

Here's a sampling of available representative GPS apps for your phone or tablet. Note that some apps are for Android, some are for Apple iOS, and some may be available for both.

- **Commander Compass Lite:** An app that determines your position and then turns your mobile device into a compass for navigation
- **Cyclemeter GPS:** An app designed for cyclists that finds your position and keeps track of your distance, riding time, and speed

- **Golf GPS & Scorecard—Swing By Swing Golf:** An app designed for golfers that determines your location and allows you to view golf courses and their features through aerial imagery
- **GPS Essentials:** An app that contains several GPS utilities (and bills itself as the "Swiss army knife" of GPS navigation)
- **Sygic: GPS Navigation:** An app that uses GPS, maps, and 3D visualization for mobile navigation
- **Travel Altimeter Lite:** An app that determines your position and elevation and also serves as a compass
- **Wikiloc Free:** An app that uses GPS to find your position for use in outdoor activities such as hiking

GPS and GNSS in Social Media

On Facebook, check out the following:

- DeLorme GPS:
  www.facebook.com/DeLormeGPS
- European GNSS Agency:
  www.facebook.com/EuropeanGnssAgency
- Garmin:
  www.facebook.com/Garmin
- Geocaching.com:
  www.facebook.com/geocaching
- GPS World Magazine:
  www.facebook.com/GPSWorldMag
- Magellan GPS:
  www.facebook.com/MagellanGPS
- TomTom:
  www.facebook.com/TomTom

Become a Twitter follower of:

- DeLorme GPS: @DeLormeGPS
- EGNOS Portal: @EGNOSPortal
- European GNSS Agency: @EU_GNSS

- **Garmin Blog Team:** @Garmin
- **Galileo GNSS:** @GalileoGNSS
- **Geocaching.com:** @GoGeocaching
- **GPS World Magazine:** @GPSWorld
- **Magellan GPS:** @magellangps
- **Tom Tom:** @TomTom
- **Trimble:** @trimbleGPS

Become an Instagram follower of:

- **Garmin:** @garmin_pics
- **Geocaching:** @geocaching

For further up-to-date info, read up on this blog:

- **Garmin Blog** (the official blog from Garmin with news and updates about GPS products and applications): **http://garmin.blogs.com**

Key Terms

3D trilateration (p. 95)
almanac (p. 93)
C/A code (p. 93)
channels (p. 93)
Compass (p. 103)
constellation (p. 91)
control segment (p. 92)
CORS (p. 101)
DGPS (p. 100)
dual frequency (p. 93)
EGNOS (p. 103)
ephemeris (p. 93)
Galileo (p. 103)
geocaching (p. 105)
geocoin (p. 106)
GLONASS (p. 102)
GNSS (p. 90)

GPS (p. 89)
GPX (p. 104)
MSAS (p. 103)
multipath (p. 99)
NAVSTAR GPS (p. 90)
NDGPS (p. 101)
P code (p. 93)
PDOP (p. 98)
pseudorange (p. 94)
SBAS (p. 101)
selective availability (p. 97)
single frequency (p. 93)
space segment (p. 90)
trilateration (p. 94)
user segment (p. 93)
WAAS (p. 101)
Y code (p. 93)

GNSS Applications

This chapter's lab will examine the use of the Global Positioning System (and other global navigation satellite systems) in a variety of activities. The big caveat is that the lab can't assume that you have access to a GPS receiver unit and that you're able to use it in your immediate area (and unfortunately, we can't issue you a receiver with this book). It would also be useless to describe exercises involving you running around (for instance) Boston, Massachusetts, collecting GPS data, when you may not live anywhere near Boston.

This lab will therefore use the free Trimble GNSS Planning Online software for examining GNSS satellite positions and other planning factors for the use of GNSS for data collection. In addition, some sample Web resources will be used to examine related GNSS concepts for your local area.

Important note: Though this lab is short, it can be significantly expanded if you have access to GPS receivers. If you do, some of the geocaching exercises described in Section 4.5 of this lab can be implemented. Alternatively, rather than just examining cache locations on the Web (as in Section 4.4), you can use the GPS equipment to hunt for the caches themselves.

Objectives

The goals for you to take away from this lab are:

▶ To examine satellite visibility charts to determine how many satellites are available in a particular geographic location and to find out when during the day these satellites are available

▶ To read a DOP chart to determine what times of day have higher and lower values for DOP for a particular geographic location

▶ To explore some Web resources to find publicly available geocaches or permanent marker sites that you can track down with a GPS receiver

Using Geospatial Technologies

The concepts you'll be working with in this lab are used in a variety of real-world applications, including:

▶ Precision agriculture, in which farmers utilize GNSS to find more efficient means of maintaining crops through having access to very accurate crop locations

▶ Geology, which uses GNSS for collecting location data while in the field, like the precise coordinates of groundwater wells

[Source: © Richard Hamilton Smith/Corbis]

Obtaining Software

No software or data is necessary to download for this lab—everything you will be working with can be done through a Web browser. However, the Trimble GNSS Planning Online software requires you to have Microsoft Silverlight installed on your computer in order to work properly. You can download and install Silverlight (for free) from here: **www.microsoft.com /getsilverlight**.

Important note: Software and online resources can change fast. This lab was designed with the most recently available version of the software at the time of writing. However, if the software or Websites have significantly changed between then and now, an updated version of this lab (using the newest versions) will be available online at **www.macmillanhighered.com /shellito/catalog.**

Lab Data

There is no data to copy in this lab. All data comes as part of the Trimble GNSS Planning Online software.

Settings

Latitude:	N 28.5635°
Longitude:	W 81.3724°
Height:	0m
Cutoff:	10°
Day:	1/23/2015
Visible Interval:	12:00 AM ▾ Time Span [hours]: 24 ▾
Time Zone:	(UTC-05:00) Eastern Time (US & Canada) ▾

Pick...

Obstructions...

Today

Apply

[Source: Trimble]

4.2 GNSS Satellite Availability Information

1. Trimble GNSS Planning Online allows you to examine satellite visibility (and other factors) from a local area at various times throughout the day. To choose which group of satellites you're examining, press the Satellite Library button. Answer Question 4.1.

> **Question 4.1** How many GNSS satellites are currently being tracked by the GNSS Planning Online? How many of them are considered healthy?

2. You'll see in the window all of the various satellites being tracked by Trimble GNSS Planning Online. By clicking on the button that corresponds with the name of the GNSS you'll see the active and inactive satellites that corresponds with the particular GNSS. Click on Galileo and answer Question 4.2.

> **Question 4.2** How many Galileo satellites are currently active?

3. Next, click on BeiDou (this is the name that Trimble GNSS Planning Online uses for Compass) and answer Question 4.3.

> **Question 4.3** How many Compass satellites are currently active?

4. For this lab application, we will only be working with GPS satellites. Remove the checkmarks from the GLONASS, Galileo, BeiDou, and

QZSS boxes, and then click on the GPS box so that only the GPS satellites are selected (and the GPS button is indented).

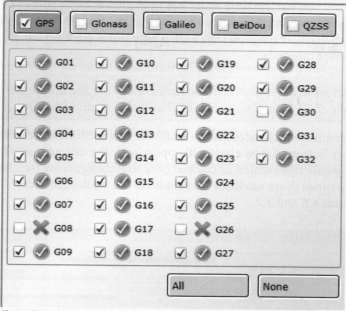

4.3 GNSS Satellite Visibility Information

1. With only the GPS satellites chosen, we'll now look at the number of satellites available for the chosen location on the chosen date. Press the **Number of Satellites** button and the view will change. A graph will appear, showing the number of visible satellites on the y-axis and a 24-hour time frame on the x-axis. Answer Questions 4.4 and 4.5.

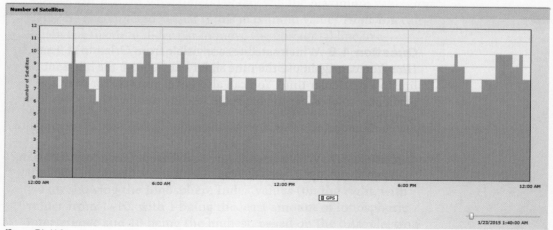

> **Question 4.4** At what specific times of day are the maximum
> number of GPS satellites visible? What is this maximum number?

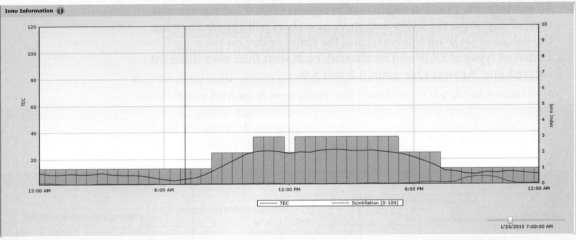

[Source: Trimble]

> **Question 4.10** At what time(s) of day will the highest value of the
> ionospheric index be encountered? What is this value? *Hint*: You can
> again use the time slider bar to move the horizontal line to a
> particular time.

> **Question 4.11** At what time(s) of day will the lowest value of the
> ionospheric index be encountered? What is this value?

5. Take another look at the three graphs you've used and answer
   Questions 4.12 and 4.13.

> **Question 4.12** Just based on number of visible satellites and the
> ionospheric and DOP calculations, what would be the "best" (i.e., peak)
> time range to perform GPS field measurements in Orlando and why?

> **Question 4.13** Just based on number of visible satellites and the
> ionospheric and DOP calculations, what would be the "worst" time
> range to perform GPS field measurements in Orlando and why?

4.4 GPS and Geocaching Web Resources

With some preliminary planning information available from Trimble GNSS
Planning Online, you can get ready to head out with a GPS receiver. For
some starting points for GPS usage, try the following:

1. Go to **www.geocaching.com**. This Website is home to several hundred
   thousand geocaches (small hidden items) at locations throughout the
   world and lists their coordinates so that you can use a GPS to track
   them down.

2. In order to access the features of geocaching.com, you will have to create an account. The basic account is free—click on Sign In, then follow the steps to create your account. The Website may send you an email with a link to use to verify your account information. Once your account is ready, log in to it using the username and password you just created and you'll be moved to the main Web page.

[Source: geocaching.com]

3. On the main page of the Website, enter 32801 (one of Orlando's many zip codes) in the box that asks for a ZIP code. Several potential geocaches should appear on the next Web page.

4. You can also search for benchmarks (permanent survey markers set into the ground) by going to **www.geocaching.com/mark**. On this benchmark page, input 32801 for the postal code. Answer Questions 4.14 and 4.15.

> **Question 4.14** How many benchmarks are available in the database for this postal code? How many of them are located within one mile or less of the Orlando 32801 postal code?

> **Question 4.15** What are the coordinates of the closest benchmark to the origin of the postal code?

4.5 GPS and Geocaching Applications

As we touched on back in the introduction, there's a lot more that can be done application-wise if you have access to a GPS receiver. For starters, you can find the positions of some nearby caches or benchmarks by visiting **www.groundspeak.com** or **www.geocaching.com** and then tracking them down using the receiver and your land navigation skills. However, there are a number of different ways that geocaching concepts can be adapted so that they can be explored in a classroom setting.

The first of these is based on material developed by Dr. Mandy Munro-Stasiuk and published in her article "Introducing Teachers and University Students to GPS Technology and Its Importance in Remote Sensing Through GPS Treasure Hunts" (see the online chapter references for the full citation of the article). Like geocaching, the coordinates of locations have been determined ahead of time by the instructor and given to the participants (such as university students or K–12 teachers attending a workshop) along with a GPS receiver. Having only sets of latitude/longitude or UTM coordinates, the participants break into groups and set out to find the items (such as statues, monuments, building entrances, or other objects around their campus or local area). The students are required to take a photo of each object they find, and all participants must be present in the photo (this usually results in some highly creative picture taking). The last twist is that the "treasure hunt" is a timed competition, with the team that returns first (and having found all the correct items) earning some sort of reward, such as extra credit or—at the very least—bragging rights. In this way, participants learn how to use the GPS receivers, how to tie coordinates to real-world locations, and they also get to improve their land navigation skills by reinforcing their familiarity with concepts such as how northings and eastings work in UTM.

A similar version of this GPS activity is utilized during the OhioView SATELLITES Teacher Institutes (see Chapter 15 for more information on this program). Again, participants break up into groups with GPS receivers, but this time each group is given a set of small trinkets (such as small toys or plastic animal figures) and instructed to hide each one, registering the coordinates of the hiding place with their GPS receiver and writing them on a sheet of paper. When the groups reconvene, the papers with the coordinates are switched with another group, and each group now has to find the other's hidden objects. In this way, each group sets up its own "geocaches" and then gets challenged to find another group's caches. Like before, this is a timed lab with a reward awaiting the team of participants that successfully finds all of the hidden items at the coordinates they've been given. This activity helps to reinforce not only GPS usage and land navigation, but also the ties between real-world locations and the coordinates being recorded and read by a GPS receiver. Both of these activities have proven highly successful in reinforcing these concepts to the participants involved.

However you get involved with geocaching, a helpful utility for managing geocached data and waypoints from different software packages is the Geocaching Swiss Army Knife, available for free download at **www.gsak .net.** This utility comes with a free trial.

Closing Time

This lab illustrated some initial planning concepts and some directions in which to take GPS field work (as well as some caches and benchmarks that are out there waiting for you to find) to help demonstrate some of the concepts delineated in the chapter. The geocaching field exercises described in section 4.5 can also be adapted for classroom use to help expand the computer portion of this lab with some GPS field work as well.

Starting with Chapter 5, you'll begin to integrate some new concepts of geographic information systems to your geospatial repertoire.

5

Working with Digital Geospatial Data and GIS

Geographic Information Systems, Modeling the Real World, Vector Data and Raster Data, Attribute Data, Joining Tables, Metadata, Esri, ArcGIS, and QGIS

Once you've measured and collected your geospatial data, it's time to do something useful with it. The nature of geospatial data lends itself to projects like analyzing where one location lies in relation to another or the appeals of one location versus another. When a new library is to be built, it should be placed at the optimal location to serve the greatest number of people without duplicating the function of other libraries in the area. When a company is deciding where to open a new retail store, they'll need information about the location—the city, the town, the neighborhood, the block, who lives there, why, and how many. When a family is planning its vacation to Florida, they'll want to know the best way to get there and the best places to stay, not too far from the theme parks they're planning to visit. All of these concepts involve the nature of geospatial data and the ability to analyze or compare particular locations and their attributes. Geospatial technology is used to address these ideas and solve these types of problems (and many, many more). Whenever examination, manipulation, or analysis of geospatial data is involved (for instance, if you need to tie non-spatial data to a location or if you want to use this information to create a map of your analysis), **geographic information systems (GIS)** are essential to getting these tasks done.

GIS is a powerful tool for analyzing geospatial data (the "geographic information" of the title). This ability to explicitly utilize geospatial data is what separates GIS from other information systems or methods of analysis. For

geographic information system (GIS) a computer-based set of hardware and software used to capture, analyze, manipulate, and visualize geospatial information

instance, a spreadsheet enables you to tabulate housing values and the amount of money paid for a house at a particular address, but even though you're using data that has something to do with a location, there's nothing in the spreadsheet that allows you to examine that data in a spatial context (such as translating the housing values into a map of where these houses are).

The ability to map the real-world locations of the houses, to examine trends in housing values in spatial terms, and to search for significant patterns or clusters in the geospatial data are what makes GIS unique. Using the coordinate and measurement concepts discussed in the previous chapters allows us to create, map, and analyze digital geospatial data in a wide variety of geospatial technology applications (see **Figure 5.1** and *Hands-On Application 5.1: Using GIS Online* for an example of GIS data used by a county government).

GIS is by its nature 100% computer-based—whether it involves software running on a desktop, laptop, mobile device, or served off the Web, the "system" being referred to is a computer system. The information handled by GIS is geospatial information—something that has direct ties to a location on Earth's surface. GIS can also utilize non-spatial data and link it directly to a location. For instance, if you have a set of data points that represent the locations of local car washes (that is, geospatial data) and a separate spreadsheet of data of how many cars use a particular car wash (that is, non-spatial data), you can link the usage data in the spreadsheet to the car wash location so that the location contains that data. From there, you can do analyses of local population demographics or road conditions about how they relate to the usage statistics of the car wash at that particular location. This wouldn't be possible without GIS being able to link non-spatial data to a location. Lastly, GIS can also perform multiple types of operations related to geospatial data—it can be used to analyze, capture, create, manipulate, store, or visualize all manner of geospatial information, not just simply design maps (though it can do that too).

FIGURE 5.1 The online GIS utility for Mahoning County, Ohio. *[Source: Mahoning County GIS Department]*

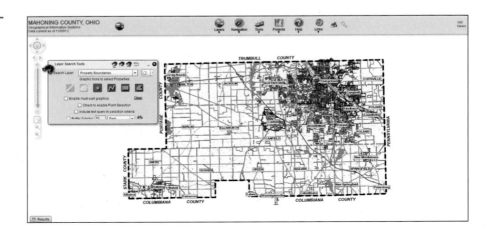

HANDS-ON APPLICATION 5.1

Using GIS Online

There are numerous online GIS resources for viewing and analyzing geospatial data. An example of a local GIS is from Mahoning County, Ohio. Open your Web browser and go to **http://gis.mahoningcountyoh.gov/mahoningpublicviewer**. Under the Layers icon, select the option for Layer Search Tools then select the Text Search tab. Use Property Boundaries as the Search Layer, PARCEL ID as the Search Layer field, and look for 29-042-0-002.00-0 (this is the parcel number for the Boardman Township Park). The map will zoom to the parcel of the park and a pop-up will appear with attribute information about that parcel (such as the sale date, acreage, and land value).

Next, under the Layers icon, select the option for Map Layers. Here, you can select which GIS layers are displayed on the map. Place a checkmark in the box of the most recent date for Orthophoto to look at overhead aerial photography of the area. You can use the mouse wheel or the options under the Navigation icon to zoom in and out of the view and get a closer look at the map and the surroundings of the park.

Expansion Questions

- From examining the aerial photos and parcels, what types of land uses are adjacent to the park boundaries (commercial, residential, etc.)?

- What attributes have been recorded for this parcel of land?

- Check your local state or county government's Websites to see if a similar kind of online GIS data viewer has been set up for your area (it may be part of your county auditor's Website or a separate GIS department for the state or county). If a GIS is available, what types of layers are available for you to view and interact with?

GIS isn't a development that just popped up over the last couple of years. Large-scale mapping efforts were done using computer-based cartography in the 1950s and 1960s by a number of different government agencies and other labs (including Harvard Laboratory). The term "GIS" first appeared in the early 1960s with the implementation of **CGIS** (the acronym for the Canadian Geographic Information System), which was designed to provide large-scale mapping of land use in Canada. The development of CGIS is attributed to the late Roger Tomlinson, who has become known as "the father of GIS."

GIS developments have continued over the years through the efforts of government agencies and the private sector, and these days GIS technology and concepts are extremely widespread (see *Hands-On Application 5.2: GIS Current Events Maps* for another example of using GIS concepts online).

> **CGIS** the Canadian Geographic Information System—a large land inventory system developed in Canada, and the first system of this type to use the name "GIS"

How Does GIS Represent Real-World Items?

GIS has a multitude of applications, ranging from mapping natural gas pipelines to determining which parts of the natural landscape are under the greatest pressure to be developed. No matter the application, the data (such as the locations and lengths of pipelines or the area and extent of land cover) needs to be represented somehow. GIS provides a means of representing (or modeling) this data in a computer environment so that it can be easily analyzed or manipulated. A model is a kind of representation that simplifies or generalizes real-world information. For instance, if you were tasked to make a map of all of the stop signs in your city, you'd need to be able to represent the locations of all the signs on a map at the scale of the entire city. By laying out a map of the city and placing a dot at each sign location, you would have a good representation of them. The same holds true if you had to come up with a map of all the roads in your town. You could attempt to draw the roads at their correct width, but the roads' locations can be represented simply and accurately by lines drawn on the map.

Whenever you're considering how to model real-world phenomena with a GIS, you first have to decide on the nature of the items you want to represent. In the GIS, there are two ways of viewing the world. The first is called the **discrete object view**. In the discrete object view, the world is made up of a series of objects, each of which has a fixed location, or a fixed starting and stopping point, or some sort of fixed boundary. A street starts at one point and ends at another, or there is a definite boundary between your property and your neighbor's. Around a university campus, there will be numerous objects representing buildings, parking lots, stop signs, benches, roads, trees, and sidewalks. By viewing the world in this way, real-world

discrete object view a conceptualization of the world in which all reality can be represented by a series of separate objects

---

HANDS-ON APPLICATION 5.2

GIS Current Events Maps

There are a lot of GIS applications available online that are used for mapping and analysis of geospatial data. For some interactive examples, go to **www.esri.com/esri-news /maps**. The Website will show you different categories of Map Galleries related to current events, such as public information or public policy issues. These interactive Web maps and applications were created using GIS software from a company called Esri. We'll talk a lot more about Esri and its software platform, ArcGIS, later in this chapter, but for now we'll be examining some of their online GIS applications.

On the main page, select one of the categories of Web maps to examine and you'll be presented with several different maps and applications to view. When you open a Web map, you'll be placed into a GIS environment where you can pan around the map, zoom in and out, and access different features of the maps. Some Web maps will allow you to measure distances and areas and get information back (as a pop-up) about layers on the map, while others will use pre-made templates to convey geospatial and attribute information to you.

Expansion Questions

- What kinds of features are being mapped in the Web maps available online? (You may want to choose one of the available categories and investigate several available Web maps.)

- How are the Web maps able to convey non-spatial information to the user (hint: click on some of the features in a Web map itself)?

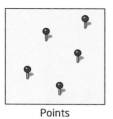 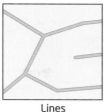 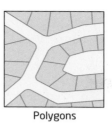

Points   Lines   Polygons

FIGURE 5.2 Discrete objects in GIS: basic points, lines, and polygons.

items can be modeled in the GIS as a series of objects (like a collection of lines to represent a road system, or a series of points to represent fire hydrant locations).

When adapting the discrete object view of the world to a data model, items in the real world are represented in the GIS by one of three objects (**Figure 5.2**):

▶ **Points**: These are zero-dimensional objects, a simple set of coordinate locations.

▶ **Lines**: These are one-dimensional objects, created from connecting starting and ending points (and any points in between that give the line its shape).

▶ **Polygons**: These are two-dimensional objects that form an area from a set of lines (or having an area defined by a line forming a boundary).

These three items are referred to as **vector objects**, as they make up the basis of the GIS **vector data model**. Basically, the vector data model is a means of representing and handling real-world spatial information as a series of vector objects—items are realized in the GIS as points, lines, or polygons. For instance, locations of fire hydrants, crime incidents, or trail-heads can be represented as points; roads, streams, or power conduits are represented as lines; and land parcels, building footprints, and county boundaries are represented as polygons. When dealing with a much larger geographic area (for instance, a map of the whole United States), cities may be represented as points, interstates as lines, and state boundaries as polygons (see **Figure 5.3** on page 128).

A lot of GIS data has already been created (such as road networks, boundaries, utilities, etc.), but in order to update or develop your own data, there are a few steps that need to be taken. **Digitizing** is a common way to create the points, lines, and polygons of vector data. With digitizing, you are in essence "tracing" or "sketching" over the top of areas on a map or other image to model features in that map or image in the GIS. For instance, if you have an aerial photograph of your house being shown on the screen of your GIS software, you could create a polygon by using the mouse to sketch the outline of your house and save it in the GIS as a representation of your house. If you sketch outlines of four of your neighbors' houses as well, you'll have several polygons stored in a single layer in the GIS. When you examine the

points zero-dimensional vector objects

lines one-dimensional vector objects

polygons two-dimensional vector objects

vector objects points, lines, and polygons that are used to model real-world phenomena using the vector data model

vector data model a conceptualization of the world that represents spatial data as a series of vector objects (points, line, and polygons)

digitizing the creation of vector objects through sketching or tracing representations from a map or image source

FIGURE 5.3 Points (city locations), lines (interstates), and polygons (state borders) as represented in GIS.
[Source: Esri]

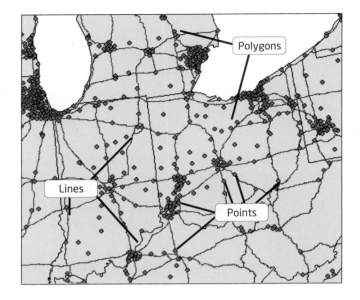

data, you'll find you have five polygon entries in that data layer (your house and four others). Similarly, you could sketch each house's driveway as a line object by tracing the mouse over the appropriate place in the image, which would result in a second data layer that's comprised of five lines. With digitizing, each of your sketches can be translated into a vector object.

Digitizing is commonly done through a process referred to as "heads-up digitizing" or "on-screen digitizing," which is similar to the sketching just described. In heads-up digitizing, a map or other image (usually an aerial photograph or satellite image) is displayed on the screen as a backdrop. Decide if you'll be sketching points, lines, or polygons and create an empty data layer of the appropriate type (to hold the sketches you'll make). From there, you use the mouse to sketch over the objects, using as much detail as you see fit (**Figure 5.4**). Keep in mind that despite your best efforts, it may likely be impossible or unfeasible to sketch every detail in the image. For instance, when you're tracing a river, it may not be possible to capture each bend or curve by digitizing straight lines.

When objects are created in GIS, they can be set up with coordinates, but the GIS still needs to have knowledge of how these objects connect to each other and what relation each object has to the others. For instance, when two lines are digitized that represent crossing streets, the GIS needs to have some sort of information that an intersection should be placed where the two lines cross. Similarly, if you have digitized two residential parcels, the GIS would require some information that the two parcels are adjacent to one another. This notion of the GIS being able to understand how objects connect to one another independent of their coordinates is referred to as **topology**. With topology, geometric characteristics aren't changed, no matter how the dataset may be altered (by projecting or transforming the data).

topology how vector objects relate to each other (in terms of their adjacency, connectivity, and containment) independently of the objects' coordinates

Topology establishes adjacency (how one polygon relates to another polygon, in that they share a common boundary), connectivity (how lines can intersect with one another), and containment (how locations are situated inside of a polygon boundary).

The layers of the National Hydrography Dataset (**NHD**) are examples of vector data applications in GIS. The NHD is a set of GIS data compiled and distributed by the USGS that maps out various water bodies across the United States. Lakes, rivers, streams, canals, ponds, and coastlines are among the features accessible in GIS through NHD. NHD layers are represented by line objects (such as rivers) or polygon objects (such as lakes) with numerous attributes attached to each feature about these water features (see **Figure 5.5** on page 130). NHD serves as the basis for water resources within The National Map (see Chapter 1) and can be freely downloaded from the Web for use in GIS. (See *Hands-On Application 5.3: The National Hydrography Dataset* for viewing this data online as well as how to obtain it.)

There's also a second way of viewing the world, in which not everything has a fixed boundary or is an object. For instance, things such as temperature, atmospheric pressure, and elevation vary across Earth and are not best represented in GIS as a set of objects. Instead, they can be thought of as a surface that's made up of a near-infinite set of points. This way of viewing the world is called the **continuous field view**. This view implies that real-world phenomena continuously vary; rather than a set of objects, a surface filled with values is used to represent things. For instance, elevation gradually varies across a landscape from high elevations to low elevations filled with hills, valleys, and flatlands, and at every location we can measure the height of the land. Similarly, if you're standing at a ranger station in a

FIGURE 5.4 An example of heads-up digitizing—tracing the boundary of the football field on Youngstown State University's (YSU's) campus to create a polygon object. *[Source: Ohio Geographically Referenced Information Program (OGRIP), Ohio Statewide Imagery Program (OSIP), April 2006/Esri]*

NHD The National Hydrography Dataset—a collection of GIS data of water resources for the United States

continuous field view a conceptualization of the world in which all items vary across Earth's surface as constant fields, and values are available at all locations along the field

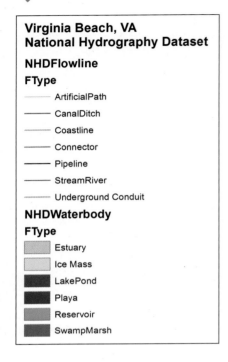

Virginia Beach, VA
National Hydrography Dataset

NHDFlowline

FType

———— ArtificialPath

———— CanalDitch

———— Coastline

———— Connector

———— Pipeline

———— StreamRiver

········ Underground Conduit

NHDWaterbody

FType

▨ Estuary

▨ Ice Mass

▨ LakePond

▨ Playa

▨ Reservoir

▨ SwampMarsh

FIGURE 5.5 Several different features from the National Hydrography Dataset for Virginia Beach, VA. *[Source: NHD/USGS]*

HANDS-ON APPLICATION 5.3

The National Hydrography Dataset

The National Hydrography Dataset (NHD) layers can be downloaded from The National Map like other datasets. To see how the NHD is used, go to **http://viewer.nationalmap.gov/viewer** to open The National Map Viewer. From the available options, place a checkmark in the Hydrography (NHD) box (this will display the NHD features and allow you to query them). Next, search for a location (such as West Palm Beach, Florida). The Viewer will zoom in to the chosen location and you'll see the various water bodies appearing in blue. Choose the Identify tool from the toolbar (the white "i" inside of the blue circle) and click on a water body to load all of the appropriate water data about it (give the Viewer a minute to do this). Once

you see the data load on the left-hand side, use the Identify tool on the water body once more to get information back about it. Clicking on the All Results button will clear your results and let you make a fresh query of the NHD once more.

Expansion Questions

- What is the area (in square kilometers) of Lake Worth in West Palm Beach, FL? According to the NHD, what type of water body is it classified as?

- What is the area (in square kilometers) of Lake Mangonia in West Palm Beach, FL? According to the NHD, what type of water body is it classified as?

park and preparing a rescue plan for stranded hikers, you'll want to know the distance from the station to every location in the park. Thus, you could have two layers in your GIS—one showing the elevation at every place and another showing the distance to the ranger station at each location.

How Can You Represent the Real World as Continuous Fields?

When these kinds of continuous fields are represented in GIS, the three vector objects (and thus, the vector data model) are often not the best way of representing data. Instead, a different conceptualization called the **raster data model** is usually used. In the raster data model, data is represented using a set of evenly distributed square **grid cells**, with each square cell representing the same area on the Earth's surface (see **Figure 5.6** for the layout of the grid cells). For instance, to represent an entire section of the landscape, each grid cell might represent 10 square feet, or 30 square meters, or 1 square kilometer, depending on the grid cell size being used for the model. Regardless of the grid cell size, each cell has the same resolution (that is, the same grid dataset cannot mix cells of 30 square feet together with cells of 5 square feet). Also, each grid cell contains a single value representing the data being modeled. For instance, in an elevation raster, each grid cell could contain a value for the elevation at that area on the ground. Datasets such as land cover, soils, or elevation are all commonly represented using the raster data model.

The National Land Cover Database (**NLCD**) is an example of a raster dataset used in GIS. Developed by a consortium of United States agencies (including the USGS and the Environmental Protection Agency), NLCD provides 30-meter raster data covering the entire United States, wherein each grid cell is coded with a value that corresponds to a specific land cover type at that location. NLCD designations include categories such as "Open Water," "Developed, High Intensity," "Deciduous Forest," "Pasture/Hay," and

> **raster data model** a way of representing spatial data that utilizes a series of equally spaced and sized grid cells
>
> **grid cell** a square unit, representing some real-world size, which contains a single value

> **NLCD** the National Land Cover Database is a raster-based GIS dataset that maps the land-cover types for the entire United States at 30-meter resolution

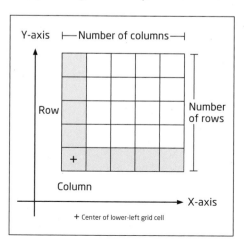

FIGURE 5.6 Grid cells of raster data model.

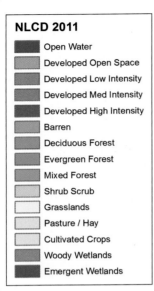

NLCD 2011

- Open Water
- Developed Open Space
- Developed Low Intensity
- Developed Med Intensity
- Developed High Intensity
- Barren
- Deciduous Forest
- Evergreen Forest
- Mixed Forest
- Shrub Scrub
- Grasslands
- Pasture / Hay
- Cultivated Crops
- Woody Wetlands
- Emergent Wetlands

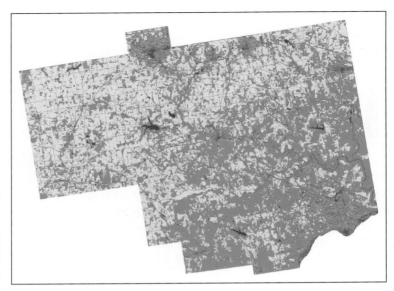

FIGURE 5.7 A section of the 2011 NLCD showing land cover in Columbiana County, Ohio, modeled with raster grid cells.
[Source: National Land Cover Database]

"Woody Wetlands." NLCD provides a means of broad land-cover classification at a state, regional, or national scale (**Figure 5.7**). NLCD products are available for land cover circa 1992, 2001, 2006, and 2011. A separate dataset is available that details the change in land-cover types between the 1992 and 2001 datasets (and datasets showing change between the other dates as well), enabling researchers not only to measure land-cover types, but also to see how the landscape is changing (like where agricultural fields or forests are converting over to urban land uses). NLCD data is distributed free of charge via Web download in a format that can be easily read or converted by GIS software. See *Hands-On Application 5.4: The National Land Cover Database (NLCD)* for how to access NLCD data.

HANDS-ON APPLICATION 5.4

The National Land Cover Database (NLCD)

The USGS has an online tool for viewing the 1992, 2001, 2006, and 2011 NLCD (and some products derived from them), as well as letting you download sections of the dataset for use in GIS. The viewer is available at **www.mrlc.gov/viewerjs**. The default view will be the 2011 NLCD data. From the Data option on the right side of the screen, you can select different NLCD data to view using the Layer Manager. If you want to download the data, the tools in the viewer will allow you to download the raster datasets directly.

Expansion Questions

- Zoom in to your own local area and examine the NLCD—what do the various colors of the grid cells represent (in terms of their land cover)?

- How are the 1992 and 2011 datasets different from one another (in terms of the land-cover types they are mapping)?

Real-world items can be represented in GIS using raster data or vector data, although each data model lends itself to different types of features. For instance, to create a precipitation map of an area, scientists may record rainfall amounts at dozens of different points. A map would then show the rainfall levels only at these scattered locations, which would be best represented with points (vector objects). However, it might be more useful to take that data and create an entire continuous surface that would predict the level of rainfall at all areas in the region, which could then be shown with raster data in which every grid cell would contain the value of the amount of predicted rainfall.

However, some objects are best represented with vector data, especially those having a definite boundary (such as building footprints or land parcels) or a fixed starting and ending point (such as a road or power line). No matter how these items are represented in GIS, it's important to note that both vector and raster data structures are only modeling the spatial features of real-world items—but GIS is also capable of handling all of the non-spatial data that goes along with them.

 ## How Is Non-Spatial Data Handled by GIS?

Whether you're mapping schools ranked by average SAT scores or the locations of the United States' top tourist destinations, all of this numeric data needs somehow to be represented in GIS. For instance (going back to the digitizing example), if you have two data layers (one containing polygons representing housing footprints and one containing lines representing the driveways), each would contain multiple objects. The GIS will have information about the spatial properties of the objects in that layer, but what it does not have is any other information about those objects (such as the tax-assessed value of the house, the house's address, the name of the owner, the number of people living in the house, the material the driveway's made of, or how old the driveway is). All of these **attributes** represent other non-spatial information associated with each of the objects. When representing attribute data in GIS, these values can take one of four forms: nominal, ordinal, interval, or ratio data.

Nominal data are values that represent some sort of unique identifier. Your Social Security number or telephone number would both be examples of nominal data—both of these values are unique to you. Names or descriptive information that are associated with a location would be nominal data, as they're unique values, the same as a value from a land-cover classification scheme. Also, the difference between numerical nominal values is not significant—you can't add your Social Security number to a friend's number and come up with a relative's number—in the same way you can't subtract

attributes the non-spatial data that can be associated with a spatial location

nominal data a type of data that is a unique identifier of some kind—if numerical, the differences between numbers are not significant

your phone number from one friend's number and come up with another friend's phone number.

Ordinal data is used to represent a ranking system of data. If you have data that is placed in a hierarchy where one item is first, another is second, and another is third, that data is considered ordinal. For instance, if you have a map of the city and you're going to identify the locations of the homes of the most successful competitors in a local car-racing event, the points on the map will be tagged as to the location of the first-place winner, the second-place winner, etc. Ordinal data deals solely with the rankings themselves, not with the numbers associated with these ranks. For instance, the only values mapped for the car-race winners is their placement in the race, not their winning times, the cars' speeds, or any other values. The only thing ordinal data represents is a clear value of what item is first, which is second, and so on. No measurements can be made of how much better the winning driver's time was than the time of the driver who took second place, only that one driver was first and one driver was second.

Interval data is used when the difference between numbers is significant, but there is no fixed zero point. With interval data, the value of zero is just another number used on the scale. For instance, temperature measured in degrees Celsius would be considered interval data since values can fall below zero—the value of zero only represents the freezing point of water, not the bottom of the Celsius temperature scale. However, since there is no fixed zero point, we can make differences between values (for instance, if it was 15 degrees yesterday and 30 degrees today, we can say it was 15 degrees warmer), but dividing numbers wouldn't work (since a temperature of 30 degrees is not twice as warm as a temperature of 15 degrees).

Ratio data are values with a fixed and non-arbitrary zero point. For instance, a person's age or weight would be considered ratio data since a person cannot be less than zero years in age or weigh less than zero pounds. With ratio data, the values can be meaningfully divided and subtracted. If we want to know how much time separated the car-racing winners, we could subtract the winning driver's time from the second-place driver's time and get the necessary data. Similarly, a textbook that costs $100 is twice as expensive as a textbook that costs $50 (as we divide one number by the other). Ratio data is used when comparing multiple sets of numbers and looking for distinctions between them. For instance, when mapping the locations of car wash centers, the data associated with the location is the number of cars that use the service. By comparing the values, we can see how much one car wash outsells the others, or calculate the profit generated by each location.

Each layer in the GIS has an associated **attribute table** that stores additional information about the features making up that layer. The attribute table is like a spreadsheet where objects are stored as **records** (rows) and the information associated with the records—the attributes—is stored as **fields**

ordinal data a type of data that refers solely to a ranking of some kind

interval data a type of numerical data in which the difference between numbers is significant, but there is no fixed non-arbitrary zero point associated with the data

ratio data a type of numerical data in which the difference between numbers is significant, but there is a fixed non-arbitrary zero point associated with the data

attribute table a spreadsheet-style form where the rows consist of individual objects and the columns are the attributes associated with those objects

records the rows of an attribute table

fields the columns of an attribute table

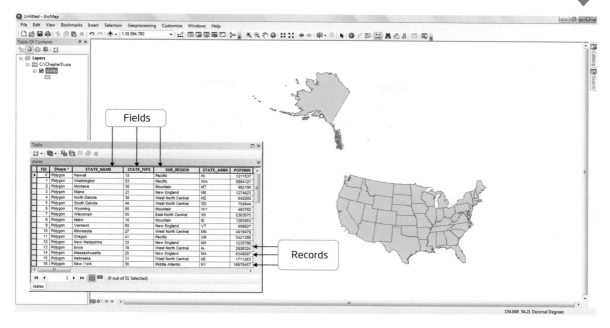

(columns). (See **Figure 5.8** for an example of an attribute table in conjunction with its GIS objects.) To use the housing example again, the houses' attribute table would consist of five records, each representing a house polygon. New fields could be created, having headings such as "Owner" and "Appraised Value" so that this type of descriptive, non-spatial data could be added for each of the records. These new attributes can be one of the four data types (for instance, "Owner" would be nominal data, while "Appraised Value" would be ratio data). In this way, non-spatial data is associated with a spatial location.

A raster attribute table can't be handled the same way as one for vector objects since a raster dataset is composed of multiple grid cells (for instance, a 30-meter resolution grid of a single county could take millions of cells to model). Rather than having separate records for each grid cell, raster data will often be set up in a table with each record featuring the value for a raster cell and an attribute showing the count of how many cells comprise the dataset. For instance, a county-land-use dataset may contain millions of cells, but only consist of seven values. Thus, the raster would have seven records (one for each value) and another field whose value would be a summation of how many cells have that value. New attributes would be defined by the raster values (for instance, the type of land use that value represents).

Besides creating new fields and populating them with attributes by hand, a **join** is another way GIS allows non-spatial data to be connected to spatial locations. This works by linking the information for records in one

FIGURE 5.8 Attribute table associated with a GIS data layer.
[Source: ArcGIS]

join a method of linking two (or more) tables together

table to corresponding records in another table. This operation is managed by both tables having a field in common (this field is also referred to as a **key**). For instance, **Figure 5.9** shows two tables, the first from a shapefile's attribute table that has geospatial information about a set of points representing the states of the United States, while the second table has non-spatial Census information about housing stock for those states. Because both tables have a common field (that is, the state's abbreviation), a join can be performed on the basis of that key.

After the join, the non-spatial Census information from the second table will be associated with the geospatial information from the states' table. For instance, you can now select the record for Alaska and have access to all of its housing information since the tables have been related to each other. Joining tables is a simple method for linking sets of data together, and especially for linking non-spatial spreadsheet data to geospatial locations. With all of this other data available (and linked to a geospatial location), new maps can be made. For instance, you could now create maps of the United States showing any of the attributes (total number of houses, number of vacation homes, or percentage of the total housing that are vacation homes) linked to the geospatial features.

FIGURE 5.9 A joining of two attribute tables, based on their common field (the state's abbreviation attribute).

[Source: Esri]

What Other Kind of Information Do You Need to Use GIS Data?

Once you have your geospatial and attribute data within the GIS, there's one more important piece of data you're going to need—descriptive information about your data. When creating data or obtaining it from another source, you should have access to a separate file (usually a "readme" text file or an XML document) that fully describes the dataset. Useful things to know would be information about the coordinate system, projection, and datum of the data; when the data was created; what sources were used to create it; how accurate the data is; and what each of the attributes represents. All of this kind of information (and more) is critical to understanding the data in order to properly use it in GIS. For instance, if you have to reproject the data, you're going to have to know what projection it was originally in; if you're going to be mapping certain attributes, then you'll have to know what each one of those strangely named fields in the attribute table actually represents. In GIS, this "data about your data" is referred to as **metadata**.

The Federal Geographic Data Committee (FGDC) has developed some standards for the content and format that metadata should have. These include:

▶ Identification of the dataset (a description of the data)

▶ Data quality (information about the accuracy of the data)

▶ Spatial data organization information (how the data is represented—whether it's vector or raster, for instance)

▶ Spatial reference information (the coordinate system, datum, and projection used)

▶ Entity and attribute information (what each of the attributes means)

▶ Distribution information (how you can obtain this data)

metadata descriptive information about geospatial data

THINKING CRITICALLY WITH GEOSPATIAL TECHNOLOGY 5.1

What Happens When You Don't Have Metadata?

Say you're trying to develop some GIS resources for your local community, and you need data about the area's sewer system. You get the data in a format that your GIS software will read, but it doesn't come with any metadata—meaning you have no information on how the sewer data was created, its accuracy, what the various attributes indicate, when it was created, or the datum, coordinate system, or projection the data is in. How useful is this dataset for your GIS work? When metadata for a GIS layer is unavailable (or incomplete), what sorts of issues can occur when you're trying to work with that data?

▶ Metadata reference information (how current the metadata is)

▶ Citation information (if you want to cite this data, how it should be done)

▶ Time period information (what dates the data covers)

▶ Contact information (whom to get in touch with for further information)

 ## What Kinds of GIS Are Available?

There are many different types of GIS software packages available, everything from free lightweight "data viewers," to open source programs, to expensive, commercially available software packages. There are numerous companies producing a wide variety of GIS software. According to a 2011 GIS Salary Survey conducted by the Urban and Regional Information Systems Association (URISA), GIS products from the company called Esri were found to be among the most popular and widely used.

Esri a key developer and industry leader of GIS products

ArcGIS Esri's GIS platform

Located in Redlands, California, **Esri** was founded by Jack and Laura Dangermond in 1969. By 1982, Esri had released the first of its "Arc" products, the initial version of Arc/Info, a key GIS software package. Esri's GIS software has gone through many changes and evolutions over the years to reach its current form: **ArcGIS**, an entire GIS software platform. The concept of a platform means multiple pieces are brought together—software on your desktop computer, cloud-based software, mobile apps, and more that all use the same common techniques and tie together. This is the best way to think of ArcGIS: a GIS platform that you can use in a variety of different ways, whether on a desktop computer, a Web browser, an app on a tablet, or a toolkit for developers.

ArcGIS for Desktop the component of ArcGIS that runs on a desktop computer

ArcGIS for Desktop is the part of the platform that runs on a personal computer. At the time of this writing, the most recent iteration of the software was ArcGIS for Desktop 10.3. ArcGIS for Desktop is available in three different varieties: ArcGIS Basic, ArcGIS Standard, and ArcGIS Advanced. These can best be thought of as different levels of the same software package— all of them are considered ArcGIS for Desktop, but the Basic version has the smallest number of functions, the Standard version has the mid-range of functions, and the Advanced version has the largest number of functions. No matter which of the three versions of ArcGIS for Desktop you're using, the main component is referred to as **ArcMap**, which is used for viewing and analyzing data (see **Figure 5.10**).

ArcMap the component of ArcGIS for Desktop used for viewing and analyzing data

Within ArcMap, GIS datasets can be added and treated as different map layers. ArcGIS offers multiple sets of tools for analysis and data manipulation and enable you to make maps from your data (things you'll be doing in the labs for Chapters 6 and 7). In addition to ArcMap, there is also ArcCatalog, a separate utility for managing available GIS data. In ArcGIS for

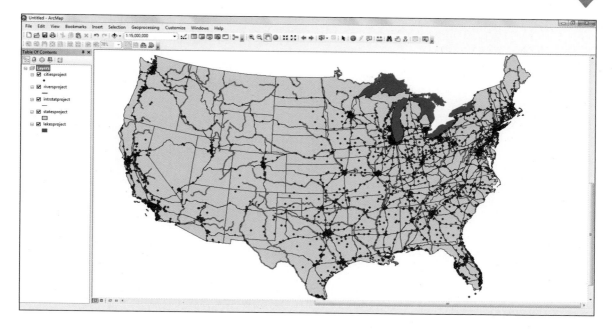

FIGURE 5.10 The basic layout of Esri's ArcMap.
[Source: Esri]

Desktop 10.3, these functions are contained in the **Catalog** in ArcMap itself. ArcGIS for Desktop 10.3 also saw the debut of **ArcGIS Pro**, a new stand-alone GIS program that can run side by side with ArcMap.

While ArcGIS is a powerful GIS program that contains a wide variety of tools and features, computer programming can be used to create customizations of the software or help automate tasks or perform analysis. The **Python** programming language is used for writing computer code (called a script) for working with ArcGIS. Python (named after the Monty Python comedy team) is commonly used with ArcGIS and is simple to use for programming tasks. The tools needed to work with writing Python scripts are available for free online.

ArcGIS Online is the cloud-based part of the ArcGIS platform (see **Figure 5.11** on page 140). ArcGIS Online allows you to build Web maps and applications using only your Web browser as well as transfer your data and maps from ArcGIS for Desktop to the cloud. You'll work with ArcGIS Online's capabilities later in Chapter 15. The ArcGIS platform gives you a full range of GIS capabilities and applications. See *Hands-On Application 5.5: Esri News and 80-Second Videos* for more information about Esri software and applications.

With the proliferation of numerous Esri products on the market and in use today, it's no surprise that a lot of the available GIS data is in an Esri data format. Raster data is widely available in (or can be easily imported into) Esri Grid format, which can be directly read by Esri software. Vector data can usually be found in one of the following three different varieties: coverage, shapefile, and geodatabase.

Catalog the component of ArcGIS for Desktop used for managing data (which contains the functionality of the previous ArcCatalog)

ArcGIS Pro a new stand-alone GIS program that comes as part of ArcGIS for Desktop 10.3

Python a free computer programming language used for writing scripts in ArcGIS

ArcGIS Online a cloud-based resource for creating and sharing GIS Web maps and applications

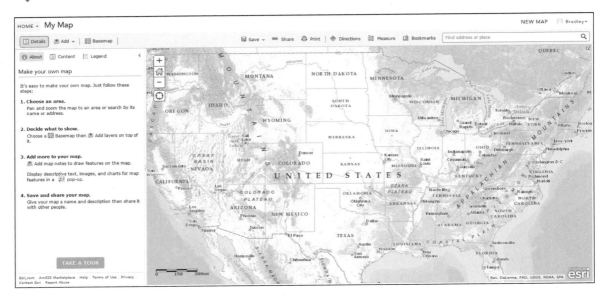

FIGURE 5.11 The basic layout of ArcGIS Online.
[Source: Esri]

A **coverage** consists of multiple files inside of a directory structure. The folder that a coverage is stored in is referred to as a workspace, and a workspace may contain multiple coverages. A workspace will also contain a separate folder called the INFO directory, which contains extra files as needed by each coverage in the workspace. Because a coverage is structured to be capable of holding multiple files, Esri has created a method for taking all the necessary files that make up one coverage and exporting

---

HANDS-ON APPLICATION 5.5

Esri News and 80-Second Videos

Esri data and software products are used in numerous applications across the globe. To keep informed about new Esri updates, software, and real-world uses of GIS, Esri publishes a pair of print publications (and makes articles and content available free online) as well as an online magazine. Check the following sites for their publications and look for articles related to fields of interest to you:

1. ArcNews: **www.esri.com/esri-news/arcnews**

2. ArcUser: **www.esri.com/esri-news/arcuser**

3. ArcWatch: **www.esri.com/esri-news /arcwatch**

In addition, Esri also produces a series of short informative videos about GIS that will only take 80 seconds of your time to watch. They pack a lot of rich content into a short amount of time. Check them out at **http://video.esri.com/channel/15 /80_dash_second-videos**

Expansion Questions

- How is Esri software enabling current developments in environmental monitoring, national security, urban planning, and law enforcement?

- What new developments are currently underway in the GIS field?

them to a single file for portability. This exported file (which is called an "interchange file," and ends with the file extension .e00) can then be moved to a different location and reimported to rebuild the coverage structure on another folder or computer. Coverages are much less common today for storing and working with geospatial data, though you may still encounter them.

A more common file type is the **shapefile**. It can hold only one type of vector object; thus there are point shapefiles, line shapefiles, and polygon shapefiles. Despite the name, a shapefile actually consists of multiple files that have to be present together to properly represent the data. The files all start with the same prefix (that is, if the shapefile is called "roads," then all the files that it consists of will be called "roads") but will have different extensions, including .shp, .shx. .dbf, and others. When copying a shapefile, you need to move all files with the same prefix to their new location for the shapefile to function properly.

A **geodatabase** is the format developed for ArcGIS for Desktop. It consists of a single item that contains all of the geospatial information for a dataset. Its object-oriented structure is set up so that multiple layers can be stored in a single geodatabase, each layer being its own **feature class**. For example, a geodatabase called "Yellowstone" could store the park boundaries as a polygon feature class, the roads running through the park as line feature classes, and the hiking trails as another line feature class. Geodatabases can also contain feature datasets, or subgroups of feature classes organized together. In addition, geodatabases can also store other data used in GIS, such as tables and rasters, and they can have GIS data in other formats (such as shapefiles) imported into them.

Geodatabases come in two varieties: the first is a **personal geodatabase**, which is stored as a single file that can hold a maximum of 2 GB of data. It is also the same file format (and thus directly compatible) as Microsoft Access. The second type is a file geodatabase, which is stored as a folder filled with files. A **file geodatabase** is essentially unlimited storage—each feature class can hold a maximum of 2 TB of data, but there is no limit on how many feature classes a file geodatabase can contain.

ArcGIS can also import data from other formats into an Esri format to use. For example, CAD (computer-aided design) data produced by programs such as AutoCAD can be imported into ArcGIS to use alongside other layers. Sometimes geospatial data comes available in a format called **SDTS** (the Spatial Data Transfer Standard), a system established by the U.S. government. The intent with SDTS was to deliver data in a file format that's "neutral," rather than software-specific. Agencies like the USGS, NOAA, and the Census Bureau sometime utilize SDTS as the format for delivering their geospatial data to you. Software programs often have a converter that enables the import of data in SDTS format to a form directly readable by the software (for instance, ArcGIS has a utility to import SDTS layers into a coverage format that you can use).

---

coverage a data layer represented by a group of files in a workspace consisting of a directory structure filled with files and also associated files in an INFO directory

shapefile a series of files (with extensions such as .shp, .shx, and .dbf) that make up one vector data layer

geodatabase a single item that can contain multiple layers, each as its own feature class

feature class a single data layer in a geodatabase

personal geodatabase a single file that can contain a maximum of 2 GB of data

file geodatabase a single folder that can hold multiple files, with nearly unlimited storage space

SDTS the Spatial Data Transfer Standard, a "neutral" file format for geospatial data, which makes the data capable of being imported into various geospatial software programs

Esri isn't the only company making GIS software—there are several other software packages and GIS companies providing a wide range of products that are used throughout the industry. No list of such things can hope to be comprehensive, but here are some of the more notable commercial GIS software products (and their vendors):

▶ Geomedia (from Intergraph/Hexagon Geospatial)

▶ Manifold System (from Manifold)

▶ MapInfo Pro (from Pitney Bowes Business Insight)

▶ Maptitude (from Caliper)

▶ TerrSet (from Clark Labs)

There are also several open source GIS products available for use and download via the Internet. An example is GRASS (Geographic Resources Analysis Support System), a long-running free (and open source) GIS software that can handle vector, raster, and image data with many different types of analysis. Quantum GIS (**QGIS**) is another prominent open source GIS program that you'll have the opportunity to use in *Geospatial Lab Application 5.1*. QGIS is freely available for download, can use many types of GIS data formats (including shapefiles and rasters), and has a layout similar to that of ArcMap (see **Figure 5.12**). QGIS features multiple types of analytical capabilities as well as mapping and data creation tools.

QGIS an open source GIS program

FIGURE 5.12 The basic layout of QGIS. *[Source: QGIS]*

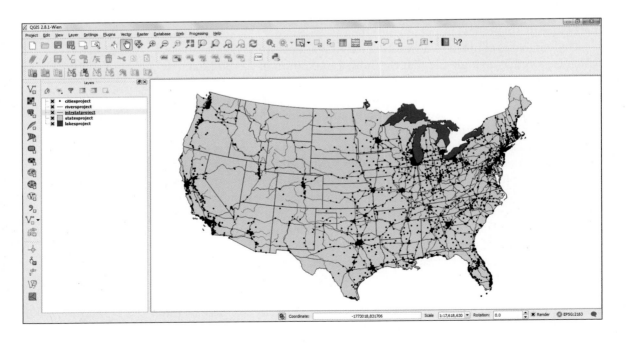

Chapter Wrapup

No matter the provider, the software, or the data representation format, GIS has a wide variety of uses, ranging from public utility mapping to law enforcement analysis, fire tracking, landscape planning, and many, many more. The next chapter will examine a number of different types of GIS analysis and address how to take this geospatial data and start doing things with it. Check out *GIS Apps* for some downloadable apps that use GIS concepts (including one for ArcGIS). Also take a look at *GIS in Social Media* for information from some blogs, Facebook pages, Twitter accounts, and Esri's YouTube channel.

The lab for this chapter provides an introduction to GIS and some basic handling of geospatial data. Two versions of the lab have been designed—the first one, *Geospatial Lab Application 5.1: GIS Introduction: QGIS Version* uses the free QGIS software package, and the second one, *Geospatial Lab Application 5.2: GIS Introduction: ArcGIS Version* uses ArcGIS for Desktop 10.3. Both labs use the same data and concepts, but implement them differently depending on which software package you have available to you.

Important note: The references for this chapter are part of the online companion for this book and can be found at **www.macmillanhighered.com/shellito/catalog**.

GIS Apps

Here's a sampling of available representative GIS apps for your phone or tablet. Note that some apps are for Android, some are for Apple iOS, and some may be available for both.

- **ArcGIS:** A mobile app version of ArcGIS, which allows access to many features of ArcGIS Online
- **ArcNews Magazine:** An app that allows you to read ArcNews on your mobile device
- **ArcUser Magazine:** An app that allows you to read ArcUser on your mobile device
- **Collector for ArcGIS:** An app designed for obtaining field data and mapping on a mobile device
- **iGIS:** An app for working with GIS (and shapefiles) on a mobile device
- **Wolf-GIS:** An app that allows for mobile GIS use of shapefiles

GIS in Social Media

Here's a sampling of some representative Facebook, Twitter, and Instagram accounts, along with some YouTube videos and blogs related to this chapter's concepts.

On Facebook, check out the following:

- **ArcGIS for Desktop:**
  www.facebook.com/ArcGISForDesktop
- **Esri:**
  www.facebook.com/esrigis
- **GIS users:**
  www.facebook.com/GISusers
- **QGIS:**
  www.facebook.com/pages/QGIS-Quantum-GIS -/298112000235096

Become a Twitter follower of:

- **ArcGIS Pro: @ArcGISPro**
- **Esri: @Esri**
- **Esri Press: @EsriPress**
- **GIS Geospatial News: @gisuser**
- **Learn ArcGIS (from Esri): @LearnArcGIS**
- **Python Software Foundation: @ThePSF**
- **QGIS: @qgis**
- **What is GIS: @GISdotcom**

Become an Instagram follower of:

- **Esri: @esrigram**

On YouTube, watch the following video:

- **Esri TV** (Esri's YouTube channel featuring numerous videos of ArcGIS applications and resources): **www.youtube.com/user/esritv**

For further up-to-date info, read up on these blogs:

- **AnyGeo Blog** (a blog about GIS, maps, and location-based technologies): **http://blog.gisuser.com**
- **ArcGIS blog: http://blogs.esri.com/esri/arcgis**
- **Esri Careers** (a blog about working and interning at Esri): **http://blogs .esri.com/esri/careers**
- **Esri Insider** (a blog about multiple geospatial and Esri software applications): **http://blogs.esri.com/esri/esri-insider**
- **QGIS Planet** (a blog about QGIS developments and applications): **http://plugins.qgis.org/planet**

Key Terms

ArcGIS (p. 138)
ArcGIS for Desktop (p. 138)
ArcGIS Online (p. 139)
ArcGIS Pro (p. 139)
ArcMap (p. 138)
attribute table (p. 134)
attributes (p. 133)
Catalog (p. 139)
CGIS (p. 125)
continuous field view (p. 129)
coverage (p. 141)
digitizing (p. 127)
discrete object view (p. 126)
Esri (p. 138)
feature class. (p. 141)
fields (p. 134)
file geodatabase (p. 141)
geodatabase (p. 141)
geographic information systems
  (p. 123)
grid cells (p. 131)
interval data (p. 134)

join (p. 135)
key (p. 136)
lines (p. 127)
metadata (p. 137)
NHD (p. 129)
NLCD (p. 131)
nominal data (p. 133)
ordinal data (p. 134)
personal geodatabase (p. 141)
points (p. 127)
polygons (p. 127)
python (p. 139)
QGIS (p. 142)
raster data model (p. 131)
ratio data (p. 134)
records (p. 134)
SDTS (p. 141)
shapefile. (p. 141)
topology (p. 128)
vector data model (p. 127)
vector objects (p. 127)

GIS Introduction: QGIS Version

This chapter's lab will introduce you to some of the basic features of GIS. You will be using a free open source program to navigate a GIS environment and begin working with geospatial data. The labs in Chapters 6 and 7 will utilize several more GIS features; the aim of this chapter's lab is to familiarize you with the basic GIS functions of the software. This lab uses the free QGIS software package.

Objectives

The goals for you to take away from this lab are:

▶ To familiarize yourself with the QGIS software environment, including basic navigation and tool use with both the Map and Browser

▶ To examine characteristics of geospatial data, such as their coordinate system, datum, and projection information

▶ To familiarize yourself with adding data and manipulating data layer properties, including changing the symbology and appearance of geospatial data

▶ To familiarize yourself with data attribute tables in QGIS

▶ To make measurements between objects in QGIS

Using Geospatial Technologies

The concepts you'll be working with in this lab are used in a variety of real-world applications, including:

▶ Public works, where GIS is used for mapping features of the infrastructure like sewer lines, water mains, and street light locations

▶ Archeology, where GIS is used for mapping boundaries and the spatial characteristics of dig sites, as well as the locations of artifacts found there

Obtaining Software

The version of QGIS used in this lab is 2.8, and available for free download at **http://qgis.org/downloads**.

Important note: Software and online resources can change fast. This lab was designed with the most recently available version of the software at the time of writing. However, if the software or Websites have significantly changed between then and now, an updated version of this lab (using the newest versions) will be available online at **www.macmillanhighered .com/shellito/catalog**.

[Source: Kenneth Garrett/National Geographic/Getty Images]

Lab Data

Copy the folder Chapter5QGIS—it contains a folder called "usaproject" in which you'll find several shapefiles that you'll be using in this lab. This data comes courtesy of Esri and was formerly distributed as part of their free educational GIS software package ArcExplorer Java Edition for Educators (AEJEE). It has already been projected for you to the US National Atlas Equal Area projection.

Localizing This Lab

The dataset used in this lab is Esri sample data for the entire United States. However, starting in Section 5.6, the lab focuses on Ohio and the locations of some of its cities. With the sample data covering the state boundaries and city locations for the whole United States, it's easy enough to select your city (or cities nearby) and perform the same measurements and analysis using those cities more local to you than ones in northeast Ohio.

5.1 An Introduction to Quantum GIS (QGIS)

1. Start **QGIS** (the default install folder is called QGIS Wein). QGIS will open in its initial mode. The left-hand column (where you'll see the word Layers) is referred to as the Map Legend—it's where you'll have a list of available data layers. The blank screen that takes up most of the interface is the Map View, where data will be displayed.

    Important note: If you don't see the Map Legend on the screen (or if other panels are open where the Map Legend should be), you can control which panels are available to view. From the **View** pull-down menu, select **Panels,** and then you can place an x in whichever panels you want to see on the screen. To begin, make sure that the only panel with an x next to its name is **Layers**.

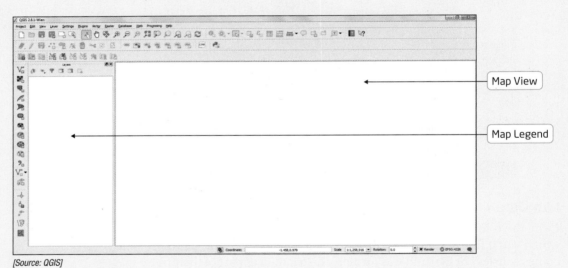

Map View

Map Legend

[Source: QGIS]

Important note: Before you begin adding and viewing data, the first thing to do is examine the data you have available to you. To do so in QGIS, you'll use the QGIS Browser—a utility designed to enable you to organize and manage GIS data.

2. Right click on the QGIS toolbar and a new menu of options will appear. From this menu, place a checkmark in the **Browser** option. The Browser dialog box will open in a new window above the Map Legend. Double-click on the word **Browser** and the dialog box will detach itself and become a new floating window that you can move and resize.

3. The Browser can be used to manage your data and to get information about it. In the Browser tree (the section going down the left-hand side of the dialog box), navigate to the **C:** drive, open the **Chapter5QGIS**

4. QGIS also allows you to specify the projection information used by the entire project. Since your layers are in the US National Atlas Equal Area projection, you'll want to tell QGIS to use this projection information while you're using this data in this lab. To do this, select the **Project** pull-down menu, and then choose **Project Properties**.

5. In the **Project Properties** dialog box, select the **Coordinate Reference System (CRS)** tab.

6. Place a checkmark in the box labeled **Enable "on the fly" CRS transformation**.

7. In the filter box, type **US National Atlas**—of the hundreds of supported projections, this will limit your search to find the US National Atlas Equal Area projection.

8. In the bottom panel, select the **US National Atlas Equal Area** option.

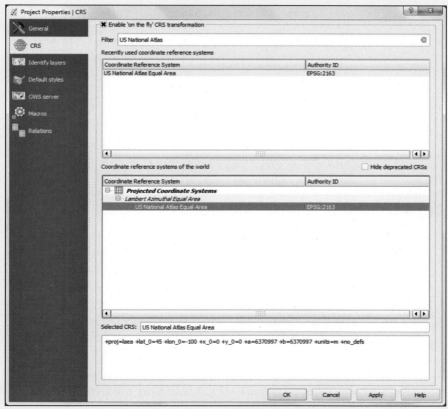

[Source: QGIS]

Question 5.4 What "family" of projections does the US National Atlas Equal Area projection belong to?

9. Click Apply to make the changes, and then click OK to close the dialog box.

5.5 Navigating the Map View

1. QGIS provides a number of tools for navigating around the data layers in the View. Clicking on the magnifying glass with the four arrows icon will zoom the Map View to the full extent of all the layers. It's good to use if you've zoomed too far in or out in the Map View or need to restart.

2. The other magnifying glass icons allow you to zoom in (the plus icon) or out (the minus icon). You can zoom by clicking in the window or clicking and dragging a box around the area you want to zoom into.

3. The hand icon is the Pan tool that allows you to "grab" the Map View by clicking on it and dragging the map around the screen for navigation.

4. Use the Zoom and Pan tools to center the Map View on Ohio so that you're able to see the entirety of the state and its cities and roads.

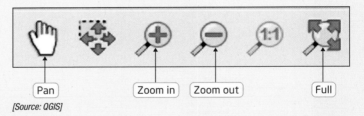

[Source: QGIS]

5.6 Interactively Obtaining Information

1. Even with the Map View centered on Ohio, there are an awful lot of point symbols there, representing the cities. We're going to want to identify and use only a couple of them. QGIS has a searching tool called Look for that allows you to locate specific objects from a dataset.

2. Right-click on the citiesproject shapefile and select Open Attribute Table. The attribute table for the layer will appear as a new window (we'll do more with attribute tables in the next section).

3. What we'll first want to do is find out which one of these 3557 records represents the city of Akron. To begin searching the attribute table, click on the button at the bottom that says Show all features and choose the option for Column Filter. A list of all of the attributes will appear; choose the one called NAME.

4. You will now be able to search through the attribute table on the basis of the NAME field. At the bottom of the dialog box, type Akron.

	NAME	CLASS	ST	STFIPS	PLACEFIP	CAPITAL	AREALAND	AREAWATER	PO
0	College	Census Designat...	AK	02	16750	NULL	18.670	0.407	
1	Fairbanks	City	AK	02	24230	NULL	31.857	0.815	
2	Kalispell	City	MT	30	40075	NULL	5.458	0.004	
3	Post Falls	City	ID	16	64810	NULL	9.656	0.045	
4	Dishman	Census Designat...	WA	53	17985	NULL	3.378	0.000	
5	Opportunity	Census Designat...	WA	53	51515	NULL	6.690	0.000	
6	Spokane	City	WA	53	67000	NULL	57.758	0.764	
7	Coeur d'Alene	City	ID	16	16750	NULL	13.129	0.467	
8	Anchorage	Municipality	AK	02	03000	NULL	1697.205	263.878	
9	Hilo	Census Designat...	HI	15	14650	NULL	54.289	4.147	
10	Kahului	Census Designat...	HI	15	22700	NULL	15.162	1.175	
11	Kihei	Census Designat...	HI	15	36500	NULL	10.159	1.731	
12	Wailuku	Census Designat...	HI	15	77450	NULL	5.066	0.374	
13	Ewa Beach	Census Designat...	HI	15	07450	NULL	1.417	0.449	
14	Pearl City	Census Designat...	HI	15	62600	NULL	4.984	0.825	
15	Waipahu	Census Designat...	HI	15	79700	NULL	2.570	0.061	
16	Waipio	Census Designat...	HI	15	79860	NULL	1.203	0.000	
17	Mililani Town	Census Designat...	HI	15	51050	NULL	3.909	0.021	
18	Schofield Barracks	Census Designat...	HI	15	69050	NULL	2.747	0.001	
19	Wahiawa	Census Designat...	HI	15	72650	NULL	2.113	0.266	
20	Makakilo City	Census Designat...	HI	15	47750	NULL	3.141	0.000	
21	Nanakuli	Census Designat...	HI	15	53900	NULL	2.522	3.224	
22	Waianae	Census Designat...	HI	15	74450	NULL	3.404	1.692	
23	Halawa	Census Designat...	HI	15	10000	NULL	2.325	0.000	
24	Kaneohe Station	Census Designat...	HI	15	28400	NULL	4.386	1.437	
25	Waimalu	Census Designat...	HI	15	77750	NULL	5.907	0.203	
26	Honolulu	Census Designat...	HI	15	17000	State	85.702	19.351	

NAME | Akron   Apply ✕ Case sensitive

[Source: QGIS]

5. Click on the Apply button. The record where the NAME attribute is "Akron" will appear as the only record in the attribute table. Click on the tab next to the NAME field (the tab will be labeled 2640) and you'll see the record highlighted in dark blue. This indicates that you've "selected" the record.

6. Drag the dialog box out of the way, so that you can see both it and the Map View at the same time.

7. You'll also see that the point symbol for the city of Akron has changed to a yellow color. This means it has been "selected" by QGIS—any actions performed on the cities layer will only affect the selected items, not all of the cities.

   However, all that you've done so far is locate and select an object. To obtain information about it, you can use the Identify Features tool (the icon on the toolbar is the white "i" and a white cursor in a blue circle). Note that Identify Features will only work with the layer that is selected in the Legend, so make sure that citiesproject (i.e. the layer you want to identify items in) is highlighted in the Legend before choosing the tool.

[Source: QGIS]

8. Select the Identify Features tool, and then click on the point representing Akron (zooming in as necessary). A new panel will appear called Identify Results (you may have to grab the top of the panel and drag it up to expand it so you can see what is inside the panel), listing all of the attributes for the citiesproject layer. By clicking on the View feature form button, you'll be able to see all of the field attribute information that goes along with the record/point representing Akron.

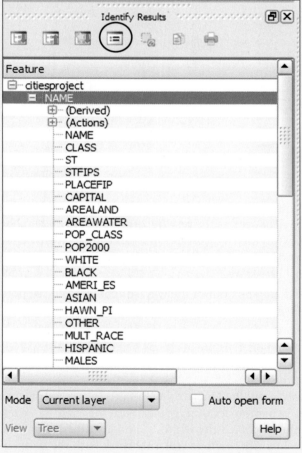

[Source: QGIS]

9. Return to the citiesproject attribute table and repeat the process to find a particular name of a city, but this time search for a city called Youngstown. Use the Identify Features tool for Youngstown as you did for Akron.

Question 5.5 According to QGIS, what was the population of Youngstown in the year 2000? (Carefully examine the attributes returned from Identify Features.)

10. Another method of selecting features interactively is to use the **Select Features** tool. To select features from the cities layer, click on citiesproject in the Map Legend. The icon for the Select Features tool is a white cursor overtop a yellow outlined polygon (note that if a different icon—i.e., a different Select tool—is present on the toolbar instead, choose the arrow to its right to activate a pull-down menu where you will be able to choose the proper Select Features icon).

[Source: QGIS]

11. The **Select Features** tool allows you to select several objects at once by defining the shape you want to use to select them with: rectangle, polygon, freehand, or radius. For now, use the rectangle (the default option) and select the four cities to the immediate northwest and west of Youngstown— they will become highlighted in yellow instead of Youngstown.

5.7 Examining Attribute Tables

1. As we've seen, each of the layers has an accompanying attribute table. To open a layer's attribute table, right-click on the citiesproject layer in the Map Legend, and then select **Open Attribute Table**.

   Important note: At the top of the cities' attribute table, you'll see that four features have been selected (the four cities near Youngstown). That's all well and good, but you'll also see that there are 3557 records in the entire attribute table (which means there are 3557 point objects in the cities' data layer). To find those four records would mean a lot of digging through the data. However, QGIS has some helpful sorting features, including one that will separate the selected records from the rest.

2. At the bottom of the attribute table, click on the **Show All Features** button, and from the list of options that appear, select the option for **Show Selected Features**. QGIS will only show the records of the four cities you selected, all highlighted in dark blue (as they have already been manually selected by you).

Question 5.6 Without using Identify Features, which four cities did you select?

Question 5.7 How many women (the attribute table lists this statistic as "Females") live in these four cities combined?

3. To clear the selected features, select the Unselect all icon at the top of the attribute table.

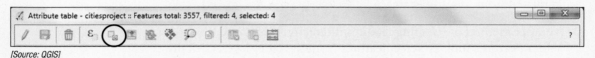

4. Close the attribute table when you're done.

5.8 Labeling Features

Rather than dealing with several points on a map and trying to remember which city is which, it's easier to simply label each city so that its name appears in the Map View. QGIS gives you the ability to do this by creating a label for each record and allowing you to choose the field with which to create the label.

1. Right-click on the citiesproject layer in the Map Legend and select Properties.
2. Click on the Labels tab.
3. Put a checkmark in the box next to Label this layer with.
4. From the pull-down menu next to the Label this layer with, choose NAME.

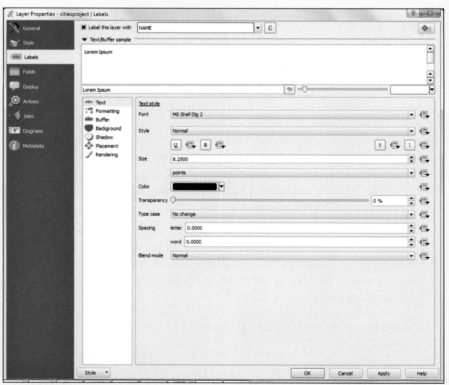

5. For now, accept all the defaults and click Apply and then OK. Back in the Map View, you'll see that labels of all the city names have been created and placed near each point.

6. Change around any of the options—font size, color, label placement, and even the angle—to set up the labels however you feel is best for examining the map data.

5.9 Measurements on a Map

With all the cities now labeled, it's easier to keep track of all of them. With this in mind, your next task is to make a series of measurements between points to see what the Euclidian (straight line) distance is between cities in northeast Ohio.

1. Zoom in tightly so that your Map View contains the cities of Youngstown and Warren, and the other cities between them.

2. Select the Measure Line tool from the toolbar—it's the one that resembles a grey ruler with a line positioned overtop.

[Source: QGIS]

3. You'll see that the cursor has turned into a crosshairs. Place the crosshairs on the point representing Youngstown and left-click the mouse. Drag the crosshairs north and west to the point representing Girard and click the mouse there. The distance measurement will appear in the Measure box. If you click on the Info text in the dialog box, you'll see that you are measuring "ellipsoidal" distance, referred to as the "surface" distance, as measured on a sphere. Note that each line of measurement you make with the mouse counts as an individual "segment" and the value listed in the box for "Total" is the sum of all segments.

4. Continue by measuring another line from Girard to Niles.

> **Question 5.8** What is the (ellipsoidal) distance from Girard to Niles? What is the total (ellipsoidal) distance from Youngstown to Niles (via Girard)?

5. You can clear all of the lines of measurement by clicking on New in the Measure box.

> **Question 5.9** What is the (ellipsoidal) distance from Boardman to Warren, and then from Warren to Niles?

> **Question 5.10** What is the (ellipsoidal) distance from Austintown, Ohio, to New Castle, Pennsylvania, and then from New Castle to Hermitage, Pennsylvania?

5.10 Saving Your Work (and Working on It Later)

When you're using QGIS, you can save your work at any time and return to it. When work is saved in QGIS, a "QGS" file is written to disk. Later, you can re-open this file to pick up your work where you left off.

1. Saving to a QGS file is done by selecting the Project pull-down menu, then selecting Save.
2. Files can be re-opened by choosing Open from the Project pull-down menu.
3. Exit QGIS by selecting Exit QGIS from the Project pull-down menu.

Closing Time

This lab was pretty basic, but it served to introduce you to how QGIS operates and how GIS data can be examined and manipulated. You'll be using either QGIS or ArcGIS in the next two labs, so the goal of this lab was to get the fundamentals of the software down. The lab in Chapter 6 takes this GIS data and starts to do spatial analysis with it, while the lab in Chapter 7 will have you starting to make print-quality maps from the data.

5.2 Geospatial Lab Application

GIS Introduction: ArcGIS Version

This chapter's lab will introduce you to some of the basic features of GIS. You will be using one of Esri's GIS programs to navigate a GIS environment and begin working with geospatial data. The labs in Chapters 6 and 7 will utilize several more GIS features; the aim of this chapter's lab is to familiarize you with the basic functions of the software. The previous *Geospatial Lab Application: GIS Introduction: QGIS Version* asked you to use the free Quantum GIS (QGIS) software; however, this lab provides the same activities but asks you to use ArcGIS for Desktop 10.3.

Objectives

The goals for you to take away from this lab are:

▶ To familiarize yourself with the ArcGIS for Desktop software environment, including basic navigation and tool use with both ArcMap and Catalog

▶ To examine characteristics of geospatial data, such as their coordinate system, datum, and projection information

▶ To familiarize yourself with adding data and manipulating data layer properties, including changing the symbology and appearance of Esri data

▶ To familiarize yourself with data attribute tables in ArcGIS for Desktop

▶ To make measurements between objects in ArcGIS for Desktop

Using Geospatial Technologies

The concepts you'll be working with in this lab are used in a variety of real-world applications, including:

▶ Public works, where GIS is used for mapping infrastructure and things such as sewer lines, water mains, and street light locations

▶ Archeology, where GIS is used for mapping boundaries and the spatial characteristics of dig sites, as well as the locations of artifacts found there

Obtaining Software

The current version of ArcGIS is not freely available for use. However, instructors affiliated with schools that have a campus-wide software license may request a 1-year student version of the software online at **www.esri.com/industries/apps/education/offers/promo/index.cfm**.

Important note: Software and online resources can change fast. This lab was designed with the most recently available version of the software at the time of writing. However, if the software or Websites have significantly changed between then and now, an updated version of this lab (using the newest versions) will be available online at **www.macmillanhighered.com/shellito/catalog**.

[Source: Kenneth Garrett/National Geographic/Getty Images]

Lab Data

Copy the folder Chapter5—it contains a folder called "usa," in which you'll find several shapefiles that you'll be using in this lab. This data comes courtesy of Esri and was formerly distributed as part of their free educational GIS software package ArcExplorer Java Edition for Educators (AEJEE).

Localizing This Lab

The dataset used in this lab is Esri sample data for the entire United States. However, starting in Section 5.6, the lab focuses on Ohio and the locations of some Ohio cities. With the sample data covering the state boundaries and city locations for the whole United States, it's easy enough to select your city (or cities nearby) and perform the same measurements and analysis using those cities more local to you than ones in northeast Ohio.

5.1 An Introduction to ArcMap

1. Start ArcMap (the default install folder is called ArcGIS). ArcMap will begin by asking you questions about "maps." In ArcMap, a "map document" is the means of saving your work—since you're just starting, click on Cancel in the Getting Started dialog box. The left-hand column (where you'll see the word Layers) is referred to as the Table of Contents (TOC)—it's where you'll have a list of available data layers. The blank screen that takes up most of the interface is the View, where data will be displayed.

[Source: ArcGIS]

Before you begin adding and viewing data, the first thing to do is examine the data you have available to you. To examine data layers, you can use the Catalog window in ArcMap—a utility designed to allow you to organize and manage GIS data.

Important note: ArcGIS for Desktop 10.3 also contains a separate application called ArcCatalog that can be used to perform the same data management tasks. However, ArcGIS for Desktop 10.3 incorporates the same functionality into ArcMap's Catalog window, so you'll be using that in this lab, rather than a separate program.

2. By default, the Catalog window is "pinned" to the right-hand side of ArcMap's screen as a tab. Clicking on the tab will expand the Catalog window. You can also open the Catalog window by clicking on its icon on the main toolbar.

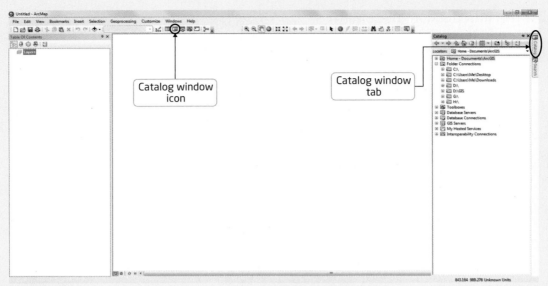

[Source: ArcGIS]

Important note: Various folders (representing network locations, hard drives on a computer, or external USB drives) can be accessed by using the Connect to Folder button. For instance, this lab assumes the Esri data you're using in this lab is stored in a folder on the C: drive of your computer.

3. Press the Connect to Folder button on the Catalog's toolbar.

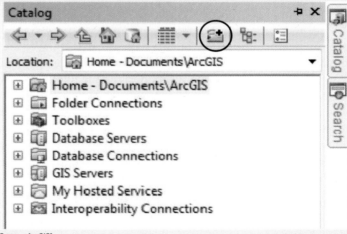

[Source: ArcGIS]

4. In the Connect to Folder dialog box, choose your computer's C: drive (or the drive or folder where the lab data is stored) from the available options and click OK.

5. Back in the Catalog, whichever folder you chose (such as C:) will become available by expanding the Folder Connections option.

Important note: This lab assumes the lab data is stored on the C:\Chapter5\ path. Navigate to that folder (or its equivalent on your computer) and open the usa folder. Several layers will be available, including cities, lakes, and states.

6. The Catalog can be used to preview each of your available data layers as well as manage them. To see a preview of the layers you'll be using in this lab, do the following:

    a. Right-click on one of the layers in the Catalog (for instance, states).

    b. Select the option for Item Description.

    c. A new window will open (called "Item Description—states"). Click on the Preview tab in this window to see what the dataset looks like. Tools will be available in the window to zoom in and out (the magnifying glasses), pan around the data (the hand), or return to the full extent of the dataset (the globe).

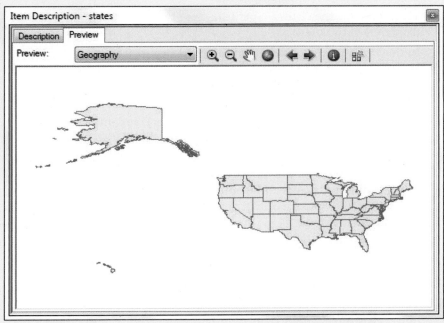

[Source: ArcGIS]

Preview each of the shapefiles (selecting a layer's Item Description will put it into the new window). Note that once the Item Description window is open (and the Preview of the Geography option is selected), you can click once on the name of the file in the Catalog and it will display a preview of that file in the Item Description window.

Question 5.1 What objects do each of the following datasets consist of: cities, rivers, and states?

5.2 Adding Data to ArcMap and Working with the TOC

1. Back in the main window of ArcMap, click on the yellow and black "plus" button to start adding the data you've previewed in the Catalog to the map.

[Source: ArcGIS]

2. In the Add Data dialog box, navigate to the Chapter5 folder, then the usa folder, and select the states.shp shapefile. Click Add (or double-click on the name of the shapefile).

3. You'll now see the states shapefile in the TOC and its content will be displayed in the View.

4. In the TOC, you'll see a checkmark in the box next to the states shapefile. When the checkmark is displayed, the layer will be shown in the View, and when the checkmark is removed (by clicking on it) the layer will not be displayed.

5. Now add two more layers: cities (a point layer), and intrstat (a line layer). All three of your layers will now be in the TOC.

6. You can manipulate the "drawing order" of items in the TOC by grabbing a layer with the mouse and moving it up or down in the TOC. Whatever layer is at the bottom is drawn first, and the layers above it in the TOC are drawn on top of it. Thus, if you move the states layer to the top of the TOC, the other layers will not be visible in the View because the states' polygons are being drawn over them.

5.3 Symbology of Features in ArcMap

1. You'll also notice that the symbology generated for each of the three objects is simple—points, lines, and polygons, each one assigned a random color. You can significantly alter the appearance of the objects in the View if you want to customize your maps.

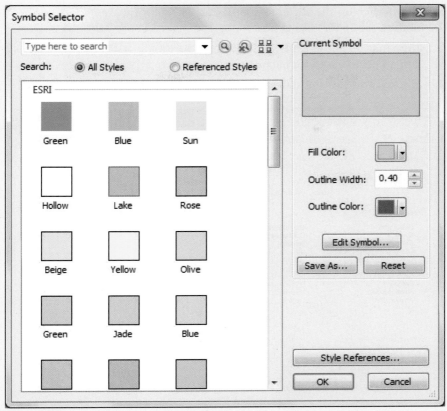

[Source: ArcGIS]

2. Right-click on the states layer and select Properties.

3. Select the tab for Symbology.

4. Press the large colored rectangle in the Symbol box—in the new Symbol Selector dialog box, you can alter the appearance of the states by changing their color (choose one of the defaults or select from the Fill Color options), as well as the outline color and thickness of the outline itself.

5. Change the states dataset to something more appealing to you, and then click OK. In the Layer Properties dialog box, click Apply to make the changes, then OK to close the dialog. Do the same for the cities and interstates shapefiles. Note that you can alter the points for the cities into shapes like stars, triangles, or crosses, and change their size as well. Several different styles are available for the lines of the interstates file as well.

5.4 Obtaining Projection Information

1. Chapter 2 discussed the importance of projections, coordinate systems, and datums, and all of the information associated with these concepts can be accessed using ArcMap.

2. Right-click on the states shapefile and select **Properties**. Click on the **Source** tab in the dialog box.

3. Information about units, projections, coordinate systems, and datums can all be accessed for each layer.

4. Click OK to exit the states **Layer Properties** dialog box.

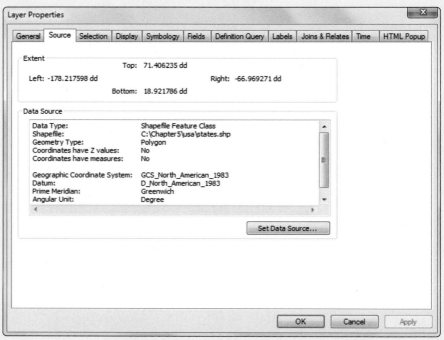

[Source: ArcGIS]

> **Question 5.2** What units of measurement are being used in the interstates dataset? What type of coordinate system (UTM, SPCS, etc.) is being used?

> **Question 5.3** What datum and projection are being used for the cities dataset?

Important note: ArcMap also allows you to change the projected appearance of the data in the View from one projected system to another. In ArcMap, all of the layers that are part of the View are held within a Data Frame—the default name for a Data Frame is "Layers," which you can see at the top of the TOC. By altering the properties of the Data Frame, you can change the appearance of layers in the View. Note that this does not alter the actual projection of the layers themselves, simply how you're seeing them displayed.

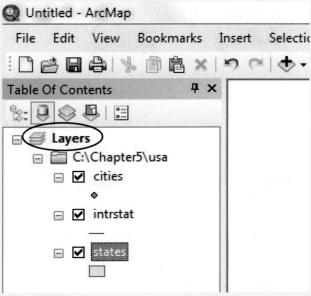

[Source: ArcGIS]

5. Right-click on the Data Frame in the TOC (the Layers icon) and select Properties.

6. In the Data Frame Properties dialog box, select the Coordinate System tab.

7. From the available options, select Projected Coordinate Systems, then Continental, and then North America. Lastly, choose US National Atlas Equal Area.

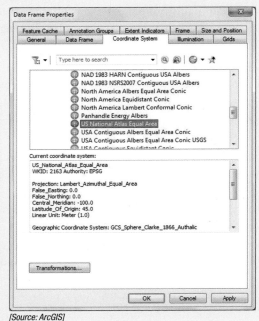

[Source: ArcGIS]

> **Question 5.4** What linear units and datum are used by the US National Atlas Equal Area projection?

8. Click Apply to make the changes, and then click OK to close the dialog box.

9. Back in the View, you'll see that the appearance of the three layers has completely changed from the flat GCS projection to the curved US National Atlas Equal Area projection. You can select other projections for the Data Frame if you wish. Try out some other appearances—when you're done, return to the US National Atlas Equal Area projection for the remainder of the lab.

5.5 Navigating the View

1. ArcMap provides a number of tools for navigating around the data layers in the View. By default, these items are on a separate toolbar docked under the pull-down menus. ArcMap has a variety of separate toolbars to access—the one with the main navigation tools is simply called Tools. If it is not present, select the Customize pull-down menu, select Toolbars, and then select Tools from the available choices.

[Source: ArcGIS]

Important note: The magnifying glass icons allow you to zoom in (the plus icon) or out (the minus icon). You can zoom by clicking in the window or clicking and dragging a box around the area you want to zoom into.

2. The hand icon is the Pan tool that allows you to "grab" the View by clicking on it, and to drag the map around the screen for navigation.

3. The globe icon will zoom the View to the extent of all the layers. It's good to use if you've zoomed too far in or out in the View or need to restart.

4. The four pointing arrows allow you to zoom in (arrows pointing inward) or zoom out (arrows pointing outward) from the center of the View.

5. The blue arrows allow you to step back or forward from the last set of zooms you've made.

6. Use the Zoom and Pan tools to center the View on Ohio so that you're able to see the entirety of the state and its cities and roads.

Database Query and Selection, Buffers, Overlay Operations, Geoprocessing Concepts, and Modeling with GIS

Chapter 5 described the basic components of GIS and how it operates, so now it's time to start doing some things with that GIS data. Anytime you are examining the spatial characteristics of data or how objects relate to one another across distances, you're performing **spatial analysis**. A very early (pre-computer) example of this is Dr. John Snow's work during an 1854 epidemic of cholera in Soho, central London—a poor, overcrowded square mile of narrow streets and alleyways. His analysis was able to identify the infected water from one pump as the source of the outbreak. At some point, Snow mapped the locations of the cholera deaths in relation to the water pumps (see **Figure 6.1** on page 178). This type of spatial analysis (relating locations of cholera deaths to the positions of water pumps) was innovative for the mid-nineteenth century, but is commonplace today—with GIS, these types of spatial analysis problems can be easily addressed.

> **spatial analysis**
> examining the characteristics or features of spatial data, or how features spatially relate to one another

Other common forms of spatial analysis questions include the following:

▶ How many objects of a particular type are within a certain distance of a particular location? Location researchers for a fast food chain might ask, "How many fast food restaurants are within one mile of this interstate exit?" A realtor might ask, "How many rental properties are within two blocks of this house that's for sale?" An environmentalist might ask, "What acreage of wetland will be threatened with removal by this proposed development?"

▶ How do you choose the most suitable locations based on a set of criteria? Many different types of questions related to this topic can be answered, such as those involving where to locate new commercial developments, new real estate options, or the best places for animal habitats.

FIGURE 6.1 A version of Dr. John Snow's 1854 map of Soho, London, showing locations of cholera deaths and water pumps. *[Source: Map 1. Published by C.F. Cheffins, Lith, Southhampton Buildings, London, England, 1854 in Snow, John. On the Mode of Communication of Cholera, 2nd Ed, John Churchill, New Burlington Street, London, England, 1855]*

This chapter examines how GIS can be used with geospatial data to answer questions like these in the context of spatial analysis. Keep in mind the two ways that GIS is used to model real-world data from the previous chapter—vector and raster. Due to the nature of some kinds of geospatial data, some types of spatial analysis are best handled with vector data and some with raster data.

How Can Data Be Retrieved from a GIS for Analysis?

One type of analysis is to find which areas or locations meet particular criteria—for instance, to take a map of U.S. Congressional Districts and identify which districts have a Democratic representative and which have a Republican representative. Alternatively, you might have a map of all residential parcels and want to identify which houses have been sold in the last month. In both of these examples, you would perform a database **query** to select only those records from the attribute table (see Chapter 5) that reflect the qualities of the objects you want to work with.

In other cases, before beginning the analysis, you might only want to utilize a subset of your GIS data. For instance, rather than dealing with all Congressional Districts in the entire United States, you might want to examine only those in Ohio. Or, if you have a dataset of all residential parcels in a county (which is likely thousands of parcels), you might only want to do analysis related to one (or a handful) of the parcels rather than examining all

query the conditions used to retrieve data from a database

of them. You'd want to go into the dataset and select only those parcels required for your study. Again, in these cases, the query would only select the records you want to deal with.

In GIS, queries are composed in the Structured Query Language (**SQL**) format, like a mathematical function. SQL is a specific format that is used for querying a layer or database to find what attributes meet certain conditions. For example, a layer made up of points representing cities in Ohio also has a number of fields representing its attributes, such as city name, population, and average income. If you just wanted to do analysis with one of these records (the city called "Boardman," let's say), then a query could be built to find all records where the field called CITY_NAME contained the characters of 'Boardman' as: CITY_NAME = 'Boardman'. This should return one record for you to use in your analysis. Another example would be to select all cities in Ohio with a population greater than or equal to 50,000 persons. The attribute table has a field called POP2010 (representing the 2010 city population) that contains this information. Thus, a query of POP2010 >= 50000 could be built, and all records with a population field containing a value of greater than or equal to the number 50,000 would be selected for your use (see **Figure 6.2** for an example of a query in GIS).

A query will use one of the following **relational operators:**

▶ Equal (=): Use this when you want to find all values that match the query. For instance, querying for CITY_NAME = 'Poland' will locate all records that exactly match the characters in the string that comprise the word 'Poland.' If the record were called 'Poland Township' it would not be located because the character string looks for an exact match.

> **SQL** Structured Query Language—a formal setup for building queries

> **relational operator** one of the six connectors (=, <>, <, >, >=, or <=) used to build a query

FIGURE 6.2 A database query for Ohio cities returning multiple records (selected records and their corresponding cities are in cyan). *[Source: Esri]*

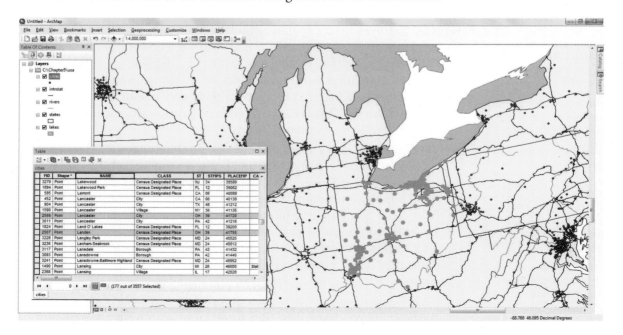

▶ Not Equal (<>): Use this when you want to find all the records that do not match a particular value. For instance, querying for CITY_NAME <> 'Poland' will return all the records that do not have a character string of the word 'Poland' (which would probably be all of the other cities with names in the database).

▶ Greater Than (>) or Greater Than Or Equal To (>=): Use this for selecting values that are more than (or more than or equal to) a particular value.

▶ Less Than (<) or Less Than Or Equal To (<=): Use this for selecting values that are less than (or less than or equal to) a particular value.

While a simple query only uses one operator and one field, a **compound query** enables you to make selections using multiple criteria. There are different ways of linking multiple criteria together for creating one of these compound queries, each using a different option of querying (referred to as a **Boolean operator**):

▶ **AND**: Suppose you want to select cities in Ohio that have a population of over 50,000 (a variable called POP2010 in the attribute table) and an average household income (a variable called AVERAGEHI) of more than $30,000. A compound query can combine these two requests by saying: POP2010 >= 50000 AND AVERAGEHI > 30000. Because "AND" is being used as the operator, only records that match both the criteria will be returned by the query. If an attribute only meets one criterion, it won't be selected. An "AND" query is referred to as an **intersection** because it returns what the two items have in common.

▶ **OR**: A different query could be built by saying: POP2010 >= 50000 OR AVERAGEHI > 30000. This would return records of cities with a population of greater than or equal to 50,000, or cities with an average household income of greater than $30,000, or cities that meet both. When an OR operator is being used, records that meet one or both criteria are selected. An "OR" query is referred to as a **union** since it returns all the elements of both operations.

▶ **NOT**: A third query could be built around the concept that you want all of the data related to one criterion, but you want to exclude what relates to the second criterion. For instance, if you want to select cities with a high population, but not those with a higher average household income, you could build a query like: POP2010 >= 50000 NOT AVERAGEHI > 30000 to find those particular records. A "NOT" query is referred to as a **negation** because it returns all the elements of one dataset, but not what they have in common with the other.

▶ **XOR**: A final type of query can be built using the idea that you want all of the data from each layer, except for what they have in common. For instance, if you wanted to find those cities with a high population or

---

compound query a query that contains more than one operator

Boolean operator one of the four connectors (AND, OR, NOT, XOR) used in building a compound query

AND the Boolean operation that corresponds with an intersection operation

intersection the operation wherein the chosen features are those that meet both criteria in the query

OR the Boolean operator that corresponds with a union operation

union the operation wherein the chosen features are all that meet the first criterion as well as all that meet the second criterion in the query

NOT the Boolean operator that corresponds with a negation operation

negation the operation wherein the chosen features are those that meet all of the first criteria and none of the second criteria (including where the two criteria overlap) in the query

XOR the Boolean operator that corresponds with an exclusive or operation

those that had high incomes, but not both types of cities, your query would be: POP1990 >= 50000 XOR AVERAGEHI > 30000. XOR works as the opposite of AND since it returns all data, except for the intersection of the datasets. An "XOR" query is referred to as an **exclusive or**, since it acts like a union but leaves out the intersection data.

Compound queries can use multiple operations to select multiple records by adhering to an order of operations, such as: (POP2010 >= 50000 AND AVERAGEHI > 30000) OR CITY_NAME = 'Boardman'. This will select cities that meet both the population and the income requirement, and would select cities named 'Boardman' as well. The result of a query will be a subset of the records that are chosen, and then those records can be used for analysis, rather than the whole data layer. Selected records can also be exported to their own dataset. For instance, the results of the previous query could be extracted to compose a new GIS layer, made up of only those records (and their corresponding spatial objects) that meet the query's criteria. Once you have your selected records or new feature layers, you can start to work with them in GIS (see *Hands-On Application 6.1: Building SQL Queries in GIS* for an example of doing queries with an online GIS utility).

> **exclusive or** the operation wherein the chosen features are all of those that meet the first criterion as well as all of those that meet the second criterion, except for the features that have both criteria in common in the query

HANDS-ON APPLICATION 6.1

Building SQL Queries in GIS

Durham, North Carolina, has an online GIS mapping utility that allows the user to do a variety of spatial analyses with the city's data, including performing queries. To work with this tool, go to **http://maps2.roktech.net/durhamnc_gomaps**. In the upper left-hand corner there are several tools (such as zooming and panning) to get oriented to the GIS data. The tool with the hammer icon allows you to build SQL queries of the data. Click on this icon and a new set of options will appear on the right—by choosing the tab for Query Builder, these will allow you to construct SQL queries of the GIS. First, select an available layer (for instance, select Stormwater Watersheds). Second, select an attribute to query (for instance, double-click on River). Third, select an operator (for instance, choose equals "=" symbol). Click on the Get Unique Values button and you'll see the available values appear to finish the SQL statement (for instance, choose CAPE FEAR). Your SQL statement will appear in the box (in this example, it should read: RIVER = 'CAPE FEAR'). To run the SQL query, click on the Execute button. All records meeting your query will be retrieved (in this case, all stormwater watersheds related to the Cape Fear river). The selected watersheds will be highlighted on the map, while the records (and the attributes that go along with them) will be displayed in a spreadsheet at the bottom of the screen. Investigate some of the other available data layers and the types of queries that can be built with them.

Expansion Questions

- How many stormwater watersheds in Durham, NC, are related to the Cape Fear river?

- How many stormwater watersheds in Durham, NC, are related to the Neuse river?

- Which of the parks in Durham, NC, have boating available? (Hint: Use the Parks layer and the boating attribute, then look for records where the boating attribute is equal to "Y.")

How Can You Perform Basic Spatial Analysis in GIS?

buffer a polygon of spatial proximity built around a feature

Once you have your selected data, it's time to start doing something with it. A simple type of analysis is the construction of a **buffer** around the elements of a data layer. A buffer is an area of proximity set up around one or more objects. For instance, if you want to know how many rental properties are within a half-mile of a selected house, you can create a half-mile buffer around the object representing the house and then determine how many rental properties were within the buffer. Buffers can be created for multiple points, lines, or polygons (**Figure 6.3**). Buffers are simple tools to utilize, but they can provide useful information when you're creating areas of analysis for various purposes.

dissolve the ability of the GIS to combine polygons with the same features together

GIS can also perform a **dissolve** operation, where boundaries between adjacent polygons that have the same properties are removed, merging the polygons into a single, larger shape. Dissolve is useful when you don't need to examine each polygon, but only regions with similar properties. For example, suppose you have a land-use map of a county consisting of polygons representing each individual parcel and how it's being used (commercial, residential, agricultural, etc.). Rather than having to examine each individual parcel, you could dissolve the boundaries between parcels and create regions of "residential" land use or "commercial" land use for ease of analysis. Similarly, if you have overlapping buffers generated from features, dissolve can be used to combine buffer zones. See **Figure 6.4** for an example of dissolve in action.

FIGURE 6.3 A 50-mile buffer constructed around a major interstate (I-70).
[Source: Esri]

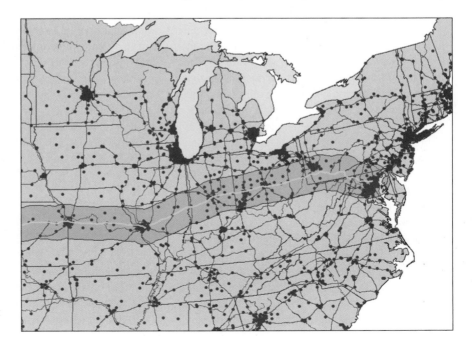

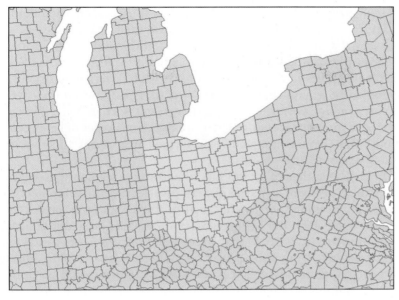

FIGURE 6.4 Before and after the dissolve operation—Ohio county boundaries are dissolved on the basis of their state's FIPS code.
[Source: Esri]

Buffers and querying are basic types of spatial analysis, but they only involve a single layer of data (such as creating a buffer around roads or selecting all records that represent a highway). If you're going to do spatial analysis along the lines of John Snow's hunt for a cholera source, you'll have to be able to do more than just select a subset of the population who died, or just select the locations of wells within the area affected by the disease—you'll have to know something about how these two things interacted with each other over a distance. For instance, you'll want to know something about how many cholera deaths were found within a certain distance of each well. Thus, you won't just want to create a buffer around one layer (the wells)—you'll want to know something about how many features from a different layer (the death locations) are in that buffer. Many types of GIS analysis involve combining the features from multiple layers together to allow examination of several characteristics at once. In GIS, when one layer has some sort of action performed to it and the result is a new layer, this process is referred to as **geoprocessing**. There are numerous types of geoprocessing methods, and they are frequently performed in a series to solve spatial analysis questions.

GIS gives you the ability not just to perform a regular SQL-style query, but to also perform a **spatial query**—to select records or objects from one layer based upon their spatial relationships with other layers (rather than using attributes). With GIS, queries can be based on spatial concepts—to determine things like how many registered sex offenders reside within a certain distance from a school, or which hospitals are within a particular distance of your home. The same kind of query would be done if a researcher wanted to determine which cholera deaths are within a buffer zone around a well. In these cases, the spatial dimensions of the buffer are being used as the selection criteria—all objects within the buffer are what will be selected. See **Figure 6.5** for an example of using a buffer as a selection tool, as well as *Hands-On Application 6.2: Working with Buffers in GIS*.

geoprocessing the term that describes when an action is taken to a dataset that results in a new dataset being created

spatial query selecting records or objects from one layer based upon their spatial relationships with other layers (rather than using attributes)

FIGURE 6.5 Effect of selecting point features (cities) by using a buffer (50 miles around I-70). *[Source: Esri]*

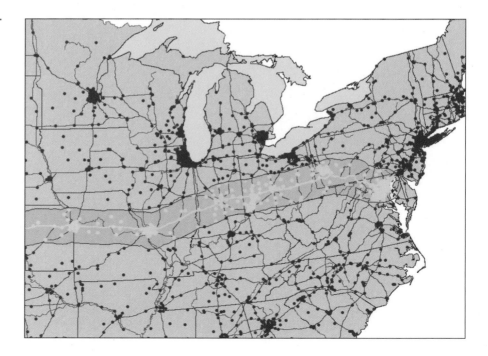

HANDS-ON APPLICATION 6.2

Working with Buffers in GIS

Honolulu has an online GIS application that looks at parcels and zoning for the area. On this Website, you can select a location, generate a buffer around it, and select all parcels touched by the buffer (and display their parcel records). To use this, go to **http://gis.hicentral.com /fastmaps/parcelzoning**, then zoom in to a developed or residential area (if you're familiar with the area, you can also search by address). To generate a buffer, select the Tools icon, and then choose the option for Circle-Select Parcels. At this point, you can specify a buffer distance (the radius of the circle) and click on a place on the map. A buffer will be generated around your placemark, and the parcels selected by the buffer will be highlighted on the map and their records returned. Try examining some areas with various buffer sizes to see how buffers can be used as a selection tool.

Expansion Questions

* Use the Search function and the Places of Interest option to locate "State Capital." How many land parcels are within a 500-foot radius of a point placed at the center of the building?

* How many land parcels are within a 1000-foot radius of the same point?

Other analyses can be done beyond selecting with a buffer or selecting by location. When two or more layers share some of the same spatial boundaries (but have different properties) and are combined together, this is referred to as an **overlay** operation in the GIS. For instance, if one layer contains information about the locations of property boundaries and a second layer contains information about water resources, combining these two layers in an overlay can help determine which resources are located on whose property. An overlay is a frequently used GIS technique for combining multiple layers together for performing spatial analysis (see **Figure 6.6** on page 186).

There are numerous ways that polygon layers can be combined together through an overlay in GIS. Here are some widely used overlay methods (see **Figure 6.7** on page 187):

▶ **Intersect**: In this operation, only the features that both layers have in common are retained in a new layer. This type of operation is commonly used when you want to determine an area that meets two criteria—for instance, a location needs to be found within a buffer and also on a plot of land used for agriculture: you would intersect the buffer layer and the agricultural layer together to find what areas, spatially, they have in common.

▶ **Identity**: In this operation, all of the features of an input layer are retained, and all the features of a second layer that intersect with them are also retained. For example, you may want to examine all of the nearby floodplain and also what portions of your property are on it.

▶ **Symmetrical difference**: In this operation, all of the features of both layers are retained except for the areas that they have in common. This

overlay the combining of two or more layers in the GIS

intersect a type of GIS overlay that retains the features that are common to two layers

identity a type of GIS overlay that retains all features from the first layer along with the features it has in common with a second layer

symmetrical difference a type of GIS overlay that retains all features from both layers except for the features that they have in common

FIGURE 6.6 An example of overlaying two layers.

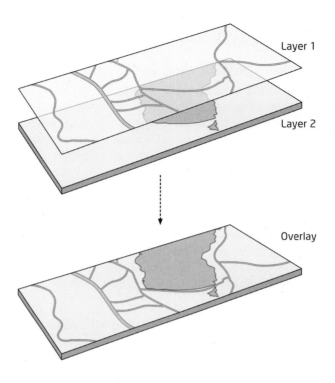

operation could show you, for example, all potential nesting areas along with the local plan for development, except where they overlap (so those areas can be taken out of the analysis).

▶ **Union (overlay):** In this operation, all of the features from both layers are combined together into a new layer. This operation is often used to merge features from two layers (for instance, in order to evaluate all possible options for a development site, you may want to overlay the parcel layer and the water resources layer together).

union (overlay) a type of GIS overlay that combines all features from both layers

Another way of performing spatial overlay is by using raster data. With raster cells, each cell has a single value, and two grid layers can be overlaid in a variety of ways. The first of these is by using a simple mathematical operator (sometimes referred to as **map algebra**), such as addition or multiplication. To simplify things, we're only going to assign our grids values of 0 or 1, where a value of 1 means that the desired criteria are met and a value of 0 is used when the desired criteria are not met. Thus, grids could be designed using this system with values of 0 or 1 to reflect whether a grid cell meets or does not meet particular criteria.

map algebra combining datasets together using simple mathematical operators

Let's start with a simplified example—say you're looking to find a plot of land to build a vacation home on, and your two key criteria are that it be close to forested terrain for hiking and close to a river or lake for water recreation. What you could do is get a grid of the local land cover and assign all

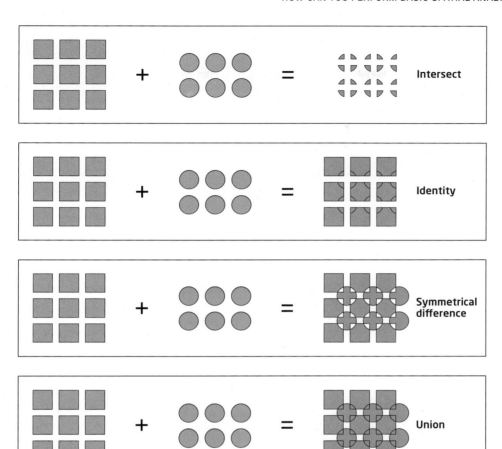

FIGURE 6.7
Various types of geoprocessing overlay operations using two sample polygon layers.

grid cells that have forests on the landscape a value of 1 and all non-forested areas a value of 0. The same holds true for a grid of water resources—any cells near a body of water get a value of 1 and all other cells get a 0. The "forests" grid has values of 1 where there are forests and values of 0 for non-forested regions, while the "water" grid has values of 1 where a body of water is present and values of 0 for areas without bodies of water (see **Figure 6.8** on page 188 for a simplified way of setting this up).

Now you want to overlay your two grids and find what areas meet the two criteria for your vacation home, but there are two ways of combining these grids. **Figure 6.9** on page 188 shows how the grids can be added and multiplied together with varying results. By multiplying grids containing values of 0 and 1 together, only values of 0 or 1 can be generated in the resulting overlay grid. In this output, values of 1 indicate where (spatially) both criteria (forests and rivers) can be found, while values of 0 indicate where both cannot be found. In the multiplication example, a value of 0 could have one of the

FIGURE 6.8 A hypothetical landscape consisting of forests, rivers, and non-forested or dry areas. *[Source: (top left) Alex L. Fradkin/Getty Images (top right) Robert Glusic/Getty Images (bottom left) Bruce Heinemann/Getty Images (bottom right) U.S. Geological Survey/photo by Stewart Tomlinson]*

FIGURE 6.9 Raster overlay examples for multiplication and addition. *[Source: (left) Alex L. Fradkin/Getty Images (right) U.S. Geological Survey/photo by Stewart Tomlinson]*

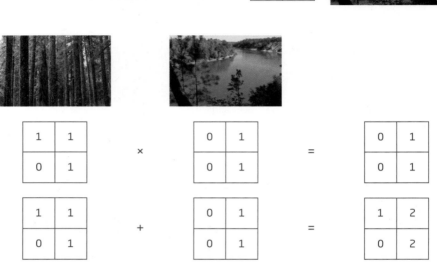

criteria, or none, but we're only concerned with areas that meet both criteria, or else we're not interested. Thus, the results either match what we're looking for or they don't.

In the addition example, values of 0, 1, or 2 are generated from the overlay. In this case, values of 0 contain no forests or rivers, values of 1 contain one or the other, and values of 2 contain both. By using addition to overlay the grids, a gradient of choices is presented—the ideal location would have a value of 2, but values of 1 represent possible second-choice alternatives, and values of 0 represent unacceptable locations.

This type of overlay analysis is referred to as **site suitability**, as it seeks to determine which locations are the "best" (or most "suitable") for meeting certain criteria. Site suitability analysis can be used for a variety of topics, including identifying ideal animal habitats, prime locations for urban developments, or prime crop locations. When using the multiplication operation, you will be left with sites classified as "suitable" (values of 1) or "non-suitable" (values of 0). However, the addition operation leaves other options that may be viable in case the most suitable sites are unavailable, in essence

site suitability the determination of the "useful" or "non-useful" locations based on a set of criteria providing a ranking of suitable sites. For instance, the best place to build the vacation home is in the location with values of 2 since it meets both criteria, but the cells with a value of 1 provide less suitable locations (as they only meet one of the criteria).

How Can Multiple Types of Spatial Analysis Operations Be Performed in GIS?

There are many criteria that play into the choice of site suitability or into determining what might be the "best" or "worst" locations for something, and often it's not as cut and dried as "the location contains water" (values of 1) or "the location does not contain water" (values of 0). Think about the vacation home example—locations were classified as either "close to water" or "not close to water." When you're choosing a vacation home location, there could be locations right on the water, somewhat removed from it but still close enough for recreation, and some that are much further away from water. Similarly, terrain is so varied that it's difficult to assign it a characteristic of "mountainous" or "not mountainous." Instead, a larger range of values could be assigned—perhaps on a scale of 1 through 5, with 5 being the best terrain for hiking or the closest to the water and 1 being furthest away from the water or the least desirable terrain for hiking.

In order to set something like this up, a distance calculation can be performed. GIS can calculate the distance from one point to all other points on a map, or calculate the distance a location is from all other features. For example, when you're examining a location for building a house in a rural locale, you may want to know the distance that spot is from a landfill or wastewater treatment plant. Alternatively, you may want to know the distance a vacation home is away from a lake or river. A way of conceptualizing distance (as is done in ArcGIS) is to think of it as a continuous surface (see Chapter 5), where each spot on that surface has a value for the distance it is away from a feature. ArcGIS makes these measurements by calculating a distance surface of raster grid cells, with each cell containing a value of the distance from the features in a layer. **Figure 6.10** on page 190 illustrates this by mapping the location of a set of cities in Ohio and calculating the distance that each cell is away from a city. Next, the distance from the cities is sliced into ten equal ranges (that is, all of the areas closest to a city are given a value of 1, the next closest are given a value of 2, and so forth).

For this example, we'll come up with a new set of grids classified along these lines and throw in a third option as well—distance away from roads (on the assumption that a vacation home would be best located well away from busy roads and the accompanying automobile noise). In all of these, the highest values represent the "best" options (the places furthest away from busy roads), and the lowest values represent the "worst" options (right next

FIGURE 6.10 An example of a raster distance surface calculation, where each grid cell contains a value representing the distance from a point (cities in Ohio), and the same grid sliced into 10 ranges. *[Source: Esri]*

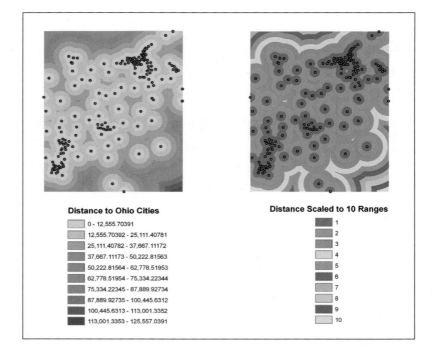

FIGURE 6.11 A raster overlay example for combining three criteria together to form a simple suitability index.

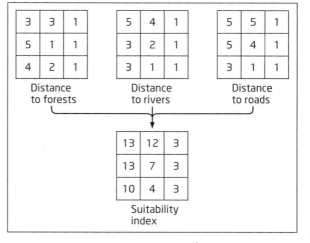

suitability index a system whereby locations are ranked according to how well they fit a set of criteria

to a busy road). **Figure 6.11** shows the new grids and how they are overlaid together through the addition operation to create a new output grid. Rather than choices of 0, 1, and 2, the output grid contains a wide range of options, where the highest number reflects the most suitable site based on the criteria and the lowest number reflects the least suitable sites. This new grid represents a very simple **suitability index**, where a range of values is generated, indicating a ranking of the viability of locations (these would be ordinal values, as we discussed back in Chapter 5). In the example in Figure 6.11, the values of 13 represent the most suitable sites, the values of 12 and 10 are possible alternative venues. The value of 7 is likely a mediocre site, and the values of 4 and 3 are probably unacceptable. The advantage of the suitability index is to offer a wider gradient of choices for suitable sites as an alternative to some of the previous options. Note that while this example is using raster data, you can also perform site suitability analysis actions using vector data.

GIS can process numerous large spatial data layers together at once. Thinking about the vacation home example, there are many additional factors that play into choosing the site of a second home—access to recreational amenities, the price of the land, the availability of transportation networks, and how far the site is away from large urban areas, for instance. All of these factors (and more) come into play when you're trying to find the optimal location for a vacation home. Similarly, if you work in the retail industry and are assessing which place is the best location for a new store, there are a lot of criteria that will influence your decision—how close the potential sites are to your existing store, how close that area is to similar stores, the population of the area, the average household income for each neighborhood, and the rental price of the property, to name just a few. GIS can handle all of these spatial factors by combining them together (see **Figure 6.12** for an example of combining several data layers together).

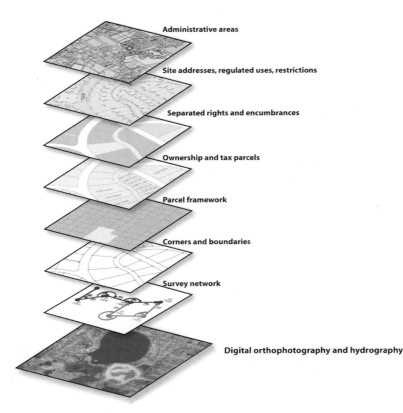

Administrative areas

Site addresses, regulated uses, restrictions

Separated rights and encumbrances

Ownership and tax parcels

Parcel framework

Corners and boundaries

Survey network

Digital orthophotography and hydrography

FIGURE 6.12 An example of combining multiple GIS data layers together. *[Source: Esri]*

GIS model a representation of the factors used for explaining the processes that underlie an event or for predicting results

A **GIS model** is a representation that combines these spatial dimensions or characteristics together in an attempt to describe or explain a process or to predict results. For instance, trying to predict where future urban developments will occur based on variables such as a location's proximity to existing urban areas, the price of the land parcels, or the availability of transportation corridors could be performed using GIS modeling, with the end result being a map of the places most likely to convert to an urban land use. There are many different types of GIS models with different features and complexities to them, and it's far beyond the scope of this book to catalog and examine all of them.

GIS models are often used to examine various criteria together to determine where, spatially, some phenomenon is occurring or could potentially occur. These are often used in decision-making processes or to answer questions (like the previous example of where to build a vacation home). Another example of this would be in determining which parcels of land surrounding Civil War battlefields are under the greatest pressure to convert to an urban land use. Battlefield land is often considered a historic monument to the men who died in the Civil War, and while the core areas of several sites are under the protection of the U. S. National Park Service, there are many other sites that have been developed or are in danger of having residential or commercial properties built on them, forever losing the "hallowed ground" of history.

GIS can be used to combine a number of factors together that influence where new developments are likely to occur (such as proximity to big cities, existing urban areas, or water bodies) in a fashion similar to site suitability in order to try and find the areas under the greatest development pressures. This type of information could be used by preservationists attempting to purchase lands with the intention of keeping them undeveloped.

The factors that contribute to a phenomenon can be constructed as layers within GIS, weighted (to emphasize which layers have a greater effect than others) and combined together. The end result will be a ranking of areas to determine their suitability. One example of this process is referred to as multi-criteria evaluation (**MCE**), which takes layers that have been standardized and initially ranked against each other, then weights and combines them together. The MCE process produces a map showing various "pressures" at each location (along the lines of the suitability index map) to determine which places have the greatest influence placed on them (based on the combination of all the input layers). This acts like a more complex form of the basic site suitability and can be used in decision-making processes—like determining which areas of battlefield land have the greatest developmental pressures on them to convert to an urban land use. Other problems, such as locating suitable animal habitats, areas of high flood potential, or places for retail sites can be approached using similar techniques.

MCE multi-criteria evaluation—the use of several factors (weighted and combined) to determine the suitability of a site for a particular purpose

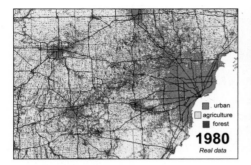

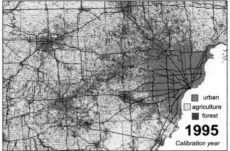

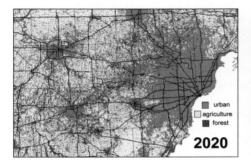

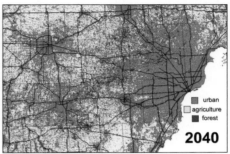

FIGURE 6.13 The use of the Land Transformation Model to examine and predict areas that will change to an urban land use in the future. *[Source: HEMA Laboratory Purdue University/ South East Michigan Council of Governments/Michigan Department of Natural Resources/Esri]*

Another example of a GIS model is the Land Transformation Model (LTM), a tool used to combine various weighted raster layers together in order to determine where areas have changed over to an urban land use, and then to predict where areas are likely to change over to an urban land use in the future. See **Figure 6.13** for an example of the output of the LTM's predictive capability, covering areas in Michigan up to the year 2040. Also see *Hands-On Application 6.3: The Land Transformation Model* for further examination of the LTM as a GIS model.

HANDS-ON APPLICATION 6.3

The Land Transformation Model

The Land Transformation Model is headquartered at Purdue University. For further background on LTM as an example of a GIS model, go to **http://ltm.agriculture.purdue .edu/default_ltm.htm**. The actual model itself (and sample datasets) can be downloaded, as well as a selection of graphics and papers related to the model, if you're feeling adventurous.

Expansion Questions

- What types of applications has the Land Transformation Model been used for?

- What kinds of GIS layers are used in this model for forecasting urban development?

THINKING CRITICALLY
WITH GEOSPATIAL TECHNOLOGY 6.1

What Are Potential Societal or Policy Impacts of GIS Models?

The output of GIS models is often used to aid in decision-making processes, and could have a significant impact in areas of concern to all of us. For instance, trying to determine which sites would make the most suitable locations for a new casino development would greatly affect the social, economic, and land use conditions of an area. If the end result of a model is to determine what areas of a residential area are under the greatest potential threat of flooding, it could have a strong impact on an area's real estate market. What kinds of impacts can these types of models have on local, state, or federal policy?

There is also this question to consider: What if the model's predictions are incorrect? If a model predicts that one area will be the optimal location for a new shopping center, and it turns out not to be, how could this kind of error affect this area and future decision making? Keep in mind that you could have a great model or design, but have inaccurate data for one (or more) of the layers. If that's the case, how can having this kind of poor-quality data in a model affect the output (and what kinds of impacts could the results of models using inaccurate data have on policy decisions)?

Chapter Wrapup

Spatial analysis is really at the core of working with geospatial data and GIS. While there are many more types of spatial analysis that can be performed with GIS, this chapter simply introduces some basic concepts of how it can be performed (and some applications of it). Check out *Spatial Analysis Apps* for some mobile apps related to this chapter's concepts, as well as *Spatial Analysis in Social Media* for some blogs and a YouTube video concerning spatial analysis. In the next chapter, we're going to look at taking GIS data (for instance, the results of an analysis or a model) and using it to make a map. This chapter's labs will return you to working with GIS software (either QGIS or ArcGIS), to start in on more functions, like querying and buffers.

Important note: The references for this chapter are part of the online companion for this book and can be found at **www.macmillanhighered.com/shellito/catalog**.

Spatial Analysis Apps

Here's a sampling of available representative GIS and spatial analysis apps for your phone or tablet. Note that some apps are for Android, some are for Apple iOS, and some may be available for both.

- **Crime Maps:** An app that contains GIS maps and location information for numerous types of crimes across the United States
- **Distance Measure:** An app that shows a buffer around your present area and also allows you to measure straight-line distance or polygon areas on a map
- **mySoil:** An app from the British Geological Survey that incorporates soils data from across Britain and Europe

Spatial Analysis in Social Media

Here's a sampling of some representative YouTube videos and blogs related to this chapter's concepts.

You Tube On YouTube, watch the following video:

- **GIS Spatial Analyst Tutorial Using John Snow's Cholera Data** (a video showing ArcGIS spatial analysis techniques using Snow's nineteenth-century data):
  www.youtube.com/watch?v=isVD8u6WrG4

For further up-to-date info, read up on this blog:

- **ArcGIS Resources—Analysis & Geoprocessing Category** (blog posts concerning spatial analysis and geoprocessing topics):
  http://blogs.esri.com/esri/arcgis/category/subject-analysis-and -geoprocessing

Key Terms

AND (p. 180)
Boolean operator (p. 180)
buffer (p. 182)
compound query (p. 180)
dissolve (p. 182)
exclusive or (p. 181)
geoprocessing (p. 184)
GIS model (p. 192)
identity (p. 185)
intersect (p. 185)
intersection (p. 180)
map algebra (p. 186)
MCE (p. 192)
negation (p. 180)

NOT (p. 180)
OR (p. 180)
overlay (p. 185)
query (p. 178)
relational operator (p. 179)
site suitability (p. 188)
spatial analysis (p. 177)
spatial query (p. 184)
SQL (p. 179)
suitability index (p. 190)
symmetrical difference (p. 185)
union (p. 180)
union (overlay) (p. 186)
XOR (p. 180)

6.1 Geospatial Lab Application

GIS Spatial Analysis: QGIS Version

This chapter's lab builds on the basic concepts of GIS from Chapter 5 by applying several analytical concepts to the GIS data that you used in that chapter's labs. This lab also expands on Chapter 5 by introducing several more GIS tools and features and explaining how they are used.

Similar to the two Chapter 5 Geospatial Lab Applications, two versions of this lab are also provided. The first version (*Geospatial Lab Application 6.1: GIS Spatial Analysis: QGIS Version*) uses the free QGIS package. The second version (*Geospatial Lab Application 6.2: GIS Spatial Analysis: ArcGIS Version*) provides the same activities for use with ArcGIS for Desktop 10.3. Although ArcGIS contains more functions than QGIS, these two programs share several useful spatial analysis tools.

Objectives

The goals for you to aim for in this lab are:

▶ To construct database queries in QGIS

▶ To create and use buffers around points, lines, and polygons

▶ To use different selection criteria (from queries) to perform simple spatial analysis

Using Geospatial Technologies

The concepts you'll be working with in this lab are used in a variety of real-world applications, including:

▶ Law enforcement, where officers use spatial analysis techniques for examining "hot spots" of crimes as well as examining where different types of crimes are occurring in relation to other locations

▶ Emergency management services, which use spatial analysis for a variety of applications, such as researching what populated areas are covered (or not covered) by tornado sirens

2. To build a query, first select the Field to use. In this case, double-click ST (which contains the state name abbreviations).

3. Next, select an Operator to use. In this case, select the = operator.

4. Lastly, select the appropriate Values. Click on the All button to show a list of all available values for the Field called ST. For this query, double-click the Value called 'SD' (which stands for South Dakota). The query will appear in the filter expression box as: "ST" = 'SD'. This will select all records whose ST fields are equal to the characters 'SD'.

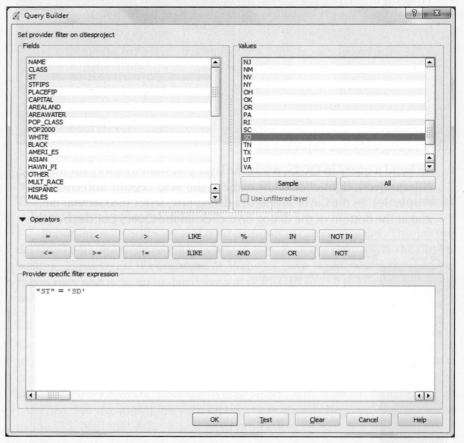

[Source: QGIS]

5. Click OK to run the query. After it runs, the query builder will close and you'll return to the Layer Properties dialog box. Close that dialog box as well.

6. Open the citiesproject layer's attribute table by right-clicking on its name in the Map Legend and selecting Open Attribute Table. The

number of filtered records (i.e., those that met the results of your query) will be shown at the top of the attribute table dialog box—you'll see that this particular query will return nine records. Close the open citiesproject attribute table as well.

7. Take a look at the Map View again and you'll see that those nine cities in South Dakota are the only points being displayed.

8. Next, we'll build a compound query to find all of the cities in South Dakota with a Year 2000 population of greater than 50,000 persons. Bring up the **Query Builder** again, this time creating a query looking for ST = 'SD' (but don't click OK yet).

9. Now click the button for the **AND** operator, then finish the second half of the compound query by selecting **POP2000** as the Field, then select the **greater than Operator** (>), and lastly, type in the number **50000** in the filter expression box. Your final query should look something like this:

[Source: QGIS]

10. Click **OK** to query the database.

> **Question 6.1** How many cities in South Dakota have a population greater than 50,000 persons? What are they?

11. Return to the citiesproject layer's attribute table and you'll see the records representing the cities that were your answer to Question 6.1. Hold down the shift key and click the mouse on the small grey box to the immediate left of the NAME field. This will select each of the records (you'll see them highlighted in dark blue in the attribute table and their corresponding points shown in yellow in the Map View). You can then close the attribute table (your points will remain selected).

12. Also, return to the Query Builder and click on the **Clear** button at the bottom of the dialog. This will clear the filtering that was done by your

query and once more show all of the points from the layer. You can then close the Query Builder. At this point, you should have no filtered results (you're able to see all of the points from the layer, not only the ones you queried for), but your query results should also be selected (displayed in yellow).

6.3 Creating and Using Buffers around Points

1. QGIS allows you to create buffers around objects for examination of spatial proximity and gives you the ability to use the spatial dimensions of the buffer as a means of selecting other objects.

2. To create a buffer, first select the layer whose selected features you wish to buffer from the Map Legend (in this case, select the citiesproject layer).

3. Next, from the Vector pull-down menu, select Geoprocessing Tools, then select Buffer(s).

4. The Buffer dialog box will appear:

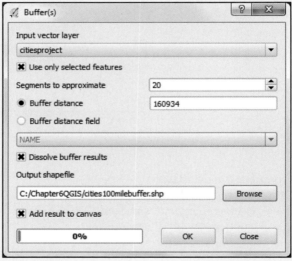

[Source: QGIS]

5. We'll start with a simple buffer around the cities, so the Input vector layer should be citiesproject.

6. Place a checkmark in the box next to Use only selected features. This will create a buffer around the cities you selected for Question 6.1.

7. Set the Segments to approximate option to 20. This is a measure of the "smoothness" of the buffer being created by QGIS.

8. Select the ratio button next to Buffer distance and type in 160934. This value is the number of meters equal to 100 miles. Since your citiesproject data is being measured in meters, QGIS will use that for the buffer's units of measurement.

9. Place a checkmark in the box next to Dissolve buffer results.

10. Click the Browse button and navigate to your C:\Chapter6QGIS folder. Name the Output shapefile (the buffer that will be created) cities100milebuffer.shp and Save it.

11. Place a checkmark in the box next to Add result to canvas.

12. Back in the Buffer(s) dialog box, click OK to create the buffer.

13. When prompted about adding the new layers to the Map Legend, click Yes. The circular 100-mile buffers will be created around the selected cities. Close the Buffer(s) dialog, then drag your citiesproject layer to the top of the layers in the Map Legend.

14. QGIS allows you to select features based on their proximity to other features. For instance, you can find out how many cities are within the buffer you just created and create a new selection of these cities. To do this, select the Vector pull-down menu, then choose Research Tools, and then choose Select by location.

15. You'll want to Select features in citiesproject.

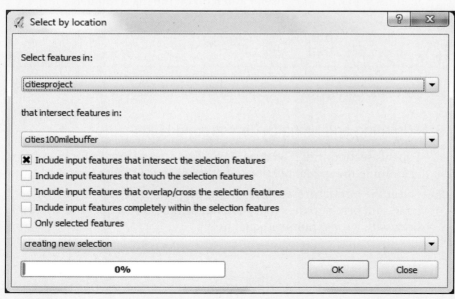

[Source: QGIS]

16. You'll want to find those features that intersect features in cities100milebuffer.

17. Place a checkmark in the box next to Include input features that intersect the selection features.

18. In the pull-down menu at the bottom of the dialog, choose Creating new selection.

19. Click OK when you're ready. This process will locate the points in the citiesproject layer that are within the boundaries of the buffers you created, and then select these points. You can close the Select by Location dialog box as well.

20. Open the attribute table and view only the selected records (see *Geospatial Lab Application 5.1: GIS Introduction: QGIS Version* for how to do this).

> **Question 6.2** How many cities are within a 100-mile radius of the cities found in Question 6.1? (Be very careful to note what actually got selected in the "select by location" procedure when you answer this question.)

21. Turn off the cities100milebuffer layer.

22. Before proceeding, unselect (clear) all of the currently selected features. Click on the Unselect all icon on the main toolbar:

[Source: QGIS]

23. All of the currently selected points will be unselected.

6.4 Creating and Using Buffers around Lines

1. For this section, you'll be creating buffers around line features and examining the spatial relations of other objects to these lines.

2. Select the intrstatproject layer in the Map Legend, open its attribute table, and bring up the Query Builder again. Build a query to select all lines with a route number equal to I94.

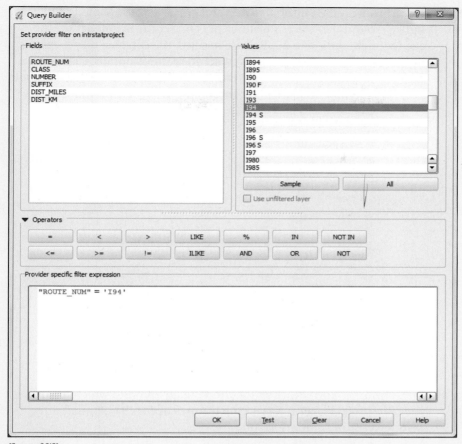

[Source: QGIS]

3. The field that corresponds with route number is ROUTE_NUM.

4. Click All to show all available values.

5. Select = for the Operator, then from the Values table choose 'I94' (note that QGIS doesn't necessarily sort all values in numerical order).

6. Click OK when the query is built (and then you can also close the Layer Properties dialog box as well). Zoom out so that you can see the length of I-94 (a major artery running east to west through the northern United States) on the screen.

7. Like you did in Step 11 of part 6.2, open the intrstatproject layer's attribute table and select all of the records that were filtered (these will be the records that correspond with the sections of I-94).

8. Also, like you did in Step 12 of part 6.2, clear the filtered records so that you can see all of the interstate lines but those that correspond with I-94 will still remain selected.

9. Next, bring up the Buffer(s) dialog and create a 32186.9-meter (20-mile) buffer around the selected features of I-94. Set the Segments to approximate option to 20. Be sure to select the Dissolve buffer results option and choose the Use only selected features option. Call the output shapefile I94_20milebuffer.shp. Also be sure to select Add result to canvas so that the buffer will automatically display on the screen.

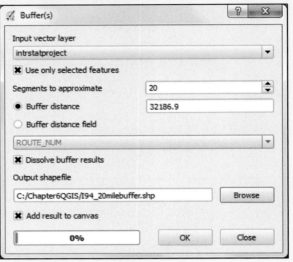

[Source: QGIS]

10. Next, bring up the Select by location dialog and find which features in the citiesproject layer intersect the features of the I94_20milebuffer layer by creating a new selection.

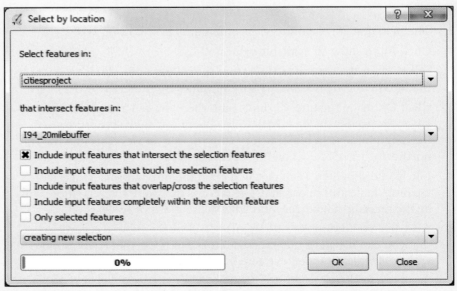

[Source: QGIS]

11. Select the **citiesproject** layer from the Map Legend, then open the attribute table for the cities and view the selected results.

> **Question 6.3** How many cities are within 20 miles of I-94?

12. What we're going to do next is redo the analysis, but this time with a smaller buffer. To do this, you're going to want to clear the results of your cities but not your interstates. To do this, click on the Unselect all icon inside the **citiesproject attribute table** (do not click the Unselect all icon on the main toolbar—that will clear the features from both the citiesproject and the intrstatproject layers).

	NAME	CLASS	ST	STFIPS	PLACEFIP	CAPITAL	AREALAND	AREAWATER	PO
0	College	Census Designat...	AK	02	16750	NULL	18.670	0.407	
1	Fairbanks	City	AK	02	24230	NULL	31.857	0.815	

Attribute table - citiesproject :: Features total: 3557, filtered: 3557, selected: 307

[Source: QGIS]

13. Redo the analysis, this time with a 16093.4-meter (i.e., 10-mile) buffer. Be sure to use **20 Segments to approximate**, and to dissolve the buffer results. Create a new output shapefile called **I94_10milebuffer** and find which cities intersect this new buffer as a new selection.

> **Question 6.4** How many cities are within 10 miles of I-94?

14. Unselect all of the selected cities (and only unselect from the citiesproject layer) and redo the analysis one more time, this time with a 8046.72-meter (i.e., 5-mile) buffer. Be sure to use **20 Segments to approximate** and to dissolve the buffer results. Create a new output shapefile called **I94_5milebuffer** and find which cities intersect this new buffer as a new selection.

> **Question 6.5** How many cities are within 5 miles of I-94?

15. Unselect all of the currently selected features and turn off the buffers.

6.5 Selecting by Location with Polygons

1. For your next question, you'll want to know how many cities are within states that have a high population. Unfortunately, the states layer only has total population values and gives no information about numbers of cities or populated places. However, this type of question can be answered spatially using the Query Builder and select by location tools that you've been using so far in this lab.

2. In the Map Legend, choose the **statesproject** layer and bring up the **query builder** dialog box.

3. Construct a query that will select all states with a year 2005 population (the Field is called POP2005) of greater than 2,000,000 persons.

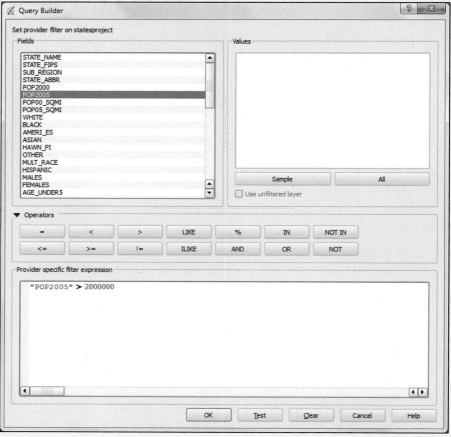

[Source: QGIS]

4. Click OK when the query is ready.

> **Question 6.6** How many states had a year 2005 population of greater than 2,000,000 persons?

5. Like you did before, open the statesproject layer's attribute table and select the records that met the query. Also, return to the query builder and clear the filtered results.

6. Next, bring up the **Select by location tool** to select cities that intersect one of these selected states (that have a population of more than 2,000,000 persons).

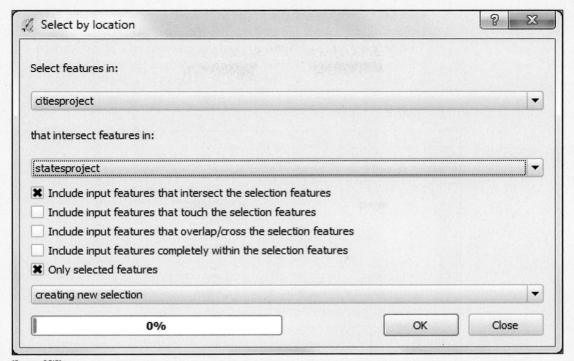

[Source: QGIS]

7. You'll want to select features from the citiesproject layer that intersect features in the statesproject layer. Also be sure to use only selected features (since you only want to locate which cities are within one of the high population states you just selected). Click OK when the settings are correct.

8. Answer Question 6.7 (this will involve examining the selected cities in the attribute table).

> **Question 6.7** How many cities are located within high-population states (defined as states that have populations of more than 2,000,000 persons)?

9. Unselect all of the currently selected features from both the citiesproject and the statesproject layers.

6.6 More Spatial Analysis

1. The next analysis that you'll perform involves analysis of the features in relation to the Great Lakes, so zoom in on the Map View so that you can see all five Great Lakes. The first thing you'll need to do is select the Great Lakes to work with.

2. Select the lakesproject shapefile in the Map Legend and bring up the Query Builder. You'll see that the two available fields that are available for query are the Name and Area of the lake records. You'll want to build a query that looks for the name of each of the five Great Lakes as follows: "NAME" = 'Lake Erie' OR "NAME" = 'Lake Huron' OR "NAME" = 'Lake Michigan' OR "NAME" = 'Lake Ontario' OR "NAME" = 'Lake Superior'.

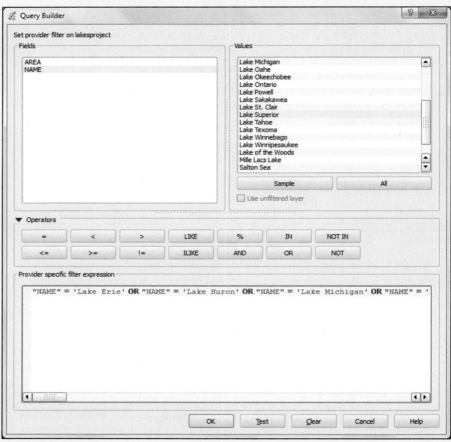

[Source: QGIS 2.8]

3. Answer Question 6.8, and then click OK. All five Great Lakes will now be selected.

> **Question 6.8** Why is the "OR" operator used in this expression rather than the "AND" operator? What would the same query give you back is you used "AND" rather than "OR" in each instance?

4. As before, open the lakesproject layer's attribute table and select the features the met the query. Then return to the Query Builder and clear the filtered records. You should have all five of the Great Lakes selected (i.e. highlighted in yellow).

5. Next, bring up the Buffer(s) dialog, and find out how many rivers are within 5 miles (8046.72 meters) of the Great Lakes by creating a buffer called GL_5milebuffer. Then use Select by location to find which features from the riversproject layer intersect this buffer. Be sure to use 20 Segments to approximate and to dissolve the buffer results. Note that even if part of the river is within the 5-mile buffer, the entire river will be selected.

> **Question 6.9** Which rivers does QGIS consider to be within 5 miles of the Great Lakes? Which river system are they part of?

6. Lastly, use the GL_5milebuffer layer to find out how many cities are within 5 miles of the Great Lakes.

> **Question 6.10** How many cities are within 5 miles of the Great Lakes?

Closing Time

You can exit QGIS by selecting Exit QGIS from the Project pull-down menu. There's no need to save any data in this lab. Now that you have the basics of working with GIS data down and know how to perform some initial spatial analysis, the Geospatial Lab Application in Chapter 7 will describe how to take geospatial data and create a print-quality map, using QGIS or ArcGIS.

6.2 Geospatial Lab Application

GIS Spatial Analysis: ArcGIS Version

This chapter's lab builds on the basic concepts of GIS from Chapter 5 by applying several analytical concepts to the GIS data you used in *Geospatial Lab Application 5.2: GIS Introduction: ArcGIS Version*. This lab also expands on Chapter 5 by introducing several more GIS tools and features and explaining how they are used.

The previous *Geospatial Lab Application 6.1: GIS Spatial Analysis: QGIS Version* uses the free QGIS software package; however, this lab provides the same activities for use with ArcGIS 10.3.

Although ArcGIS contains more functions than QGIS, the two programs share several useful spatial analysis tools.

Objectives

The goals for you to aim for in this lab are:

▶ To construct database queries in ArcGIS for Desktop

▶ To use different selection criteria (from queries and selecting by location) to perform simple spatial analysis

Using Geospatial Technologies

The concepts you'll be working with in this lab are used in a variety of real-world applications, including:

▶ Law enforcement, where officers use spatial analysis techniques for examining "hot spots" of crimes as well as examining where different types of crimes are occurring in relation to other locations

▶ Emergency management services, which use spatial analysis for a variety of applications, such as researching what populated areas are covered (or not covered) by tornado sirens

[Source: © ZUMA Press, Inc./Alamy]

Obtaining Software

The current version of ArcGIS is not freely available for use. However, instructors affiliated with schools that have a campus-wide software license may request a 1-year student version of the software online at **www.esri .com/industries/apps/education/offers/promo/index.cfm**.

  Important note: Software and online resources can change fast. This lab was designed with the most recently available version of the software at the time of writing. However, if the software or Websites have significantly changed between then and now, an updated version of this lab (using the newest versions) will be available online at **www.macmillanhighered.com /shellito/catalog**.

Lab Data

Copy the folder Chapter6—it contains a folder called "usa," in which you'll find several shapefiles that you'll be using in this lab. This data comes courtesy of Esri and was formerly distributed as part of their free educational GIS software package ArcExplorer Java Edition for Educators (AEJEE).

Localizing This Lab

The dataset used in this lab is Esri sample data for the entire United States. However, starting in Section 6.2, the lab focuses on North Dakota and South Dakota and the locations of some cities and roads in the area, and then switches to the Great Lakes region in Section 6.6. With the sample data covering the state boundaries and city locations for the whole United States, it's easy enough to select your city (or cities nearby), as well as major roads or lakes and perform the same measurements and analysis using those places more local to you than the ones in these areas.

6.1 Some Initial Pre-Analysis Steps

1. Start ArcMap.

2. From the usa folder, add the following shapefiles: states, cities, lakes, rivers, and intrstat.

3. Change the color schemes and symbology of each layer to make them distinctive from each other when you'll be doing analysis. (See *Geospatial Lab Application 5.2* for how to do this.)

4. Position the layers in the Table of Contents (TOC) so that all layers are visible (that is, the states layer is at the bottom of the TOC and the other layers are arranged on top of it).

5. When the layers are set how you'd like them, zoom in on South Dakota, so the state (and the nearby data) fills up the screen.

6. You'll also be changing the projection and coordinate system of the data frame. This will allow you to work with your data in a projected format. Just as you did in *Geospatial Lab Application 5.2: GIS Introduction: ArcGIS Version*, change the Data Frame properties coordinate system to the U.S. National Atlas Equal Area projection.

6.2 Database Queries

1. A database query is a standard feature in ArcMap. We'll start off with a simple query (using only one expression). To build a query in ArcMap, choose Select by Attributes from the Selection pull-down menu.

2. The Select by Attributes dialog box will appear.

3. You'll be building a simple query to find which points (representing cities) are in South Dakota (SD).

4. To build a query, first select the Layer to use. In this case, choose cities (since it's the layer that we'll be querying).

5. Next, select the Field to query. In this case, choose "ST" by double-clicking it with the mouse.

6. Next, select an operator to use. In this case, select the = operator by clicking on the symbol with the mouse.

7. Lastly, select the appropriate values. Press the Get Unique Values button to get a list of all the possible values associated with the "ST" Field. In this case, choose 'SD.'

8. The query will appear in the dialog box ("ST" = 'SD'). This will select all records whose ST field is equal to the characters 'SD'.

9. Click OK to run the query.

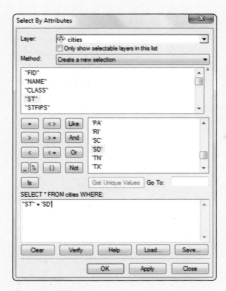

[Source: ArcGIS/Esri]

10. In the View, you'll see several points selected as a result of the query.

11. Open the cities layer's attribute table and show only the selected records (see *Geospatial Lab Application 5.2* for how to do this).

12. This particular query will return nine records (listed as the nine records shown selected in the attribute table).

13. Next, we'll build a compound query to find all the cities in South Dakota with a population of greater than 50,000 persons. Bring up the **Select by Attributes** dialog box again, this time creating a query looking for ("ST" = 'SD') (it may still be there from the last time the query was created—make sure to put in the parentheses).

14. Now click the button for the **And** operator, then finish the second half of the compound query by selecting the **( )** symbol to place the new query in parentheses.

15. Choose **POP2000** as the Field (don't display its unique values), then select the **greater than (>)** operator, and lastly, type the number **50000** in the box. Your final query should look something like this:

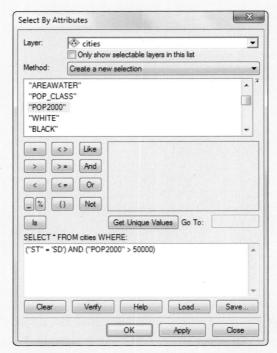

[Source: ArcGIS/Esri]

16. Click OK to query the database.

> **Question 6.1** How many cities in South Dakota have a population of greater than 50,000 persons? What are they?

6.3 Selecting by Location with Points

1. ArcMap allows you to select features based on their proximity to other features. To do so, choose Select by Location from the Selection pull-down menu. This will allow you to select features that are nearby your selected cities.

2. In Select by Location, you'll specify one or more layers as the Target layer (the layer from which you'll be selecting features), and a second layer as the Source (the layer whose spatial dimensions you'll be using to perform the selection). There are several different ways these two layers can spatially intersect with each other—however, in this case, you'll be finding which of the features from the Target layer (the cities) are within the boundaries of the Source layer (the selected cities).

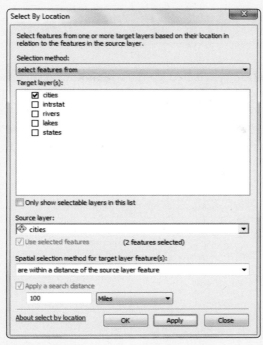

[Source: ArcGIS/Esri]

3. For the Selection method, choose Select features from.

4. Choose cities for your Target layer.

5. Choose cities as your Source layer. You'll notice that the two features you selected back in Question 6.1 remain selected—thus, you'll be finding other cities based on their relation to only these two.

6. Choose are within a distance of the Source layer feature as the Spatial selection method for target layer feature(s).

7. Click Apply a search distance and use 100 Miles.

8. Leave the other options unused, and then click OK.

9. Open the cities attribute table, and examine the selected cities (see the lab in Chapter 5 for how to do this).

> **Question 6.2** How many cities are within a 100-mile radius of the cities you selected in Question 6.1? (Be very careful to note what actually got selected in the Select by Location procedure when you answer this question.)

6.4 Selecting by Location with Lines

1. To create some new selections, you should first clear (unselect) the currently selected features. Click on the Clear Selected Features icon on the toolbar:

All of the currently selected points will no longer be unselected.

2. For this section, you'll be using Select by Location with line features and examining the spatial relations of other objects to these lines.

3. Bring up the Select by Attributes dialog box again, and this time build a simple query to select all lines in the intrstat layer with a route number equal to I94 (I-94 is a major cross-country east–west road that runs through the northern United States).

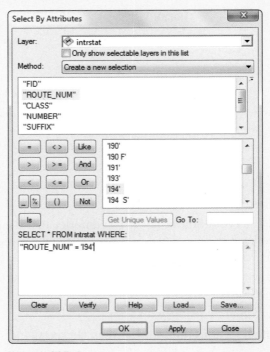

4. Use intrstat as the Layer to select from.

5. Use Create a new selection as your selection method.

6. Use "ROUTE_NUM" as the Field to select from.

7. Choose = as your operator.

8. Press the Get Unique Values button and choose 'I94' as the value to use.

9. Your query should read "ROUTE_NUM" = 'I94'

10. Click OK when the query is built. Zoom out so that you can see the length of I-94 on the screen.

11. Next, use Select by Location to find all cities that lie within 20 miles of I-94.

12. Open the cities attribute table and examine the selected records.

> **Question 6.3** How many cities are within 20 miles of I-94?

13. Clear the selected results and redo the analysis as follows:

    a. Use Select by Location to find all cities that lie within 10 miles of I-94.

    b. Open the cities attribute table and examine the selected records.

> **Question 6.4** How many cities are within 10 miles of I-94?

14. Clear the selected results and redo the analysis one last time as follows:

    a. Use Select by Location to find all cities that lie within 5 miles of I-94.

    b. Open the cities attribute table and examine the selected records.

> **Question 6.5** How many cities are within 5 miles of I-94?

15. Clear all of the currently selected features.

6.5 Selecting by Location with Polygons

1. For your next question, you'll want to know how many cities are within states that have a high population. Unfortunately, the states layer only has total population values, and gives no information about numbers of cities or populated places. However, this type of question can be answered spatially using the Select by Attribute and Select by Location tools that you've been using so far in this lab.

2. Bring up the Select by Attributes dialog box.

3. Construct a query that will select all states with a year 2005 population (the Field is called POP2005) of greater than 2,000,000 persons.

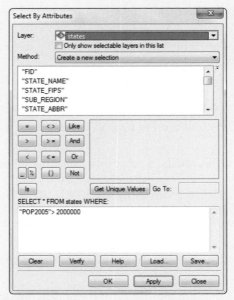

[Source: ArcGIS/Esri]

4. Click OK when the query is ready.

> **Question 6.6** How many states had a year 2005 population of greater than 2,000,000 persons?

5. Next, bring up the Select by Location tool to select cities that fall within one of these selected states (that have a population of more than 2,000,000 persons).

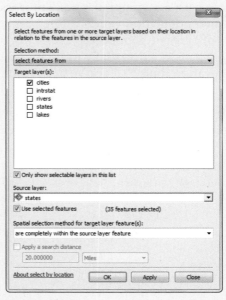

[Source: ArcGIS/Esri]

6. You'll want to select features from the cities layer (the target layer) that are **completely within the source layer feature** (the states). Click Apply when the settings are correct and OK to close the dialog box.

7. Answer Question 6.7 (this will involve examining the selected cities in the attribute table).

> **Question 6.7** How many cities are within a state that has a population of more than 2,000,000 persons?

8. Clear all of the currently selected features.

6.6 More Spatial Analysis with the Selection Tools

1. The next analysis that you'll perform involves analysis of the features in relation to the five Great Lakes, so zoom in on the View so that you can see all five Great Lakes. The first thing you'll need to do is select the Great Lakes to work with.

2. Bring up the Select by Attributes dialog box. Choose lakes as the Layer to select from. You'll see that three fields are available to you—FID, AREA, and NAME. Select "NAME" and press the Get Unique Values button.

3. You'll want to build a query that looks for the name of each of the five Great Lakes as follows: ("NAME" = 'Lake Erie' OR "NAME" = 'Lake Huron' OR "NAME" = 'Lake Michigan' OR "NAME" = 'Lake Ontario' OR "NAME" = 'Lake Superior').

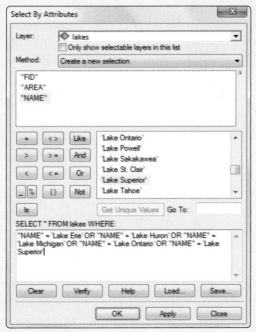

[Source: ArcGIS/Esri]

4. Answer Question 6.8, and then click OK. All five Great Lakes will now be selected. Close the Select by Attributes dialog box.

> **Question 6.8** Why is the "OR" operator used in this expression rather than the "AND" operator? What would the same query give you back if you used "AND" rather than "OR" in each instance?

5. Use the Select by Location tool to find out how many rivers are within 5 miles of the Great Lakes. Note that even if part of the river is within the 5-mile distance, the entire river will be selected.

> **Question 6.9** Which rivers are within 5 miles of the Great Lakes? Which river system are they part of?

6. Lastly, use Select by Location to find out how many cities are within 5 miles of the Great Lakes.

> **Question 6.10** How many cities are within 5 miles of the Great Lakes?

7. Exit ArcMap by selecting Exit from the File pull-down menu. There's no need to save any data in this lab.

Closing Time

Now that you have the basics of working with GIS data down and know how to perform some initial spatial analysis, the Geospatial Lab Application in Chapter 7 will describe how to take geospatial data and create a print-quality map using QGIS or ArcGIS.

7 Using GIS to Make a Map

Scale, Map Elements, Map Layouts, Type, Thematic Maps, Data Classification Methods, Color Choices, and Digital Map Distribution Formats

One thing that makes spatial data unique is that it can be mapped. Once you've used GIS to create the boundaries of a Civil War battlefield and analyzed its proximity to new housing and commercial developments, the next step is to make a map of the results of your work. However, making a good map with GIS involves more than taking the final product of the analysis, throwing on a title and legend, clicking on "print," and then calling it a day. There are numerous design elements and considerations that should be taken into account when you're laying out a map. In fact, there's a whole art and science of mapmaking (called **cartography**), involving things like color selection, the positioning of items on the map, or what kind of message the final map product is intended to convey to the reader.

A **map** is a representation of spatial data that is designed to convey information to its reader. Making digital maps with GIS offers a wide variety of options for creating an end product, but there are a few considerations to take into account before you start the mapping process. For example, say there's a park nearby—a municipal wooded area that's used by the local populace for walking trails, biking, picnicking, or walking dogs. The only map of note is an old hand-drawn one you can pick up at a kiosk at the park entrance, and the local government wants to update it. As you're the local geospatial technology expert, they come to you to do the job. Before you break out your GIS tools, there are some basic questions to ask that are going to define the map you design.

First, what's the purpose of the map, and who will be using it? If the park map you've been asked to design is intended for the park's visitors to use, then some of the important features the map should contain are locations of park entrances, parking, comfort stations, and designated picnic areas. It should be at an appropriate scale, the trails should be accurately mapped, and trailhead locations and points of interest throughout the park should be

> **cartography** the art and science of creating and designing maps
>
> **map** a representation of geographic data

Why Is Map Design Important?

Why is a map's appearance so important? After all, you could argue that a poorly put together map of New Zealand is still (at the end of the day) a map of New Zealand. However, a well-designed map is going to be more useful than a poorly designed one—but why? Why are things like the choice of colors, lettering, and placement of items on the map critical to the usefulness of a map? Consider what kind of effect a poorly designed map might have, and what sort of impact bad design could make on the usefulness of, say, a map of a local county fair, a real estate map of a new subdivision, a promotional map showing the location of a small business, or a map of a proposed new urban development readied for promotional purposes. How can the design of a map influence the way the map's information is perceived by its reader?

---

highlighted. If the map is to be used for zoning purposes, things like exact boundaries, road systems, and parcel information will likely be some of the most important factors. This is part of the **visual hierarchy** of the map, which determines those elements of the map that are most important (and thus, most prominently displayed) and what information is less important (and therefore not so visually conspicuous).

> **visual hierarchy** how features are displayed on a map to emphasize their level of prominence

Second, is the information on the map being effectively conveyed to the map reader? For instance, if you're designing a trail map, should all trails be named or marked, or will that cause too much congestion and clutter? Will the colors and symbols be easily understood by novice map readers who just want to find their way around the park? The trail map will likely not need an inset map of the surrounding area to put the park into a larger context, nor should the map reader be left guessing the meanings of the symbols used on the map.

Third, is the map well designed and laid out properly as a representation of the park? A good map should be well balanced with regard to the placement of the various map elements, for best ease of use by the reader. This chapter will examine several of these cartographic design and data display functions, and by the time we reach the lab, you'll be designing a professional-looking map of your own using GIS.

How Does the Scale of the Data Affect the Map (and Vice Versa)?

> **geographic scale** the real-world size or extent of an area

A basic map item should be information about the scale of the map—and there are a couple of different ways of thinking about scale. First, there's the **geographic scale** of something—things that take up a large area on the ground (or have large boundaries) would be considered large in terms of geographic scale. Study of a global phenomenon occurs on a much larger geographic scale than study of something that exists at a city level, which would be a much smaller geographic scale. Something different is **map scale**, a value representing that *x* number of units of measurement on the map equals *y* number of units in the real world. This relationship between the real world and the map can be expressed as a representative fraction (**RF**). An example of an RF would be a map scale of 1:24000—a measure of one unit on the map would be equal to 24,000 units in the real world. For instance, measuring one inch on the map would be the same as 24,000 inches in the real world, or one foot on the map is equal to 24,000 feet in the real world (and so on).

Maps are considered **large-scale maps** or **small-scale maps** depending on that representative fraction. Large-scale maps show a smaller geographic area and have a larger RF value. For instance, a 1:4000-scale map is considered a large-scale map—due to the larger scale, it shows a smaller area. The largest-scale map you can make is 1:1—where one inch on the map is equal to one inch of measurement in the real world (that is, the map will be the same size as the ground you're actually mapping—a map of a classroom will be the same size as the classroom itself). Conversely, a small-scale map has a smaller RF value (such as 1:250,000) and shows a much larger geographic area.

For instance, on a very small-scale map (such as one that shows the entire United States), cities will be represented by points, and likely only major cities will be shown. On a slightly larger-scale map (one that shows all of the state of New York), more cities are likely to be shown as points, along with other major features (additional roads can be shown as lines, for example). On a larger-scale map (one that shows only Manhattan), the map scale allows for more detail to be shown—points will now show the locations of important features and many more roads will be shown with lines. On an even larger-scale map (one that shows only a section of lower Manhattan) buildings may now be shown as polygon shapes (to show the outline or footprint of the buildings) instead of points, and additional smaller roads may also be shown with lines.

The choice of scale will influence how much information the map will be able to convey and what symbols and features can be used in creating the map in GIS. **Figure 7.1** on page 226 shows a comparison between how a feature (in this case, Salt Lake City International Airport) is represented on large-scale and small-scale maps. The actual sizes of the maps greatly vary, but you can see that more detail and definition of features is available on the larger-scale map than on the smaller-scale one (also see *Hands-On Application 7.1: Powers of 10—A Demonstration of Scale* for a cool example of visualizing different scales).

The same holds true for mapping of data—for instance, the smaller-scale map of the whole state of New York could not possibly show point locations of all of the buildings in Manhattan. However, as the map scale grows larger, different types of information can be conveyed. For instance, in

map scale a metric used to determine the relationship between measurements made on a map and their real-world equivalents

RF representative fraction—a value indicating how many units of measurement in the real world are equivalent to how many of the same units of measurement on a map

large-scale map a map with a higher value for its representative fraction; such maps will usually show a small amount of geographic area

small-scale map a map with a lower value for its representative fraction; such maps will usually show a large amount of geographic area

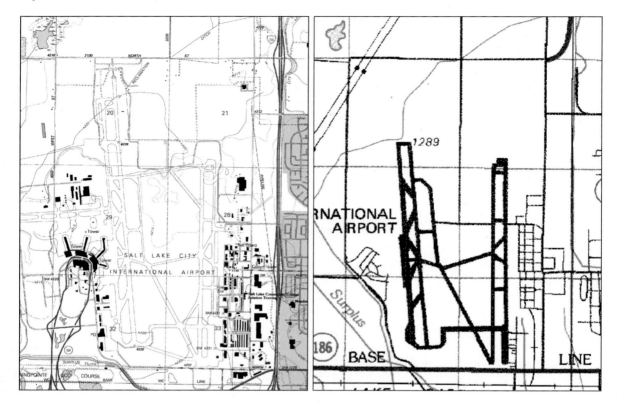

FIGURE 7.1 Salt Lake City International Airport—a comparison of its representation on a 1:24,000 scale map (left) and a 1:100,000 scale map (right). *[Source: Utah State Geographic Information Database (SGID)]*

---

cartographic generalization
the simplification of representing items on a map

Figure 7.1, the large-scale map can convey much more detail concerning the dimensions of the airport runways, while the smaller-scale map has to represent the airport as a set of simplified lines. This kind of **cartographic generalization** (or the simplification of representing items on a map) is going to affect the kind of GIS data you're able to derive from it. For instance, if you digitize the lines of the airport runways, you'll end up with two very different datasets (one more detailed, one very generalized).

This relationship between the scale of a map and data that can be derived from a map can be a critical issue when you're dealing with geospatial data. For instance, say you're using GIS for mapping a university campus. At this small geographic scale, you're going to require detailed information that fits your scale of analysis. If the hydrologic and transportation data you're working with is derived from 1:250,000-scale maps, it's likely going to be way too coarse to use. Data generated from smaller-scale maps is probably going to be incompatible with the small geographic scale you're working at. For instance, digitized features on a small-scale map (like a 1:250,000 scale) are going to be much more generalized than data derived from larger-scale maps (like a 1:24,000 scale), or from sources such as aerial photos taken at a larger scale (for example, 1:6000). (See Chapter 9 for more information on using aerial photos for analysis.)

HANDS-ON APPLICATION 7.1

Powers of 10—A Demonstration of Scale

Though it's not a map, an excellent demonstration of scale and how new items appear as the scale changes is available at **http://micro.magnet.fsu.edu/primer/java/scienceopticsu/powersof10.**

This Website (which requires Java to run properly) shows a view of Earth starting from 10 million light years away. Then the scale changes to a view from 1 million light years away, and then 100,000 light years away (a factor of 10 each time). The scale continues changing until it reaches Earth—and then continues all the way down to sub-atomic particles. Let it play through, and then use the manual controls to step through the different scales for a better view of the process.

Expansion Questions

• Examine the video at the following scales: 1000 km, 100 km, 10 km, 1 km, and 100 m. The imagery at each of these scales could be used as a basemap for digitizing GIS data—what kinds of features and phenomena would be most appropriate to map at each scale?

• Using the same five scales as the previous question, what kinds of features and phenomena would be the least appropriate to map at each scale?

What Are Some Design Elements Included in Maps?

There are several elements that should show up on a good map. For instance, someone reading a map should be able to quickly figure out what scale the map is. A map element might simply be text of the RF (such as 1:24,000, 1:100,000, or whatever the map scale might be). A **scale bar** is a graphical representation of equivalent distances shown on a map (**Figure 7.2**). The scale bar provides a means of measuring the map scale itself, except that a

> **scale bar** a graphical device used on a map to represent map scale

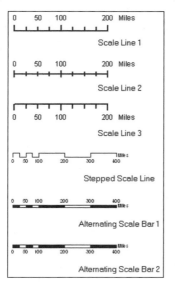

FIGURE 7.2 Examples of various scale bars. *[Source: Esri]*

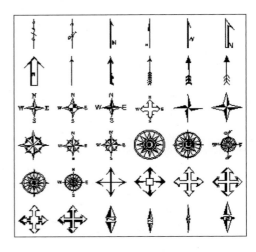

FIGURE 7.3 Examples of various north arrows. [*Source: Esri*]

measurement of x distance on the scale bar is equal to y units of distance in the real world.

A second map element is a **north arrow**, a graphical device used to orient the direction of the map. However the map is oriented, the north arrow is used to point in the direction that is due north. A north arrow may sometimes be drawn as a compass rose, which shows all cardinal directions on the map. North arrows can be as simple as an arrow with the letter "N" attached, or as complex as a work of art. See **Figure 7.3** for examples of north arrows used in map design.

Another item is the map's **legend**—a guide to what the various colors and symbols on the map represent. A good legend should be a key to the symbology of the map (see **Figure 7.4** for an example of a map legend). Since the legend is the part of the map where you can explain things to the map's reader, you may want to call it something other than "legend." Instead, consider calling the legend something more useful like "park trail guide" or "county population change."

The choice of **type** used on the map is also important, as it's used for such things as the title, the date of the map's creation, the name of the map's creator, the origin of the data sources used to make the map, and the

north arrow a graphical device on a map used to show the orientation of the map

legend a graphical device used on a map that explains what the various map symbols and colors represent

type the lettering used on a map

FIGURE 7.4 Examples of various legend items. [*Source: Esri*]

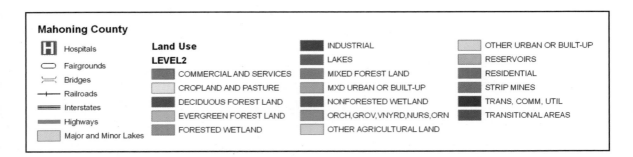

Map Source: Esri Data	Map Source: Esri Data
Map Source: Esri Data	**Map Source: Esri DATA**
Map Source: Esri Data	**Map Source: Esri Data**
Map Source: Esri Data	Map Source: Esri DATA
Map Source: Esri Data	Map Source: Esri Data

FIGURE 7.5 Examples of various fonts available for type in map layouts.

lettering attached to the map's features. GIS programs will usually allow you to **label** map features (adding things like the names of rivers, roads, or cities to the map) using an automatic placement tool or allowing you interactively to select, move, and place map labels.

When you're selecting the type to use for particular map elements, a variety of different **fonts** (or lettering styles) are available. GIS programs (like word-processing programs such as Microsoft Word) will often have a large number of fonts to select from—everything from common fonts (like Arial or Times New Roman) to much flashier fonts (such as Mistral or Papyrus). See **Figure 7.5** for some examples of different versions of type fonts available from Esri. Often, a map will contain only two different fonts, carefully selected to fit well with each other without clashing. Using too many fonts, or crafting a map's type out of several of the more elaborate fonts, is likely to make a map difficult to read and reduce its usefulness. See *Hands-On Application 7.2: TypeBrewer Online* for a tool you can use as an aid in selecting appropriate type for a map.

In GIS, a map is put together by assembling all of the elements together in a **layout**. A good way to think of a layout is as a digital version of a blank piece of paper on which you'll arrange the various map elements together in a

label text placed on a map to identify features

fonts various styles of lettering used on maps

layout the assemblage and placement of various map elements used in constructing a map

HANDS-ON APPLICATION 7.2

TypeBrewer Online

TypeBrewer is a very neat (and free) online utility that allows you to examine different font styles and combinations used on maps. With TypeBrewer, you can interactively change font size, density, or appearance. Go to **www.typebrewer.org** to get started (you'll need to have Adobe Flash Player installed on your computer for it to work properly). With TypeBrewer, you can select from several different pre-set styles of type on a map, then alter aspects of the type (such as its size, density, and tracking) to see the effects of those changes on the map. The aim of TypeBrewer is to examine different forms of map type so you can apply them to your own maps. Try looking at several different formats of the type, then decide what the most appropriate ones would be and why.

Expansion Questions

- Examine the three options for the Formal, Informal, Classic, and Contemporary mapping templates. What are the fonts being used for each of them?

- Adjust the size, density, and tracking of the fonts being used for the 12 different options. How does the presentation of the map change when you adjust these to different settings? What does it mean for map appearance to change size, density, or tracking?

FIGURE 7.6 Example of balancing the placement of items on a map layout.
[Source: Esri]

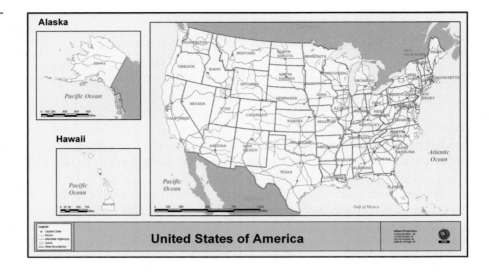

United States of America

cartographic design to create a version of what will ultimately be your map. Sometimes, the software (such as ArcGIS) will include several **map templates**, which provide pre-created designs for your use. Using a template will take your GIS data and place things like the title, legend, and north arrow at predetermined locations and sizes on the layout. Templates are useful for creating a quick printable map layout, but GIS will also allow you to design a map the way you want it with regard to size, type, and placement of elements.

When you're designing a layout, it's important to not simply slap a bunch of elements on the map and decide that the map's ready. Balancing the placement of items is important, partly so as not to overload the map with too much information, but also to provide a useful, readable map. Good map balance provides a means of filling in the "empty" spaces on the map, so it doesn't have all its information crammed into the top or bottom and leave other parts of the map blank. Items should be of uniform size (instead of, say, making the north arrow gigantic just to fill in some empty space) and placed in proportion to one another (see **Figure 7.6** for an example of trying to balance out the placement, size, and proportion of items in a map layout with a template).

map templates pre-made arrangements of items in a map layout

How Is Data Displayed on a GIS Map?

There are several different types of maps that can be made using GIS tools. The purpose of a **reference map** (like you'd see in an atlas or a road map) is to give location information and to highlight different features. A map of park trails, a Manhattan restaurant guide map, a map of the casinos on the Las Vegas Strip, or a zoning map of your neighborhood are all examples of reference maps. Topographic maps (that also show landforms) are another example of reference maps (we'll deal with them in Chapter 13).

reference map a map that serves to show the location of features, rather than thematic information

Another main type of map is the **thematic map**, which is geared toward conveying one particular theme to the reader. Thematic maps can be used to show things such as the state-by-state increase in U.S. population over a particular period, or presidential election results county by county (see *Hands-On Application 7.3: Presidential Election Thematic Maps* for examples of types of thematic maps online). In GIS, a layer's attributes are used for displaying information on the map. For instance, to create a map of the 2012 U.S. presidential election results by county, each polygon representing that county will have an attribute designating whether Barack Obama or Mitt Romney had a higher number of votes in that county (colored red for counties won by Romney and blue for counties won by Obama). This attribute will then be the data being displayed on the map (see **Figure 7.7** on page 232). Other thematic maps may use **graduated symbols** for display. In this case, points (or other symbols) of differing sizes are plotted to represent the thematic factors.

Of course, many thematic maps don't rely on a simple two-choice attribute like the election map (in which each county will be marked Obama or Romney—or else contain no data). Attributes such as the percentage of colleges and universities per state that are using this textbook in a class (or some other values) are mapped using a type of thematic map called a **choropleth map**. In order to best display this information on a map, the data will have

thematic map a map that displays a particular theme or feature

graduated symbols the use of different-sized symbology to convey thematic information on a map

choropleth map a type of thematic map in which data is displayed according to one of several different classifications

HANDS-ON APPLICATION 7.3

Presidential Election Thematic Maps

The results of a U.S. presidential election by state can be presented as a thematic map. A state would either be colored red if its electoral college votes favored the Republican candidate or blue if they favored the Democratic candidate. Presidential election thematic maps can be seen online at **www.votenight.com**. For the 2012 presidential election between Barack Obama and Mitt Romney, you can click on a state to mark its status as red, blue, or grey (undecided). From the U.S. presidential election history pull-down menu, you can examine thematic maps of past presidential elections going back to 1932. Check out the various thematic maps (and voter distributions) over the last 80 years of elections.

Expansion Questions
- Which presidential elections have shown the greatest difference between the numbers of

electoral college votes received by each candidate?

- Under the electoral college system, the candidate with the greatest number of electoral votes is the winner, not the candidate who wins the greatest number of states. Turn on the option to display the number of electoral votes per state and re-examine the presidential election thematic maps over the years. Which elections saw the winning candidate take a smaller number of states but a larger number of electoral votes?

- From a mapping context, would it be more informative to map presidential election results as red states and blue states, or should the states be mapped according to their electoral vote count?

FIGURE 7.7 A thematic map showing the results of the 2012 U.S. presidential election by county (the vote in blue counties going for Barack Obama and the vote in red counties going for Mitt Romney).

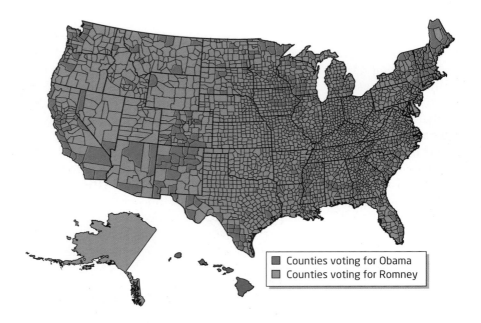

Counties voting for Obama
Counties voting for Romney

data classification various methods used for grouping together (and displaying) values on a choropleth map

to be classified or divided into a few categories. GIS gives several options for **data classification** of multiple values—each method classifies data differently, which can result in some very different results being displayed on the maps. (See **Figure 7.8** for four different maps of the percentage of the total number of houses per state that are considered seasonal homes or vacation homes by census, each created using a different data classification technique.) Note that each map has the data broken into the same number of classes (four) for comparison purposes. Also keep in mind that choropleth maps already have the boundaries of their polygons set in place (elements such as the states' boundaries being already defined) and the attributes being mapped do not define these boundaries. For instance, the states' boundaries are already delineated and the choropleth map is showing a classification of a value of seasonal home percentages utilizing all of these predetermined boundaries.

natural breaks a data classification method that selects class break levels by searching for spaces in the data values

The first of these data classification methods is called **natural breaks** (also referred to as the Jenks optimization). Like the name implies, this method takes all of the values being mapped and looks at how they're grouped together. The spaces in between the data values are used to make different classes (or ranges of data) that are displayed. This is shown in Figure 7.8a—states with the lowest percentages of seasonal home values (such as Nebraska, Oklahoma, and Texas) end up in one class, and states with the highest percentages of seasonal homes (such as Maine, Vermont, and New Hampshire) end up together in another class.

quantile a data classification method that attempts to place an equal number of data values in each class

The map in Figure 7.8b shows the results of using the **quantile** method of data classification. This method takes the total number of data values to

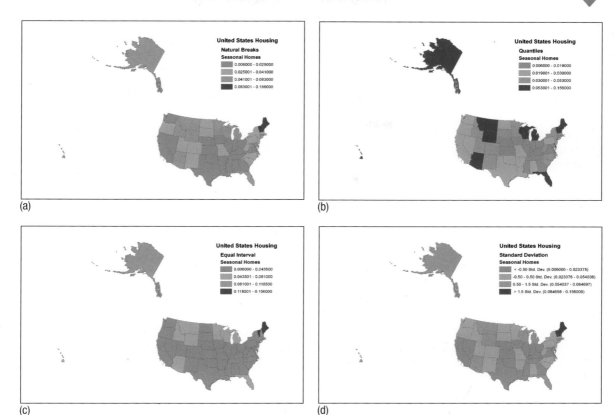

be mapped and splits them up into a number of classes. It tries to distribute values so that each range has a similar number of values in it. For instance, with 50 states being mapped (plus the District of Columbia as a 51st area), each of the four ranges will have about 13 counties' worth of data being shown in each class. Since the break points between the ranges are based on the total number of items being mapped (that is, the number of states ending up in each range), rather than the actual data values being mapped, the quantile method causes a relatively even distribution of values on the map.

The third method (shown in the map in Figure 7.8c) uses **equal intervals** for data classification. It works like it sounds—it creates a number of equally sized ranges and then splits the data values among these ranges. The size of each range is based on the total span of values to be mapped. For instance, in the seasonal home map, the data is divided into four classes, and the range of values goes from the state with the lowest seasonal home percentage (0.6% of the total housing stock in Illinois) to the state with the highest seasonal home percentage (15.6% in Maine). Equal interval takes the complete span of data values (there is 15% separating the lowest and highest values) and divides it by the number of classes (in this case, four), and that value (in this case, 3.75%) is used to compute the breaking point between classes. So the first class represents states that have a seasonal home

FIGURE 7.8 Four choropleth map examples created using the same data (the year 2000 percentage of the total number of houses that are considered seasonal or vacation homes), but employing different data classification methods, as follows: (a) natural breaks, (b) quantiles, (c) equal interval, and (d) standard deviation. [Source: Esri]

equal interval a data classification method that selects class break levels by taking the total span of values (from highest to lowest) and dividing by the number of desired classes

value of 3.75% more than at lowest end (for instance, the first class would have values between 0.6% and 4.35%). Note that this method simply classifies data based on the range of all values (including the highest and lowest) but does not take into account clusters of data or how the data is distributed. As such, only a few states end up in the upper class because their percentages of seasonal homes were greater than three-fourths of the total span of values.

The final method uses the concept of **standard deviation** (see Figure 7.8d). A standard deviation is the average distance that a single data value is away from the mean (the average) of all data values. The breakpoints for each range are based on these statistical values. For instance, the GIS would calculate the average of all United States seasonal home values (3.9%) and the standard deviation for them (3.1%). So when it comes to the percentage of the total housing stock that is seasonal, each state's percentage is an average of 3.1% away from the average state's percentage. These values for the mean and standard deviation of the values are used to set up the breakpoints. For instance, the breakpoint of the first range is of all states whose seasonal home values are less than half a standard deviation value lower than the mean—those states with a seasonal home percentage of less than the mean minus 0.5 times the standard deviation (1.55%), or 2.34%. The fourth range consists of those counties with a value greater than 1.5 times the standard deviation away from the mean. The other ranges are similarly defined by the mean and standard deviation values of the housing data.

As Figure 7.8 shows, the same data can produce choropleth maps that look very different and carry different messages, depending on which method is used to classify the data. For instance, the map in Figure 7.8d (standard deviation) shows that most states have roughly an average (or below average) percentage of seasonal homes, while the other maps show various distinctions between the states that are classified as having a higher or lower percentage of seasonal homes. Thus, the same data can result in different maps, depending on the chosen method of classification. When you're selecting a method, having information about the nature of the data itself (that is, if it is evenly distributed, skewed toward low or high numbers, or all very similar values) will help you to determine the best kind of mapping.

You should keep in mind that while different data classification methods can affect the end result shown on the choropleth map, the type of data values being mapped can also greatly affect the outcome of the choropleth map. An issue involved with choropleth mapping is displaying the values of data that can be counted (for instance, the total population values per state or the total number of housing sales per county) when the sizes of the areas being mapped are very different. If you're mapping the number of vacation homes in each state, a very large state like California is probably going to have a larger number of homes (12,214,549 homes in total, with 239,062 of them classified as seasonal, according to the 2000 Census) than a smaller state like Rhode Island (439,837 homes in total with 13,002 of them

standard deviation a data classification method that computes class break values by using the mean of the data values and the average distance a value is away from the mean

classified as seasonal). Thus, if you're making a choropleth map of the number of vacation homes in each state, California will show many more vacation homes than Rhode Island, just because it has a lot more houses.

A better measure to map is the percentage of the total number of houses that are vacation homes—in this way, the big difference in housing counts between California and Rhode Island won't be a factor in the map. Instead, you'll be mapping a phenomenon that can be comparably measured between the two states. When you map the percentages of the total housing that are considered vacation homes, California's seasonal homes only make up about 2% of the total, while Rhode Island's seasonal homes make up about 3%. To make a choropleth map of count data, the data must first be **normalized**—that's to say, it must have all count values brought onto the same level. For instance, dividing the number of seasonal homes by the total number of houses would be a way of normalizing the data. See *Hands-On Application 7.4: Interactive Thematic Mapping Online* and *Hands-on Application 7.5: The Census Data Mapper* for online tools used for creating chrropleth maps.

> **normalized** altering count data values so that they are at the same level of representation (such as using them as percentages instead of regular count values)

HANDS-ON APPLICATION 7.4

Interactive Thematic Mapping Online

It's now time to start making your own thematic maps. Open your Web browser and go to **http://thematicmapping.org/engine**. This is the Website of the Thematic Mapping Engine, which allows you to create numerous kinds of thematic maps of data from a variety of topics. First, select an indicator (start with something simple like Population) and choose a year (say, 2010). Choose Choropleth for the technique and pick a set of colors. For the Classification method, choose Quantiles. Lastly, click on the Preview button to see a thematic map of world population displayed in an interactive Google Earth interface in your Web browser (alternatively, you can select the Download option to actually download the thematic map as a new layer in Google Earth). After the Quantile map is set up, try the same settings, but using the equal interval classification method.

Try some of the other indicators (such as CO_2 emissions, GDP per capita, Infant mortality rate, or Mobile phone subscribers) over a range of years. These thematic maps can be created using graduated symbols, choropleth maps, or even 3D-style prism maps (we'll use the Thematic Mapping Engine again in Chapter 14 with these types of maps). You can specify which data classification method to use, and finally display the mapped results, using Google Earth or the online preview.

Expansion Questions

- For creating a choropleth map of 2010 world population, which classification method would be more appropriate, quantiles or equal interval, and why?

- What were the CO_2 emissions per capita in the year 2000 in Brazil, China, India, and the United States? What were the CO_2 emissions per capita in the year 1980 for the same countries?

- What was the number of Internet users per 100 persons in the year 2005 in Brazil, China, India, and the United States? What was the number of Internet users per 100 persons in 1999 for the same countries?

HANDS-ON APPLICATION 7.5

The Census Data Mapper

The U.S. Census Bureau has an online utility for mapping demographic data from the U.S. Census as choropleth maps. To get started with it, go to **http://datamapper.geo .census.gov/map.html**—this is the Website for the Census Data Mapper. When the Website completely loads, you'll go through a quick six-step process to create a choropleth map. First, select the theme you want to map, such as Age and Sex, Population and Race, or Family and Housing. Choose the Population and Race option for your first map. Next, you'll choose the data table (or the attribute) associated with that theme that you'll make a map of—choose Percent Black or African American. Third, choose an appropriate color palette for displaying the data. Fourth, select the number of classes you want to display—for instance, for this first map, choose 5. Fifth, choose the classification method you want to use—for instance, choose Quantile. Lastly, click on Make Your Map to generate the choropleth map of your choices. The choropleth map is also interactive; you can zoom in and out of areas and by clicking on a county, you'll receive the mapped data back for that place. Try some of the other themes and choices to generate choropleth maps.

Expansion Questions

- In Cook County, Illinois (which contains the city of Chicago), what is the percentage of the population that is Black or African American? What is the percentage of the population listed as White, as well as the percentage of the population listed as Latino or Hispanic?

- In Wayne County, Michigan (which contains the city of Detroit), what is the percentage of the population listed as male, as female, and as 65 years and over?

- What is the percentage of occupied housing units in Allegheny County, Pennsylvania (where Pittsburgh is located), Laramie County, Wyoming (where Cheyenne is located), and Clark County, Nevada (where Las Vegas is located)?

What Kinds of Colors Are Best to Use with GIS Maps?

RGB a color scheme based on using the three primary colors red, green, and blue

CMYK a color scheme based on using the colors cyan, magenta, yellow, and black

The choice of colors for a map is also important—first to create something pleasing for the map reader, but also in terms of graphic design and usability in a digital format. For instance, the choropleth maps in Figure 7.8 use a range of values of dark green for low percentages and dark blue for high percentages, but there are a lot of ways those classifications could have been colored. There are different color schemes for use in GIS maps—one of these is **RGB**, a setup used for design on computer screens. RGB utilizes the colors red, green, and blue as the primary colors, mixing them together as needed to create other hues. Another is **CMYK**, used in graphic design, which uses the colors of cyan, magenta, yellow, and black (the "K" of the acronym) for creating hues. Chapter 10 will deal further with generating different colors from an initial set of primary colors in the context of examining digital imagery.

With a reference map, different colors are used to represent the various objects and symbols. On a choropleth map (like those shown in Figure 7.8), a range of colors is needed to show the lowest classes of data through the highest classes. In this case, a **color ramp** is used to show a set of colors to represent the classes of values. A color ramp is a selected set of colors that will show the different color changes of a color scheme—for instance, using a single hue of a color in a color ramp might run the range of lightest to darkest blue or light green to very dark green. Other color ramps (like those shown in Figure 7.8) incorporate multiple hues ranging from green through blue. Several different examples of color ramps are shown in **Figure 7.9**.

> **color ramp** a particular range of colors applied to the thematic data on a map

Part of why color choice becomes important with a map is the wide range of visualization media that are available for viewing it. Colors of map items may appear one way on a computer monitor, another way when they're projected onto a screen with an LCD projector, and appear differently yet again when they're printed on an inkjet or laserjet printer. Just because you like the look of the shades of blue and green on the GIS map on your monitor, you shouldn't assume that the same colors will translate the same way to a printed page or to a photocopy of the printout. Your colors might easily have a washed out appearance when the map is projected onto a big screen. Check out *Hands-On Application 7.6: Color-Brewer Online* to investigate the effect that different color choices will have on a final map.

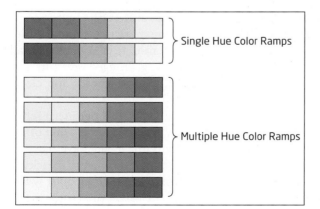

Single Hue Color Ramps

Multiple Hue Color Ramps

FIGURE 7.9 Examples of various color ramps.

HANDS-ON APPLICATION 7.6

ColorBrewer Online

To investigate further the impact of color or choices on a map, check out Color-Brewer 2.0, which is available online at **http://colorbrewer2.org**. ColorBrewer is a great tool for color choice and color ramp selection for a variety of different map data and formats. ColorBrewer 2.0 sets up a pre-determined choropleth map while allowing you to examine different color selections, see how different color choices on choropleth maps can be distinguished from one another, or examine the best choices for printing, photocopies, and color-blind–safe schemes. Try out several of the options to determine (based on the pre-set choropleth maps) what the best color scheme would be for your map. You can use ColorBrewer in a similar way

as TypeBrewer—to determine the optimum setup of colors and then apply those settings to your own maps.

Expansion Questions

- What color schemes are suggested as being photocopy-able? Why would these schemes be more useful than other schemes when you're producing a map that will be photocopied?

- If you were producing a map with five data classes and needed it to be print-friendly only, which classes does ColorBrewer recommend? Of these sequential or diverging schemes, which would you choose for your map, and why?

How Can GIS Maps Be Exported and Distributed?

Once a map has been designed and formatted the way you want it, it's time to share and distribute the results. Rather than just print a copy of the map, there are several digital formats to which maps can be quickly and easily converted for ease of distribution. A simple way is to export the map as a graphical raster file—this saves a "snapshot" of the map as a digital graphic that can be viewed like a picture, either as a file or an image placed on a Website. There are many different formats for map export—two common ones are **JPEG** (Joint Photographic Experts Group) and **TIFF** (Tagged Image File Format). Images saved in JPEG format can experience some data loss due to the file compression involved. Consequently, JPEG images usually have smaller file sizes for viewing or downloading. Images saved in TIFF format have a much larger file size but are a good choice for clearer graphics.

The clarity of an image is a function of what **DPI** (dots per inch) setting is used when the map is exported to a graphic. The lower the value of DPI (such as a value of 72), the less clarity the resultant image will have. Very low values of DPI will result in the exported image being very blocky or pixelated. Maps exported with higher values of DPI (such as 300) will be very crisp and clear. However, the higher the DPI value, the larger the file size will be, which becomes important when you're distributing map data

JPEG the Joint Photographic Experts Group image or graphic file format

TIFF the Tagged Image File Format used for graphics or images

DPI dots per inch—a measure of how coarse (lower values) or sharp (higher values) an image or map resolution will be when exported to a graphical format

online—the larger the file is, the more time will be needed to transfer or display the images. For professional-print-quality maps, the TIFF file format is used at a higher DPI value (such as 300).

A **GeoPDF** is another option for exporting and distributing maps. A GeoPDF allows the user to export a GIS map in the commonly used PDF file format, which can be opened by the Adobe Reader software using a special free plug-in. A GeoPDF differs from a regular PDF in that it allows the user to interact with map layers and get information about the coordinates of locations shown in the PDF. By using the GeoPDF format, you can click on a location and get the latitude and longitude of that point, as well as make measurements of lengths or areas on the map in real-world units. Also, a GeoPDF can contain multiple layers of data that the user can turn on and off (such as annotations for road names or a separate layer of the roads themselves) as in GIS (see Chapter 13 for more uses of GeoPDFs with other maps).

> **GeoPDF** a format that allows maps to be exported to a PDF format, yet contain multiple layers of geographic information

Chapter Wrapup

Once GIS data is created, compiled, or analyzed, the results are usually best communicated using a map, and the ability to create a map of data is a standard feature in GIS software packages such as ArcGIS or QGIS. However, for producing a useful, readable, well-balanced map, there are several choices that go into design, layout, and the presentation of data, colors, and symbols.

This chapter presented an overview of several cartographic and map-design concepts, and this chapter's Geospatial Lab Application will have you take the data you've been analyzing in Chapters 5 and 6 to create a professional-quality map. In the next chapter, we're going to examine some specific types of maps—road network maps—and see how GIS and geospatial technologies are used to design street maps that can be used to locate addresses and compute shortest paths (and directions) between locations. Before that, check out *Cartography Apps* for some downloadable apps, as well as *Cartography in Social Media* for some mapping blogs and related Twitter accounts.

Important note: The references for this chapter are part of the online companion for this book and can be found at **www.macmillanhighered .com/shellito/catalog**.

Cartography Apps

Here's a sampling of available representative cartographic apps for your phone or tablet. Note that some apps are for Android, some are for Apple iOS, and some may be available for both.

- **Atlas Lite:** A free app for displaying pre-generated maps
- **Atlas of the World:** An app for displaying pre-made maps of world population by country and also by state within the United States
- **Cartovista:** An app for displaying pre-made datasets as choropleth maps
- **Census Quickfacts Data Browser:** An app for examining population maps of Census data of the United States by state or county

Cartography in Social Media

Here's a sampling of some representative Facebook and Twitter accounts, along with some blogs related to this chapter's concepts.

On Facebook, check out the following:

- **The History of Cartography Project:**
  www.facebook.com/HistoryofCartographyProject

Become a Twitter follower of:

- **Bjorn Sandvik** (creator of Thematic Mapping Engine):
  @thematicmapping
- **British Cartographic Society:** @bcsweb

For further up-to-date info, read up on these blogs:

- **ArcGIS Resources—Mapping Category** (blog posts concerning mapping and cartography topics):
  http://blogs.esri.com/esri/arcgis/category/mapping
- **Thematic Mapping Blog** (a blog from the creator of the Thematic Mapping Engine):
  http://blog.thematicmapping.org
- **Strange Maps** (a blog dedicated to all types of weird and cool maps):
  http://bigthink.com/blogs/strange-maps

Key Terms

cartographic generalization (p. 226)

cartography (p. 223)

choropleth map (p. 231)

CMYK (p. 236)

color ramp (p. 237)

data classification (p. 232)

DPI (p. 238)

equal intervals (p. 233)

fonts (p. 229)

geographic scale (p. 224)

GeoPDF (p. 239)

graduated symbols (p. 231)

JPEG (p. 238)

label (p. 229)

large-scale maps (p. 225)

layout (p. 229)

legend (p. 228)

map (p. 223)

map scale (p. 225)

map templates (p. 230)

natural breaks (p. 232)

normalized (p. 235)

north arrow (p. 228)

quantile (p. 232)

reference maps (p. 230)

RF (p. 225)

RGB (p. 236)

scale bar (p. 227)

small-scale maps (p. 225)

standard deviation (p. 234)

thematic map (p. 231)

TIFF (p. 238)

type (p. 228)

visual hierarchy (p. 224)

GIS Layouts: QGIS Version

This lab will introduce you to the concept of taking GIS data and creating a print-quality map from it. This map should contain the following:

▶ The population per square mile of the contiguous United States (48 states without Alaska and Hawaii), set up in an appropriate color scheme

▶ The data displayed in a projection with units of measurement other than decimal degrees (the default units used by the lab data are meters—be sure that the scale bar reflects this information)

▶ An appropriate legend (make sure your legend items have regular names, and that the legend is not called "legend")

▶ An appropriate title (make sure that your map title doesn't include the word "map" in it)

▶ A north arrow

▶ A scale bar

▶ Text information: your name, the date, and the sources of the data

▶ Appropriate borders, colors, and design layout (your map should be well designed, and not look like map elements were thrown on at random)

Important note: At the end of this lab, there is a checklist of items to aid you in making sure the map you make is complete and of the best quality possible.

Like the geospatial lab applications in Chapters 5 and 6, two versions of this lab are provided in this chapter. The first version (*Geospatial Lab Application 7.1: GIS Layouts: QGIS Version*) uses the free QGIS. The second version (*Geospatial Lab Application 7.2: GIS Layouts: ArcGIS Version*) provides the same activities for use with ArcGIS for Desktop 10.3.

Objectives

The goals for you to take away from this lab are:

▶ To familiarize yourself with the map composer functions of QGIS

▶ To arrange and print professional-quality maps from geographic data using the various layout elements

Using Geospatial Technologies

The concepts you'll be working with in this lab are used in a variety of real-world applications, including:

▶ Urban planning, where GIS is used to compose maps showing data about housing and population distribution, transportation routes, land parcels, zoning information, and other matters

▶ Law enforcement, where agents use GIS to create maps for distribution among police officers and the community showing the major locations of particular types of crimes, analysis of crime hot spots, and locations of 911 calls

[Source: Monty Rakusen/© Cultura Creative (RF)/Alamy]

Obtaining Software

The version of QGIS used in this lab is 2.8, and available for free download at **http://qgis.org/downloads**.

Important note: Software and online resources can change fast. This lab was designed with the most recently available version of the software at the time of writing. However, if the software or Websites have significantly changed between then and now, an updated version of this lab (using the newest versions) will be available online at **www.macmillanhighered.com /shellito/catalog**.

Lab Data

Copy the folder Chapter7QGIS—it contains a folder called "usaproject," in which you'll find several shapefiles that you'll be using in this lab. This data comes courtesy of Esri, and was formerly distributed as part of their free educational GIS software package ArcExplorer Java Edition for Educators (AEJEE). For use in this lab with QGIS, it has already been projected for you to the US National Atlas Equal Area projection.

Localizing This Lab

The dataset used in this lab is Esri sample data for the entire United States, which you'll be using to create a map layout of the United States. However, the layout tools can be used to make a map of whatever dataset you desire—rather than creating a layout map of the entire United States, focus on your home state and create a layout of that instead (see Section 7.5 for how to focus the layout on one state instead of the whole United States). If you're only going to work with one state, use the countiesproject.shp file instead of the statesproject.shp file and make a map of county population by square mile.

7.1 Initial Pre-Mapping Tasks and Map Projections

If needed, refer back to *Geospatial Lab Application 5.1: GIS Introduction: QGIS Version* for specifics on how to do the following.

1. Start QGIS.
2. Add the statesproject shapefile from the usaproject data folder. Leave the statesproject symbology alone for now—you'll change it in the next step.
3. Pan and zoom the Map View so that the lower 48 states are shown filling up the Map View.

    Important note: The data used in this lab application has already been projected to the U.S. National Atlas Equal Area projection for you to use. However, there are many more projections to choose from if you desire—you can change the projection of a data layer by right-clicking on it, selecting Save As, and then choosing a projected coordinate system to change the layer into.

4. The next step is to set the properties of the project you're working with so that QGIS will be able to properly render some of the map elements you'll be working with (such as the scale bar). From the Project pull-down menu, choose Project Properties.

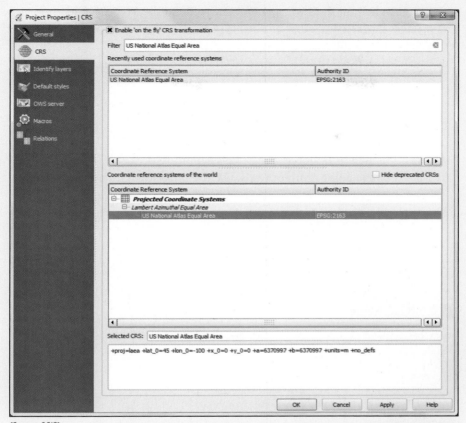

[Source: QGIS]

5. Select the Coordinate Reference System (CRS) tab.

6. Place a checkmark in the Enable 'on the fly' CRS transformation box.

7. In the box next to Filter, type US National Atlas Equal Area. This will search through all available projections and find the one you'll need.

8. In the Coordinate reference systems of the world box, click on the US National Atlas Equal Area option, then click Apply. This has set the CRS of the project environment to the chosen projection. Click OK to close the dialog box.

7.2 Setting Graduated Symbology

1. QGIS gives you the ability to change an object's symbology from a single symbol to multiple symbols or colors, and allows for a variety of different data classification methods.

2. Right-click on the **statesproject** shapefile in the Map Legend and select **Properties**. Click on the **Style** tab. To display the states as graduated symbols, use the following settings:

   a. From the pull-down menu to the right of the Style tab, select **Graduated**.

   b. The Column to use (in this lab) is the **POP05_SQMI** (the states' population per square mile from the year 2005).

   c. Use **5** for the number of Classes.

   d. Use **Quantile (Equal Count)** for the Mode.

   e. For the color ramp options, use the pull-down menu to select an appropriate choice.

3. When you have things arranged how you want them, click **Apply** to make the changes. Take a look at the map and make any further color changes you think are needed.

4. Click **OK** to close the dialog box.

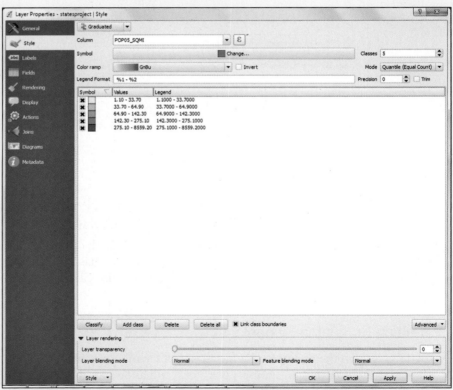

[Source: QGIS]

5. Now, the symbology of the statesproject has changed in the Map View and the values that make up each one of the breaks can be seen in the Map Legend. If the breaks and classes are not already displayed, you can show them by pressing the plus button, which you'll find to the left of the statesproject layer.

7.3 The Composer in QGIS

To begin laying out the print-quality version of the map, you'll need to begin working in the Composer. This mode of QGIS works like a blank canvas, allowing you to construct a map using various elements.

1. To begin, select **New Print Composer** from the **Project** pull-down menu.

[Source: QGIS]

2. Before the Composer opens, you'll be prompted to give your composition a title. Type in a descriptive title, such as "United States Population Density 2005" or something similar. Click **OK** after you've put in your title.

3. A new window, the Composer, will open. In the Composer, the screen represents the printed page of an 8½ × 11 piece of paper, so be careful when you're working near the edges of the page and keep all elements of the map within that border.

4. The composer has its own toolbar across the top with a new set of tools—locate and examine these navigation tools.

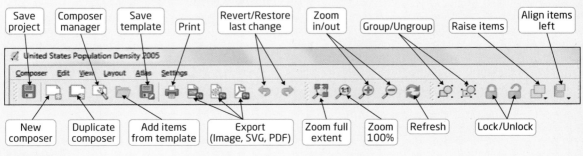

[Source: QGIS]

5. Starting at the left and moving right, the tools and their uses are as follows:

   a. The blue disk is used to save a composition.

   b. The white paper will create a new composition.

   c. The white paper over the blue paper is used to create a duplicate composition.

   d. The white paper with the wrench will open the Composer Manager.

   e. The yellow folder will allow you to add a template to the composition.

   f. The blue disk with the green bar will allow you to save a template.

   g. The printer icon is used when you're printing (see later in the lab).

   h. The next three icons will allow you export your composition to either 1) an image, 2) SVG format, or 3) PDF format.

   i. The two curved arrows allow you to either revert to the last change you made or restore the last change you made (which are useful when you need to "back up" a step in your map design).

   j. The magnifying glass with three arrows will zoom to the full extent.

   k. The magnifying glass with the 1:1 text will zoom the map to 100%.

   l. The plus and minus magnifying glasses are used to zoom in and out of the layout.

   m. The twin curved blue lines icon is used to refresh the view.

   n. The square and circle icons allow you to gather up several items and treat them as a single group (group items) or to turn a group of items back into individual items (ungroup).

   o. The lock and unlock icons allow you to fix items in place in the composition (lock) or remove this fix so they can be moved (unlock).

   p. The last two icons with the blue and yellow boxes are used for raising or aligning map elements.

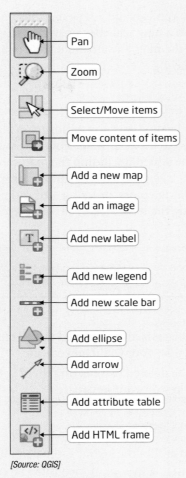

Pan

Zoom

Select/Move items

Move content of items

Add a new map

Add an image

Add new label

Add new legend

Add new scale bar

Add ellipse

Add arrow

Add attribute table

Add HTML frame

[Source: QGIS]

6. There is also a vertical toolbar down the left-hand side of the screen with additional tools:

   a. The hand is used for panning and adjusting the content of a particular window—for instance, for moving the position of what's showing on the map without changing the map itself.

   b. The magnifying glass is used to zoom in on elements on the map.

   c. The cursor pointing to the yellow box is used to move map elements to different places in the layout.

   d. The green square with the icon is used to move the content of items.

   e. The other tools are used for the addition of map elements to the layout, which we'll use later in the lab (and you'll be referred back to these icons). You can add a new map, an image, a new label, a new legend, and a new scale bar, as well as ellipse and arrow shapes for annotating the map, the attribute table of a layer, and an HTML frame for displaying Web content

7.4 Map Elements

1. Again, think of the layout (or composition) as a blank sheet of paper that you'll use to construct your map. There are numerous map elements that can be added, including the map canvas, scale bars, north arrows, and a legend. Each element has properties (such as creating borders, filling colors, or fixing size and position) that can be accessed by selecting the Item Properties tab on the right side of the composer:

[Source: QGIS]

2. Map elements can be resized and moved by selecting them with the mouse and resizing like you would any object.

3. Map elements can be deleted by selecting them with the cursor and pressing the delete key on your keyboard.

7.5 Inserting a Map Canvas

1. The first element to add to the layout is the default map canvas. This will be an element that shows all of the visible layers in the Map Legend.

2. Click on the Add new map icon from the vertical toolbar. This will allow you to draw a box on the layout that will display the current Map View. Draw a rectangular box on the layout to create the canvas.

3. Once it's drawn, you can click and drag the canvas around the map or resize it as you see fit (using the four blue tabs at its corners), and you can treat the entire canvas as if it's a single map element. When you're adjusting the size, keep in mind that you'll be adding several more map elements (such as a legend or a title), so you'll need to adjust the size, spacing, and balance of the elements accordingly.

4. You can't manipulate individual layers (for instance, you can't click and drag a state somewhere else), but using the move item content icon on the toolbar will allow you to drag the contents of the canvas around and alter what's being displayed inside the canvas.

5. To make further adjustments to the canvas itself (such as creating a border or altering the color of the border), select the Item Properties tab on the right side of the screen and scroll down to see what choices are available (such as expanding the Frame option to change the appearance of the map's border or the Background option to change the fill color of the map itself). Investigate the other choices under Item Properties (such as the options under Main Properties for adjusting the scale) to set up the canvas the way you'd like it to appear on the final printed map.

7.6 Inserting a Scale Bar

1. To add a scale bar to the map, select the Add new scale bar icon from the toolbar, then click on the composition where you want the scale bar to be added.

2. Once the scale bar appears, click on it so that the four blue boxes appear at the corners. In the Item Properties box on the right-hand side of the screen, you'll see several options for altering the scale bar. Use the options to change the following properties of the scale bar until it looks appropriate:

    a. The number of segments on the left or the right side of the "0" value on the scale bar

    b. The size of each segment

    c. The number of map units per bar units (to even out the breakpoints on the scale bar)

    d. The physical dimensions of the scale bar itself (height, line, width, etc.)

    e. The font and color used

[Source: QGIS]

7.7 Inserting a North Arrow

1. To add a north arrow to the map, select the Add Image icon from the toolbar, then click on the composition where you want the north arrow to be added and draw a box the size of how large you want to north arrow symbol to be.

2. An empty box and several different graphics options will appear in the Item Properties pane. The Add Image option allows you to add a graphic of your own to the map or to select from several pre-made

graphics, including several different north arrows. Expand the options for **Search directories** and scroll part-way down and you'll see a number of options for north arrows.

[Source: QGIS]

3. Select an appropriate north arrow and it will appear in the empty box on the composition. Use the cursor to position or resize the north arrow.

7.8 Inserting a Legend

Each layer you use in QGIS (such as statesproject) has a set of layer properties. Anything changed in the layer properties will be reflected in changes to the composition. For instance, if you change the symbology of a layer in its properties, its appearance will be transferred over to a legend in the composition. Similarly, whatever name is given to a layer in the Map Legend will carry over to a legend in the composition.

1. Return to the regular QGIS window. Currently, the statesproject layer is named just that—however, you can change this name to something more descriptive before adding a legend to the composition. In the QGIS Map Legend, right-click on the name you want to change (the **statesproject** layer) and select **Rename**.

2. In the Map Legend itself, you can type a new name for the layer (like "Population Per Square Mile"). Click the **Enter** key when you've typed in a new name.

3. Return to the composer. To add a legend to the map, select the **Add new legend** icon from the toolbar, then click on the composition where you want the legend to be added.

4. A default legend will be added to the map, consisting of all the layers in the Map Legend with whatever names are assigned to them. Use the cursor to move and resize the legend as needed.

5. To make changes to the default legend, click on the legend itself, and there will be several options available in the Item Properties. You can change the name of the legend (don't just call your legend "Legend"), its font, symbol width, and appearance, as well as spacing between items.

[Source: QGIS]

7.9 Adding Text to the Layout

1. To add text to the map (such as a title, or any other text you may want to add), select the **Add new label** icon from the toolbar, then click on the composition where you want the text to be added and draw a box of the size you want the text to be.

2. Note that when text is added, a small box filled with the word "QGIS" will be added to the composition. The lettering in the box, as well as its font, size, and color, can be changed in the Item Properties.

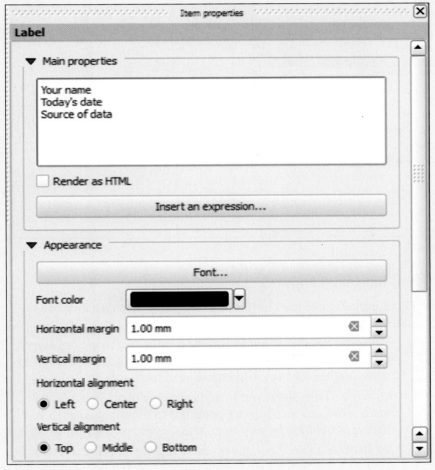

[Source: QGIS]

3. Type the text you want to appear in the Item Properties box under the Label heading. To alter the font, color, and other properties of the text, click the **Font** and **Font color** … options. More options will

appear—note that you can alter each text box separately, create larger fonts for titles and smaller fonts for type, and so on. You can also resize each text box and move it to a new position.

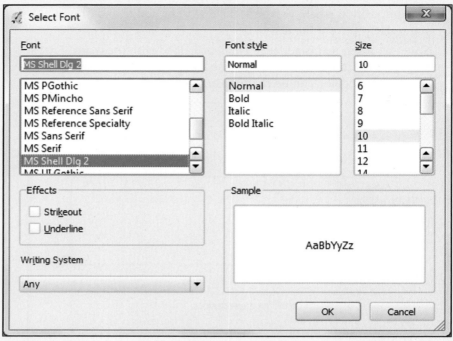

[Source: QGIS]

7.10 Other Map Elements

1. Though they're not used in this lab, four other map elements can be added:

   a. A shape: This allows you to add the outline of a shape (an ellipse, rectangle, or triangle) to the composition for additional graphics or annotation. Use the **Add ellipse** icon to do this.

   b. An arrow: This allows you to add a graphic of an arrow (not a north arrow) so that you can point out or highlight areas to add further annotation to your map. Use the **Add arrow** icon to do this.

   c. An attribute table: This allows you to add a graphic of a layer's attribute table to the map to enable the presentation of additional information. Use the **Add attribute table** icon to do this.

   d. An HTML frame: This allows you to add a frame that will contain a Web page or other Web content. In the Item properties you can specify the URL for what you want to display inside the frame. Use the **Add HTML frame** icon to do this.

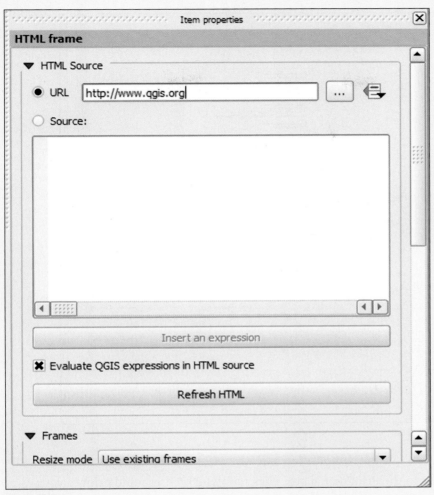

[Source: QGIS]

7.11 Printing the Composition

1. When you have constructed the map the way you want, choose the **Composer** pull-down menu and choose **Page Setup**. This will allow you to specify if you want your map printed in landscape or portrait format or add margins for printing.

[Source: QGIS]

2. When the composition is ready, select the **Print** icon from the toolbar.
3. In the Print dialog box, click **Print** and QGIS should print to your computer's default printer.

Closing Time

Close the composer by selecting the **Composer** pull-down menu and then selecting **Quit**. Exit QGIS by selecting the **Project** pull-down menu and selecting **Exit QGIS**. The geospatial lab application in Chapter 8 will return to using Google Earth Pro for analysis. You'll be using Google Earth Pro to take a series of addresses and plot them on a map, and then do some analysis of the results. You'll also use Google Maps and Google Earth to analyze the shortest paths between these plotted points.

Final Layout Checklist

_____ The contiguous 48 states (without Alaska and Hawaii) United States population per square mile, classified, and displayed with an appropriate color scheme

_____ Appropriate map legend, with all items listed with normal names (the legend should not be called "legend" and should be appropriate size, font, number of columns, and so on)

_____ Scale bar (with appropriate divisions and numerical breaks)

_____ The proper units displayed on the scale bar

_____ Appropriate title (don't use the words "map" or "title" in your title) and appropriate size and font

_____ A north arrow of appropriate size and position

_____ Type (your name, the date, the source of the data) in an appropriate size and font

_____ Overall map design (appropriate borders, color schemes, balance of items and placement, and so on)

7.2 Geospatial Lab Application

GIS Layouts: ArcGIS Version

This lab will introduce you to the concepts of taking GIS data and creating a print-quality map from it. This map should contain the following:

▶ The population per square mile of the contiguous United States (48 states without Alaska and Hawaii), set up in an appropriate color scheme

▶ The data in a different projection than the default GCS one

▶ An appropriate legend (make sure your legend items have regular names and that the legend is not called "legend")

▶ An appropriate title (make sure that your map doesn't include the word "map" in it)

▶ A north arrow

▶ A scale bar

▶ Text information: your name, the date, and the source of the data

▶ Appropriate borders, colors, and design layout (your map should be well designed instead of map elements thrown on at random)

Important note: At the end of this lab, there is a checklist of items to aid you in making sure the map you make is complete and of the best quality possible.

Geospatial Lab Application 7.1: GIS Layouts: QGIS Version uses the free Quantum GIS (QGIS); however, this lab provides the same activities for use with ArcGIS for Desktop 10.3.

Objectives

The goals for you to take away from this lab are:

▶ To familiarize yourself with the layout functions of ArcMap

▶ To arrange and print professional-quality maps from geographic data using the various layout elements

Using Geospatial Technologies

The concepts you'll be working with in this lab are used in a variety of real-world applications, including:

▶ Urban planning, where GIS is used to compose maps showing data about housing and population distributions, transportation routes, land parcels, information, and other matters

▶ Law enforcement, where agents use GIS to create maps for distribution among police officers and the community showing the major locations of particular types of crimes, analysis of crime hot spots, and locations of 911 calls

[Source: © Monty Rakusen/© Cultura Creative (RF)/Alamy]

Obtaining Software

The current version of ArcGIS for Desktop is not freely available for use. However, instructors affiliated with schools that have a campus-wide software license may request a 1-year student version of the software online at **www.esri.com/industries/apps/education/offers/promo/index.cfm**.

Important note: Software and online resources can change fast. This lab was designed with the most recently available version of the software at the time of writing. However, if the software or Websites have significantly changed between then and now, an updated version of this lab (using the newest versions) will be available online at **www.macmillanhighered.com /shellito/catalog**.

Lab Data

Copy the folder Chapter7—it contains a folder called "usa" in which you'll find several shapefiles that you'll be using in this lab. This data comes courtesy of Esri and was formerly distributed as part of their free educational GIS software package ArcExplorer Java Edition for Educators (AEJEE).

Localizing This Lab

The dataset used in this lab is Esri sample data for the entire United States, which you'll be using to create a map layout of the United States. However, the layout tools can be used to make a map of whatever dataset you desire—rather than creating a layout map of the entire United States, focus on your home state and create a layout of that instead (see Section 7.6 for how to focus the layout on one state instead of the whole United States). If you're only going to work with one state, use the counties.shp file and make a map of county population by square mile instead.

7.1 Initial Pre-Mapping Tasks and Changing Map Projections

If needed, refer back to the *Geospatial Lab Application 5.2: GIS Introduction: ArcGIS Version* for specifics on how to do the following.

1. Start ArcMap.
2. Add the states shapefile from the usa sample data folder.
3. Pan and zoom the View so that the lower 48 states are shown filling up the View.
4. Change the projection in the View from the default GCS to something more visually pleasing.

Important note: Geospatial Lab Application 5.2: GIS Introduction: ArcGIS Version used the US National Atlas Equal Area projection, but there are many more to choose from, such as the Lambert Conformal Conic projection—keep in mind that your final map will be of the United States when you select a regional or world projection. You should use a Projected Coordinate System for this lab.

5. Leave the states symbology alone for now—you'll change it in the next step.

7.2 Setting Graduated Symbology

ArcMap gives you the ability to change an object's symbology from a single symbol to multiple symbols or colors and allows for a variety of different data classification methods.

1. Right-click on the states shapefile in the TOC and select Properties. Click on the Symbology tab. To display the states as graduated symbols, use the following settings:

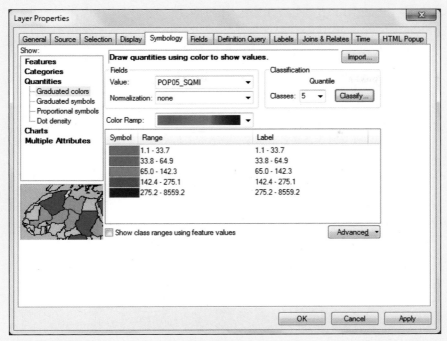

[Source: ArcGIS/Esri]

   a. Under the Show options, select Quantities, then select Graduated colors.

   b. The Field to use in this exercise is the POP05_SQMI (the states' population per square mile from the year 2005).

   c. For the color ramp, select an appropriate choice from the available options.

   d. Use 5 for the number of Classes.

   e. To change the classification method, press the Classify button. From the new options available in the pull-down menu, choose Quantile for the Method.

2. When you have things arranged how you want them, click Apply to make the changes. Take a look at the map and make any further color changes you think are needed.

3. Click OK to close the dialog box.

4. Now, the symbology of the states has changed in the View, and the values that make up each one of the breaks can be seen in the TOC.

7.3 Page Setup and Layouts

1. Before starting the layout process, first set up the properties of the map page, so that they'll fit with the printer you'll be using to print the map (to avoid any scaling or printing problems at the end).

2. From the File pull-down menu, select **Page and Print Setup** …

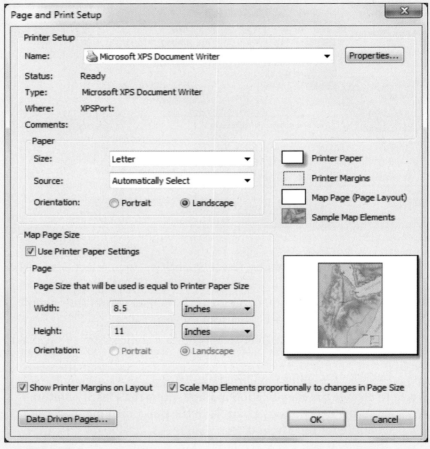

[Source: ArcGIS/Esri]

3. Under Name, select the printer you'll be using.

4. For Orientation, choose how you want to compose the layout—using portrait (a vertical orientation for the map) or landscape (a horizontal orientation).

5. Place check marks in the following boxes:

   a. Use Printer Paper Settings

   b. Show Printer Margins on Layout

   c. Scale Map Elements Proportionally to Changes in Page Size

6. Click OK when all settings are correct.

7.4 Layouts in ArcMap

1. To begin laying out the print-quality version of the map, you'll need to begin working in the Layout View. This mode of ArcMap works like a blank canvas and allows you to construct a map using various elements.

2. Note that however things appear in the Data View, they will appear that same way in the Layout View. For instance, if you change the symbology of the states in the Data View to a different color ramp, that same color ramp will be used in the Layout View.

3. To begin, select **Layout View** from the **View** pull-down menu.

    a. You can switch back to the Data View by selecting it from the View pull-down menu.

    b. You can also switch between Layout and Data Views by using the icons at the bottom left-hand corner of the View.

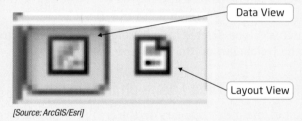

[Source: ArcGIS/Esri]

4. In the Layout View, the black border around the white page represents the border of the printed page of an 8½ × 11 piece of paper, so be careful when you're working near the edges of the page to keep all elements of the map within that border.

5. The toolbar will switch to a new set of tools—locate and examine the navigation tools.

[Source: ArcGIS/Esri]

6. Starting at the left and moving right, the tools and their uses are as follows:

    a. The plus and minus magnifying glasses are used to zoom in and out of the layout.

    b. The hand icon is used to pan around the map.

    c. The icon with the four triangles pointing outward is used to zoom the map to its full extent.

    d. The 1:1 icon is used to zoom the layout to 100%.

e. The white pages with the four black arrows are used for zooming in and out from the center of the page.

f. The white pages with the blue arrows are used for returning to previous scales.

7.5 Map Elements

1. Again, think of the layout as a blank sheet of paper that you'll use to construct your map. There are numerous map elements that can be added, including scale bars, north arrows, and a legend. Each element has properties (such as creating borders, filling colors, or fixing size and position) that can be accessed individually after you have selected the element (using the black "select elements" arrow on the Tools toolbar).

2. Map elements can be resized and moved by selecting them with the mouse and resizing as you would any object.

3. Map elements can be deleted by selecting them with the cursor and pressing the Delete key on the keyboard.

7.6 Working with the Data Frame on the Layout

1. The default element that appears on the layout is the Data Frame, which contains all of the GIS data. All of the visible layers in the TOC under the word "Layers" will appear in this frame. You can't manipulate parts of individual layers (for instance, you can't click and drag a state somewhere else), but you can treat the entire Data Frame as if it's a map element. Click and drag the Data Frame around the map or resize it as you see fit (using the eight blue tabs at its corners and midpoints). When you're placing and resizing the Data Frame, keep in mind that you'll have several other map elements on the final layout, so be mindful of the position and size of the frame, as well as the available blank white space.

2. Even though you've resized the Data Frame, you may have more than the contiguous 48 states visible (i.e., parts of Alaska may be peeking into the Frame), or you may not have all of the states and their boundaries visible. In the Layout View, the Map Scale pull-down menu on the Standard toolbar will allow you to alter the scale you're presenting on the layout:

[Source: ArcGIS/Esri]

3. You can select one of the pre-set map scales or type in your own value to adjust the scale of the map. Try some of the different values and see how they affect the map's appearance.

4. You can also access the Data Frame's properties to create (or remove) borders around the outline of the Data Frame itself. Right-click on the Data Frame element and select Properties. In the Data Frame Properties dialog box, select the Frame tab to make adjustments to the thickness and color of the border (you can also select the option for ‹None› to make the Data Frame's border transparent in the layout).

7.7 Inserting a Scale Bar

1. To add a scale bar to the map, choose the Insert pull-down menu and select Scale Bar.

2. A number of options for scale bars will appear in a new Scale Bar Selector dialog box:

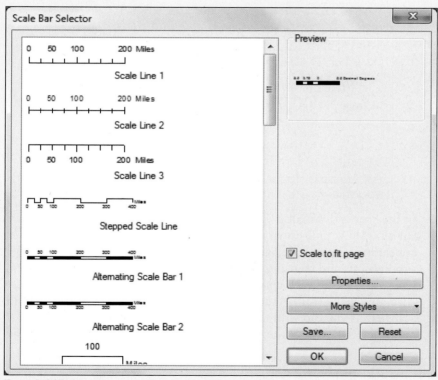

[Source: ArcGIS/Esri]

3. Additional scale bars are available by clicking the More Styles button.

4. Select an appropriate scale bar and click OK. Use the cursor to position or resize the scale bar on the layout.

5. You'll see that the scale bar will appear with its default units set to decimal degrees. To use different units of measurement (and also to change the appearance of the scale bar itself), right-click on the scale bar element and select Properties.

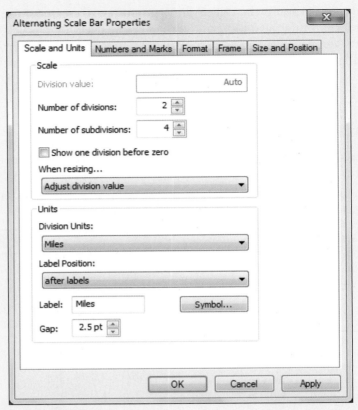

[Source: ArcGIS/Esri]

6. In the Properties dialog box, select the Scale and Units tab. From the Division Units pull-down menu, select Miles. You can also alter the appearance of the scale bar by changing the number of divisions and subdivisions and by using the option to show one division before zero. Click Apply after each change to see the effect on the scale bar element. Also, under the Frame tab, you can change the border and color of the scale bar as you did with the Data Frame. When you finally have the scale bar's appearance the way you want it, click OK to close the dialog.

7.8 Inserting a North Arrow

1. To add a north arrow to the map, choose the Insert pull-down menu and select North Arrow.

2. A number of options for north arrows will appear in a new North Arrow Selector dialog box:

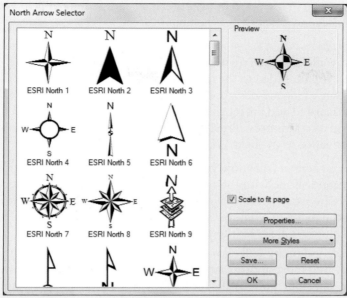

[Source: ArcGIS/Esri]

3. Additional north arrows are available by clicking the More Styles button.

4. Select an appropriate north arrow and click OK. Use the cursor to position or resize the north arrow on the layout.

5. You can alter the appearance of the north arrow (adjusting the color, rotation angle, etc.) by right-clicking on the north arrow element and selecting Properties. In the North Arrow Properties dialog box, there are a number of available options for customizing the north arrow.

7.9 Inserting a Legend

Each layer you use in ArcMap (such as the states) has a set of layer properties. Anything changed in the layer properties will be reflected in changes to the layout. For instance, if you change the symbology of a layer in its properties, it will be changed in layout elements like the map legend. However, to change the name of the layer and how it will appear in something like a map legend, the shapefile must be renamed in its Layer Properties.

1. To do this, right-click on the states layer and choose Properties. In the Layer Properties dialog box, choose General. You can type a new name for the layer (like "Population Per Square Mile"). Click OK when you've typed in a new name.

2. To add a legend to the map, choose the Insert pull-down menu and select Legend.

270

3. The Legend Wizard will open—this is a set of menus to help you to set up the legend properly for the map.

   Important note: At any step, you can see what the legend will look like on the map by clicking the Preview button in the wizard. By pressing Preview, the legend will be added to the map, and you can see what it looks like at the current stage of legend design. If you're satisfied, you can click Finish in the Legend Wizard to add the legend. If you want to keep going with the steps in the Legend Wizard, press Preview again and the legend will disappear, so you can advance to the next Wizard step.

4. The first screen will allow you to choose the layers that will appear in the legend, along with the number of columns you want the legend to have. Choose the states layer for this map (in the dialog box it will have the name of whatever you changed it to in the TOC) and select one column. Click Next to advance to the second menu.

5. The second menu will allow you to give the legend a name—give it something more appropriate than "legend." You can also adjust the color, size, font, and justification of the legend itself. Once you've set these values, click Next to advance to the third menu.

6. The third menu will allow you to adjust the color, thickness, and appearance of the legend's border, background, and drop-shadow effects. Examine various options for the legend's appearance and when you're ready, click Next to advance to the fourth menu.

7. The fourth menu will allow you to alter the symbol patch of the legend items (in this case, the states layer). When you're ready, click Next to advance to the fifth menu.

8. The fifth and final menu will allow you to adjust the appearance and spacing of items in the legend itself. When you're satisfied with how the legend's going to look, click Finish. The legend will then be added to the layout.

9. Once the legend is created, you can make several different alterations to it by right-clicking on the legend itself in the layout and selecting Properties. For instance, selecting the Items tab under Properties will allow you to access a "Style" option that you can use to make further modifications to the appearance of the legend, while the options under the Frame tab allow you to change the border of the legend.

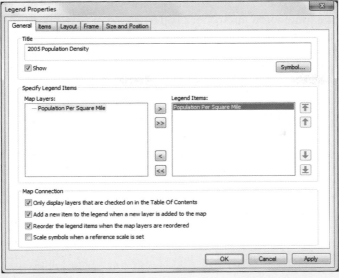

[Source: ArcGIS/Esri]

7.10 Adding Text and a Title to the Layout

1. To add text to the map (such as your name or the date), select Text from the Insert pull-down menu.

2. Note that when text is added, a small box called "Text" is placed near the center of the map. You'll have to move each text box to somewhere else in the layout.

3. Right-click the text box and select Properties from the new menu items. The Properties dialog box will appear.

[Source: ArcGIS/Esri]

4. Under the Text tab, type into the box the text you want to appear. To alter the font, color, and other properties of the text, click the Change Symbol . . . button. More options will appear—note that you can alter each text box separately, to create larger fonts for titles, smaller fonts for type, and so on.

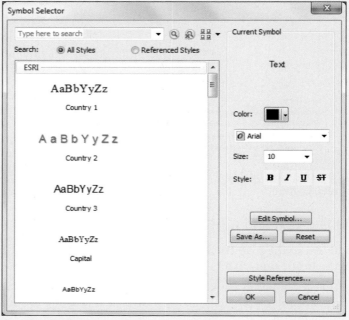

[Source: ArcGIS/Esri]

5. While you can use the Insert Text option to create a title box, ArcGIS also contains a separate option for adding a title. From the Insert pull-down menu, select Title.

6. A new dialog box will appear—type the map's title into this box and click OK.

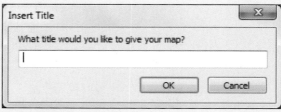

[Source: ArcGIS/Esri]

7. The title will appear in a text box at the top of the layout. Treat this text box the same way as any other—you can change its properties (by making the font size larger, or changing the justification or title) just like you did previously.

7.11 Other Map Elements

1. Though they're not used in this lab, other map elements can be added from the Insert pull-down menu:

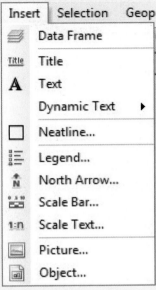

[Source: ArcGIS/Esri]

a. Dynamic Text allows you to insert information about the current date, current time, the date the map was saved, and so on.

b. Neatline allows you to draw a line around map elements.

c. Scale Text allows you to insert verbal statements of scale rather than only using a scale bar for representation.

d. Picture allows you to add graphics to your map, including downloaded images, digital camera images, or other graphics.

e. Object allows you to insert things from other software applications, including video clips, Microsoft PowerPoint slides, or sound clips.

7.12 Printing the Layout

1. When you have constructed the map the way you want, choose Print Preview from the File pull-down menu. You'll see how your map will appear on the printed page. Pay careful attention to borders, lines, and placement of objects when you're comparing how they look on the screen with how they're going to look on paper. Make whatever adjustments are necessary to the final draft of the layout before you start printing.

2. When the map's ready, select Print from the File pull-down menu—
ArcMap will use the Print dialog box to send the layout to the printer.

3. Exit ArcMap by selecting the File pull-down menu and then choosing
Exit. There's no need to save any data in this lab.

Closing Time

The geospatial lab application in Chapter 8 will return to using Google Earth
Pro for analysis. You'll be using Google Earth Pro to take a series of addresses and plot them on a map, and then do some analysis of the results.
You'll also use Google Maps and Google Earth to analyze the shortest paths
between these plotted points.

Final Layout Checklist

_____ The contiguous 48 states (without Alaska and Hawaii) United States
population per square mile, classified, and displayed with an appropriate color scheme

_____ States shown in a different (and appropriate) projection than the
default

_____ Appropriate map legend, with all items listed with normal names
(the legend should not be called "legend" and should be appropriate
size, font, number of columns, and so on)

_____ Scale bar (with appropriate divisions and numerical breaks)

_____ Appropriate title (don't use the words "map" or "title" in your title)
and appropriate size and font

_____ A north arrow of appropriate size and position

_____ Type (your name, the date, the source of the data) in an appropriate
size and font

_____ Overall map design (appropriate borders, color schemes, balance of
items and placement, and so on)

Getting There Quicker with Geospatial Technology

Satellite Navigation Systems, Road Maps in a Digital World, Creating a Street Network, Geocoding, Shortest Paths, and Street Networks Online

Satellite navigation systems (the kind of devices made by companies like Garmin, Magellan, or TomTom) are really revolutionary technologies. One small device mounted on the dashboard will find your precise location, plot it on a map, determine where the nearest gas stations are, then compute the quickest route to get you there—all with turn-by-turn directions that announce the names of streets and the distance to the next turn (see **Figure 8.1** on page 276). Many smartphone apps can do the same things, effectively turning your phone or mobile device into a satellite navigation system. The position determination is straightforward—the device has a GPS receiver in it that finds your location on Earth's surface, using the methods we discussed in Chapter 4. In fact, most people who own a device with a navigation system simply refer to it as a "GPS," as in "Punch our destination into the GPS" or "What does the GPS say is the shortest route to get there?" However, it's doing a disservice to these things to call them simply a "GPS"—they do so much more than a regular GPS receiver does.

These devices rely on a GIS-style system at their core—hence their ability to handle spatial data in the form of road-network maps, use those maps for routing, determine the shortest path between two (or more) points, and match an address to a spatial location. This same type of data is used to route emergency vehicles to the site of a 911 emergency phone call or to manage a fleet of delivery vehicles. You'll find the same sort of system at the heart of online mapping applications such as MapQuest or Google Maps, and in the location and mapping apps of a smartphone. This type of technology is changing fast and getting better at what it does.

A March 2009 article in *USA Today* describes an incident where a vehicle navigation device instructed the driver to turn off a road and follow a

> **satellite navigation system** a device used to plot the user's position on a map, using GPS technology to obtain the location

FIGURE 8.1 A Garmin GPS satellite navigation system. *[Source: Edward J Bock III/ Dreamstime.com]*

snowmobile trail toward a destination—this ended with the car stuck in the snow and the state police being called for emergency help. Similarly, a March 2009 article of the *Daily Mail* relates a story of a vehicle navigation system that directed a driver along a footpath, which ended in a sheer drop of 100 feet off the edge of a cliff (luckily, the driver stopped in time). When errors like these occur, it's usually not because the GPS receiver is finding the incorrect position from the satellite information. The problem is much more likely to be with the base network data itself. These types of systems are only as accurate as the base network data they have available to them. This chapter delves into how these types of geospatial technology applications function, how they're used, what makes them tick, and why they may sometimes lead you astray.

THINKING CRITICALLY WITH GEOSPATIAL TECHNOLOGY 8.1

What Happens When the Maps Are Incorrect?

How many times has this happened to you when you're using some sort of mapping service—either the directions take you to the wrong place, or they indicate you should turn where you can't, or they just can't find where you want to go (or the road you want to go on)? When these things happen, the GPS location is probably correct, but the maps being used for reference are likely either outdated or have errors in them. How much does the usefulness of this aspect of geospatial technology rely on the base data? For example, when Apple released its Maps component of iOS 6 in 2012, several location inaccuracies were reported, including the mislabeling of a supermarket as

a hospital. If the base maps are incomplete or not updated, then how useful is a mapping system to the user?

It should also be noted that many vehicle navigation systems will recompute a new route on the fly if you miss a turn or start to take another route to your destination. However, if the system has taken you on an improper route, how useful will the system be in getting you back to where you want to go? A key purpose of mapping systems is to navigate through unfamiliar territory—but if they don't properly fulfill that function, what can travelers rely on? Are there any sorts of liability issues that could result from inaccurate mapping and misleading instructions?

How Do You Model a Network for Geospatial Technology?

Back in Chapter 5, we discussed how real-world items are modeled, or represented, using GIS. Any type of **network** is going to involve connections between locations, whether those locations are streams, power lines, or roads. Thus (to use Esri terminology), in GIS a network in its most basic form is represented by a series of **junctions** (point locations or nodes) that are connected to each other by a series of **edges** (lines or links). For instance, in a road network, junctions might be the starting and ending points of a road, or they might be points of intersection with other roads, while the edge is the line that represents the road itself. When you're designing a road network, keep in mind that there may be many types of edges and junctions to represent. For example, a city's road network will have edges that represent streets, highways, railroads, light-rail systems, subway lines, or footpaths, while junctions may represent not only the beginnings and endings of streets, but also highway entrances and exits, freeway overpasses and underpasses, subway stops, or rail terminals.

When you're dealing with all these different types of edges and junctions, the **connectivity** of the network in GIS is essential when you come to model it. With proper connectivity, all junctions and edges should properly connect to one another, while things that should not connect do not. For example, if a freeway crosses over a road via an overpass, the network connectivity should not show a valid junction where the street is allowed access to the freeway. If your vehicle navigation system leads you to this point, then instructs you to do the impossible and "turn right onto the highway," it's because the network data is set up to make it think you can. In the same way, a railroad line may intersect with a street, but the network should not have a connection showing that the street has direct practical access to the railroad line as they cross. If this kind of connection is built into the data, you could conceivably be routed to turn onto the railroad line and continue on it toward your destination. It sounds silly to think of driving your car on the railroad tracks, but if the network data is incorrect, this line might simply represent the next road to take to get to your destination.

Thinking along these lines, other features of a road network must also be included in the model. For instance, some streets may be one way, some junctions may not allow left-hand turns, or U-turns may not be permitted. These types of features need to be properly modeled for the system to be an accurate, realistic model of the road network. You can see the "one way" street sign when you're driving, but if that feature hasn't been properly set up in the network, the system has no way of knowing it shouldn't route cars in both directions along the street. As noted, an overpass or underpass should not show up as a viable option for a turn—if a device or GIS instructs you to make a right-hand turn onto the freeway that you're currently driving under, then something's gone wrong with some aspect of the technology.

network a series of junctions and edges connected together for modeling concepts such as streets

junction a term used for the nodes (or places where edges come together) in a network

edge a term used for the links of a network

connectivity the linkages between edges and junctions of a network

line segment a single edge of a network that corresponds to one portion of a street (for instance, the edge between two junctions)

When you're modeling a network, you should consider each edge as a separate entity, though not necessarily each individual street. A long city street may be modeled in the GIS as several **line segments**, with each segment representing a section of the street. A major urban street may be made up of more than 100 line segments—each segment being a different section of the street (with each segment being delineated by roads that intersect it). For example, **Figure 8.2** shows a geospatial road network in Virginia Beach, Virginia. The road cutting through the center of the city (highlighted in blue) is Virginia Beach Boulevard, a major multi-lane city street, with numerous intersecting roads and street lights. Although we think of Virginia Beach Boulevard as one big, long street, the system models it as 129 line segments, with each line segment representing a portion of its length.

Breaking a road up into individual line segments allows the modeling of different attributes for each segment. Attributes such as the road's name, the address ranges for each side of the road, the suffix of the road's name (Drive, Avenue, Boulevard, etc.), the type of road it is (residential street, interstate, highway, etc.), and the speed limit are examples of values that can be assigned to individual segments. Thus, an attribute table of this layer in a GIS would consist of 272 records, each with multiple attributes.

street centerline a file containing line segments representing roads

Several different types of geospatial road network files are available. A **street centerline** file models each city road as a line and contains lines for the different types of roads. Street centerline files are produced as GIS line layers that represent roads. While this type of data is available from

FIGURE 8.2 Virginia Beach Boulevard (shown in blue) in a GIS network. Although it is only one road, it consists of 272 line segments as it crosses the city. [Source: US Census/Esri]

HANDS-ON APPLICATION 8.1

The U.S. Census TIGERweb

TIGER files are made freely available from the Census Bureau via the Web. Go to **http://tigerweb.geo.census.gov/tigerweb** to see the U.S. Census Bureau's online viewer for TIGER data. By zooming in on an area you can select the TIGER layers you wish to display from the choices on the left-hand side of the screen (such as roads, school districts, hydrography, and various types of boundaries).

Note that here you can view data, but if you wish to download the actual TIGER files themselves, go to **https://www.census.gov/geo/maps-data/data/tiger-line.html**. This is the part of the U.S. Census Bureau's Website for downloading TIGER/Line files. There's also full documentation of TIGER files

available in PDF format on the Website. Files for U.S. counties can be downloaded in shapefile format (see Chapter 5 for more info about shapefiles) to be used in GIS products like ArcGIS or QGIS. You can also download several other types of TIGER files besides road-network data, including census block information, hydrography, landmarks, and American Indian reference data—check to see what types of datasets are available for download.

Expansion Questions

- What kinds of transportation data are contained within TIGER/Line files besides roads?

- What kinds of TIGER data are available for your county?

numerous different sources, the U.S. Census Bureau regularly issues this type of road network data in a format usable by geospatial technology software as **TIGER/Line** files. TIGER stands for Topologically Integrated Geographic Encoding Referencing, and the files delineate different boundaries throughout the United States (such as block groups or congressional districts) in addition to containing road line data. (See *Hands-On Application 8.1: The U.S. Census TIGERweb* for more information.)

Each record in a street centerline file (such as a TIGER/Line file) represents a small segment of a road, and thus each segment (record) can have multiple attributes (fields) assigned to it. **Figure 8.3** on page 280 shows a 2014 TIGER/Line file of Virginia Beach and a portion of the attribute table of those selected segments that make up Virginia Beach Boulevard. Note how many attributes there are (information gets encoded into each road segment—the entire Virginia Beach TIGER/Line file is made up of over 19,000 segments). These attributes describe the characteristics of each segment and are also useful in referencing locations (such as address ranges) with each segment. These kinds of attributes include things such as the name of the road segment, the suffix designation of the road segment (such as a lane, avenue, drive, etc.). Different street centerline sources will use different names to represent these commonly used attributes. For instance, The National Map transportation data that is freely available from the USGS (see Chapter 1) contains street centerline data but with different names for each segment's attributes. See **Table 8.1** on page 280 for some examples on how these types of attributes are named by different sources.

> **TIGER/Line** a file produced by the U.S. Census Bureau that contains (among other items) the line segments that correspond with roads all over the United States

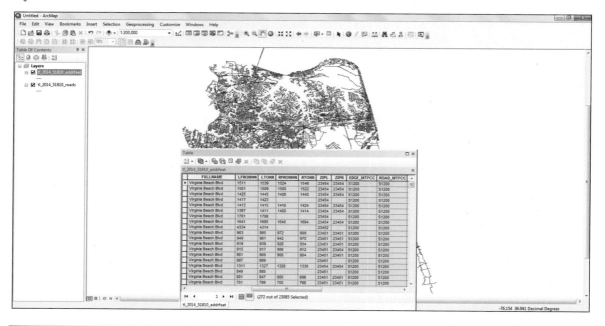

FIGURE 8.3 The attributes of the selected records making up Virginia Beach Boulevard from the 2014 Virginia Beach TIGER/Line file in ArcGIS. *[Source: US Census Bureau/Esri]*

Table 8.1 Commonly used road segment attributes and how they are named in TIGER 2000, TIGER 2014 Address Range Feature County, and The National Map transportation layers.

Attribute	TIGER 2000	TIGER 2014	The National Map
Prefix direction of road (N. Smith St.)	FEDIRP	n/a	n/a
Name of road (N. **Smith** St.)	FENAME	FULLNAME	Full_Street_Name
Type of road (N. Smith **St**.)	FETYPE	n/a	n/a
Suffix direction of road (Canal Street **E**.)	FEDIRS	n/a	n/a
Start of address ranges on left side of road	FRADDL	LFROMHN	Low_Address_Left
End of address ranges on left side of road	TOADDL	LTOHN	High_Address_Left
Start of address ranges on right side of road	FRADDR	RFROMHN	Low_Address_Right
End of address ranges on left side of road	TOADDR	RTOHN	High_Address_Right
Zip code on left side of road	ZIPL	ZIPL	Zip_Left
Zip code on right side of road	ZIPR	ZIPR	Zip_Right
Code to determine the kind of road (residential, highway, etc.)	CFCC	ROAD_MTFCC	Road_Class

These kinds of attributes define the characteristics of each road segment. Similar attributes would be found in road data, such as other street centerline files. If these attributes are incorrect, then the base network map will be incorrect. If the vehicle navigation system gives you incorrect street names or calls a road "east" when it's really west, it's likely that there are incorrect values in the base network data's attributes.

Attributes like those from the TIGER/Line files, the National Map, or similar street-network data created by others concerning specific address ranges, zip-code information, and detailed data for the names of roads can be used as a base map source for other applications, such as pinpointing specific addresses on a road. It's this source data that allows for a match of a typed street address to a map of the actual location.

How Is Address Matching Performed?

Whenever you use a program like MapQuest or Google Maps to find a map of a location, you're typing in something (like "1600 Pennsylvania Avenue, Washington, D.C.") and somehow the Website translates this string of characters into a map of a spatial location (like the White House). The process of taking a bunch of numbers and letters and finding the corresponding location that matches up with them is called **address matching** or **geocoding** (**Figure 8.4**). Although the process seems instantaneous, there are several steps involved in geocoding that are happening "behind the scenes" when you use an address-matching system (like those in GIS).

address matching another term for geocoding

geocoding the process of using the text of an address to plot a point at that location on a map

FIGURE 8.4 A map generated by querying MapQuest for "1600 Pennsylvania Avenue, Washington, D.C." *[Source: 2015 MapQuest, 2015 TomTom]*

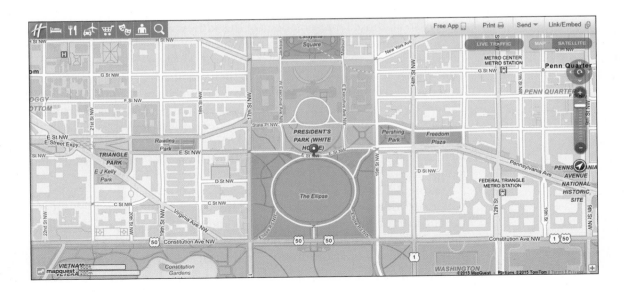

reference database
the base network data used as a source for geocoding

First, you need to have some sort of **reference database** in place—this is a road network that the addresses will be matched to. The National Map transportation data, a TIGER/Line file, or similar type of street centerline file (many are produced commercially) is needed here. What's essential is that the line segments contain attributes along the lines of those found in a TIGER/Line file—for example, street direction, name of the street, address ranges on the left and right sides of the street, street suffix, and zip codes on the left and right sides of the street. This information will be used as the source to match addresses to as well as a source for the final plotted map.

parsing breaking an address up into its component parts

Next, the address information is subjected to **parsing**, or breaking it up into its component pieces. For instance, the global headquarters of Google is located at 1600 Amphitheatre Parkway, Mountain View, CA 94043. When this address is parsed, the address number is split off from the rest of the address, along with the name of the street, the street type, the city, the state, and the zip code. The parsed version of the address would be something like: 1600 | Amphitheatre | Parkway | Mountain View | CA | 94043.

Next, the geocoding process has to have a way of handling different items in addresses that mean the same thing. For instance, if you're using Google Maps to plot the location of Google's headquarters, you might type in either "Mountain View, CA" or "Mountain View, California." Similarly, the White House is located at 1600 Pennsylvania Avenue NW in Washington D.C. However, when you're trying to find this on Google Maps, you might type in "Pennsylvania Avenue," "Pennsylvania Ave," or "Pennsylvania Av." Even though all of these things mean the same thing to us, the GIS has to also understand that they mean the same thing as well. In some cases, **address standardization** is performed to set up data in a consistent format. In this case, the geocoding process needs to standardize addresses to properly match a location using its appropriate attributes in the reference database— for instance, it might substitute the word "AVE" in place of "Avenue," "Ave," or "Av" so all those representations of the address are the same.

address standardization
setting up the components of an address in a regular format

Other geocoding systems may generate a list of multiple representations of the address, so that the GIS will treat "Avenue," "Ave," "AVE," or "Av" as the same thing. Sometimes the system will have an alternate list of road names stored in its reference database if a road is known locally by one name but is part of a larger road system—for instance, in Ohio, State Route 224 runs the length of Mahoning County, but the section of it that passes through the town of Canfield is known as "Main St." The geocoding system would use an alternate list of addresses to know that these two names mean the same place along the road.

Table 8.2 shows a number of locations in the Washington, D.C., area with their addresses, as well as these addresses parsed and standardized. For instance, in the National Gallery of Art's address, the street name is "Constitution." When the address matches, the system refers to line segments with a name attribute of "Constitution" and those segments with a street-type attribute of "AVE" (rather than ST, BLVD, LN, or any other) and a suffix

Table 8.2 **Addresses that have been parsed into their component parts and standardized.**

Location	Address	Prefix	Number	Street name	Street type	Suffix
White House	1600 Pennsylvania Avenue NW		1600	Pennsylvania	AVE	NW
National Gallery of Art	401 Constitution Avenue NW		401	Constitution	AVE	NW
U.S. Capitol	1 1st Street NE		1	1st	ST	NE
Office of the Federal Register	800 North Capitol Street NW	N	800	Capitol	ST	NW
Washington National Cathedral	3101 Wisconsin Avenue NW		3101	Wisconsin	AVE	NW
United States Holocaust Memorial Museum	100 Raoul Wallenberg Place SW		100	Raoul Wallenberg	PL	SW

direction attribute of NW (instead of some other direction). The street number, 401, is used to determine which road segments match an address range (on the left or right side of the street, depending on whether the number is odd or even).

After the address has been parsed and standardized, the matching takes place. The geocoding system will find the line segments in the reference database that are the best match to the component pieces of the address and rank them in ArcGIS. For instance, in trying to address match the National Gallery of Art, a line segment with attributes of Name = "Constitution," Type = "AVE," Suffix Direction = "NW," and an address range on the left side of "401–451" would likely be the best (or top-ranked) match. A point corresponding with this line segment is placed at the approximate location along the line segment to match the street number. For instance, our address of 401 would have a point placed near the start of the segment, while an address of 425 would be placed close to the middle. The method used to plot a point at its approximate distance along the segment is called **linear interpolation**.

Keep in mind that the plotting is an approximation of where a specific point should be. For instance, if a road segment for "Smith Street" has an address range of 302 through 318, an address of 308 will be placed near the middle. However, if the actual real-world location of house number 308 is closer to the end of the street, then the placement of the plotted point will not necessarily match up with the actual location of the house. Where streets contain only a handful of houses that correspond with the address range in the reference file, plotted locations may be estimated incorrectly. See **Figure 8.5** on page 284 for an example of plotting a geocoded point on a road network in GIS.

linear interpolation a method used in geocoding to place an address location among a range of addresses along a segment

FIGURE 8.5 Plotting an address point on a line segment in GIS.

[Source: Esri]

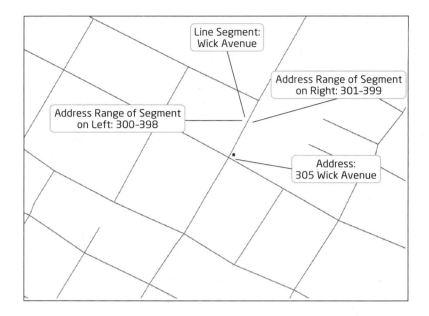

In GIS, a satellite navigation system, or a smartphone equipped with geospatial technology, whenever you specify an address, the system will match the address and fix it as a destination point to be found. The process of geocoding multiple addresses at once is referred to as **batch geocoding.** For instance, in batch geocoding, you can have a list of the addresses of all the coffee shops in Seattle, and the GIS will match every address on the list, so you won't have to input the addresses one at a time (for an example, see *Hands-On Application 8.2: Geocoding Using Online Resources*). If no match can be found, or if the ranking is so poor as to be below a certain threshold for a match, sometimes a point will not be matched at all for that address (or it may be matched incorrectly, or placed at something like the center of a zip code). You will sometimes be prompted to recheck the address, or to try to match the address interactively by going through the process manually.

Finally, when an address is plotted, the system may have the capability to calculate the x and y coordinates of that point and return those values to the user. For example, the street address of the Empire State Building in New York City is 350 5th Ave., New York, NY 10018. From address matching, using the free online **gpsvisualizer.com** utility, the GIS coordinates that fit that address are computed to be latitude 40.74827 and longitude −73.98531 (**Figure 8.6**). Thus, geocoded points have spatial reference attached to them for further use as a geospatial dataset.

Geocoding is a powerful process, but it's not infallible. There are several potential sources of error that cause an address to be plotted in an incorrect location. Since the addresses you're entering will be parsed to match up with the segments in the reference database, the result can sometimes be an error in matching. For instance, the address of the White House in Table 8.2 is

batch geocoding matching a group of addresses together at once

HANDS-ON APPLICATION 8.2

Geocoding Using Online Resources

Many online mapping resources (*see Hands-On Application 8.4: Online Mapping and Routing Applications and Shortest Paths* for details on what they are and how they're used) will geocode a single address, or a pair of addresses (so that you can calculate directions between the two). You can use GIS software to geocode multiple addresses, or you can use another resource, like the BatchGeo Website. Open your Web browser and go to **www.batchgeo.com**—this is a free online service that allows you to geocode an address (or multiple addresses in batches), and then view the plotted point or points on a map. Set up a batch of three or more addresses (such as home addresses for a group of family members, a group of friends, or several workplaces) and run the

batch geocode utility. The Website will give you examples of how the addresses need to be formatted. From there, you can view your results on the Google Map that the Website generates.

Expansion Questions

- Assemble a list of five addresses familiar to you (such as your home address, your work address, or a local park or attraction) and batch geocode them using the Website. Zoom in closely on the resulting Google Map that is produced for you and examine the geocoded points—how closely do the points match up with where the address should be?

- If any points are geocoded incorrectly, how are the locations improperly matched?

---

listed as "1600 Pennsylvania Avenue NW." The line segments that match up with this street should have the name listed with "Avenue" (or "Ave.") as a suffix. If you input something different, like "1600 Pennsylvania Street" or "1600 Penn Avenue," the location could potentially receive a lower ranking, end up plotted somewhere else, or otherwise fail to get properly matched. With more complete information (like "1600 Pennsylvania Avenue NW,

FIGURE 8.6 The geocoded result for the Empire State Building in New York City, with corresponding latitude and longitude calculation. *[Source: OpenStreetMap, Mapquest]*

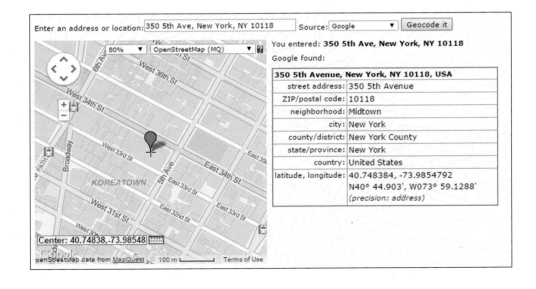

Washington, D.C., 20500"), the system will be better able to identify an accurate match.

It should also be kept in mind that the geocoding system can only properly identify locations if the line segments are in the reference database. If you're searching for an address established after the reference database network file was put together, the system will not be able to match the address properly. If the point is plotted on the correct street but at an incorrect location on the street, it's likely because of a problem with address range data in the reference database and how it reflects the real world (for instance, a house at 50 Smith Street is not necessarily halfway down a street segment that begins with 2 and ends with 100). The geocoding will usually be only as accurate as the base data it's being matched to. If the reference database doesn't contain all line segments (for instance, if it's missing subdivisions, streets, or new freeway bypasses that haven't yet been mapped and added), or if its attributes contain inaccurate address ranges or other incorrect information, or if any of its attributes are missing, the geocoding process will probably be unable to match the addresses or may plot the addresses in the wrong location.

New methods for geocoding addresses to get a more accurate match have been developed. Rather than using a line segment and interpolating the address location, point databases are being created in which a point represents the center of a parcel of land (for instance, a house or a commercial property). When geocoding with this point data, address information can be matched to the point representing the parcel, and the address location can be found for the road immediately next to the parcel.

Once locations are geocoded, the system (or GIS) can begin to examine the routes between locations to determine the shortest path from one location to another. With a vehicle navigation system, you enter the address of the destination you want to travel to, and the system will match that address. The device's current position is determined using GPS and plotted on the network map—this will be the origin. The system will then compute the shortest route between the origin and destination across the network. The same holds true for an online system to find directions—it has a matched origin and destination, and it will compute what it considers the best route for you to follow between the two points. With so many different ways to get from the origin to the destination, the system now needs a way to determine the "shortest path" between these locations.

How Are Shortest Paths Found?

When you leave your home to go to work, you probably have several different ways you can go. Some of them are pretty direct and some of them are more roundabout, but you have plenty of options available. If you want to take the "shortest" path from home to work, you'll probably focus on some of the more direct routes and eliminate some of the longer or more circuitous routes. However, a **shortest path** can mean different things. For instance, driving through city streets may be the shortest physical driving distance (in terms of mileage), but if those streets have lots of stop lights and traffic congestion, then this "shortest path" could easily take longer in time spent driving than a longer distance on a highway that doesn't have these impediments. If you want to minimize the time spent driving, the highway route will probably get you to work faster, but you'd actually be traveling a longer distance. Major city streets can take a long time to navigate at rush hour, but you'll probably sail through them much quicker if you make the same journey late at night.

All of these things have to be considered when you're figuring out the shortest path (or "best route") to take when traveling from an origin to a destination. People have their own decision-making criteria and stick to their own rules when it comes to determining what they judge to be the shortest path from one place to another—things like " always keep to the main roads," or "use highways whenever you can," or "never make a left turn." This is why vehicle navigation systems often offer multiple options, such as "shortest distance" or "shortest driving time," and sometimes apply conditions like "avoid highways" to compute the best route between points.

Within a vehicle navigation system (or GIS), each line segment has a **transit cost** (or impedance) assigned to it. The transit cost reflects how many units (of things like distance or travel time) it takes to go the length of a particular edge. The transit cost may reflect the actual distance in miles from one junction to another along the edge. The transit cost could also be the equivalent time it takes to drive that particular segment. Whatever transit cost is used, that value will be utilized in the shortest path computation. Other impedance attributes can be modeled as well—segments could have different transit costs under certain conditions (such as heavy traffic or construction). These types of impedance factors can help in making a network more realistic for use.

The shortest path is then calculated using an **algorithm**, a step-by-step mathematical procedure designed to solve a particular kind of problem—in this case, to determine the overall lowest transit cost to move along the network from a starting point to a destination. There are various types of shortest path algorithms, including **Dijkstra's Algorithm** (see *Hands-On Application 8.3: Solve Your Network Problems with Dijkstra* for more information on how to use this algorithm), which will compute the path of lowest cost to travel from a starting point to any destination in the network. For instance, say you have three different destinations you plan to travel to (work, the pizza shop, and the grocery store). Dijkstra's Algorithm will evaluate the overall transit cost from your home to each destination, and find the shortest path from your home to work, from your home to the pizza shop, and from your home to the grocery store.

shortest path the route that corresponds to the lowest cumulative transit cost between stops in a network

transit cost a value that represents how many units (of time or distance, for example) are used in moving along a network edge

algorithm a set of steps used in a process designed to solve a particular type of problem (for example, the steps used in computing a shortest path)

Dijkstra's Algorithm an algorithm used in calculating the shortest path between an origin node and other destination nodes in a network

HANDS-ON APPLICATION 8.3

Solve Your Network Problems with Dijkstra

The inner workings of Dijkstra's Algorithm are beyond the scope of this book, but there's an excellent free online resource that allows you to construct a sample network, then run the Dijkstra Algorithm to find the shortest path. The algorithm will walk through the shortest path step-by-step and describe the actions it's taking (and how the shortest paths are created). Open your Web browser and go to **www.dgp.toronto .edu/people/JamesStewart/270/9798s/Laffra /DijkstraApplet.html**. On this Website, you can set up a series of nodes (junctions) and links (edges), assign weights (transit costs) and directions to them, and run the algorithm to find the shortest path between the origin and all destinations on the network. Use the interactive interface to construct a sample network (or use the pre-made example) and use Dijkstra to set up the shortest paths for you. From the pull-down menu on the left, select Draw Nodes, then click in the large window to place the nodes (there are also options for moving nodes to

other locations or removing them altogether). Next, select the Draw Arrows option to place connections between the nodes. By selecting the Change Weights option, you can slide the arrow up and down a connector to alter the weight between two nodes. Click on the Run button on the right-hand side of the screen to execute the algorithm.

Expansion Questions

- Set up a sample network of five nodes with various connections and different weights between them, then run the algorithm. After the algorithm completes, values will be assigned to each node. What do these numbers represent?

- Clear your own sample network and press the Example button, then run the algorithm for the pre-made sample. What is the calculated shortest path to travel from node a to node e? From node a to node f? From node a to node j?

Whatever type of algorithm is used, the system will compute the shortest path, given the constraints of the network (such as transit cost or directionality of things like one-way streets). The system can then generate directions for you by translating the selected path into the various turns and names of streets that you'll take as you follow the path through to your destination. See **Figure 8.7** for an example of using an online geospatial technology utility to compute the shortest path or best route between two locations. Also check out *Hands-On Application 8.4: Online Mapping and Routing Applications and Shortest Paths* for some other Web tools for online directions and routing.

Of course, sometimes you have more than one destination to visit—say you have three places you want to stop at (shoe store, music store, and bookstore) and you want to drive the overall shortest route to hit all three places. When you're using geospatial technology to compute a shortest route between several of these **stops**, there are two types of scenario to choose from:

stops destinations to visit on a network

1. In the first scenario you need to find the shortest path when you're going to be visiting stops in a pre-defined order: This means you have to stop at the shoe store first, the music store second, and the bookstore

FIGURE 8.7 The purple line marks the shortest path from the White House to Georgetown University in Washington, D.C. (computed from MapQuest). *[Source: 2015 MapQuest, 2015 TomTom]*

HANDS-ON APPLICATION 8.4

Online Mapping and Routing Applications and Shortest Paths

There are many online applications available for creating a map of an address, and then generating a shortest path and directions from that address to another one. Examine the functionality of some of the following online mapping sites:

1. Bing Maps: **www.bing.com/maps**

2. Google Maps: **www.google.com/maps**

3. HERE: **www.here.com**

4. MapQuest: **www.mapquest.com**

5. Yahoo! Maps: **https://maps.yahoo.com**

Try inputting the same pair of locations (origin and destination) into each Web service and compare the shortest paths and routes they generate (some of them may be similar, some may be different, and some will likely give you more than one option for the "shortest path"). In addition, the services will also allow you to alter the route interactively by moving and repositioning junctions of the path, so you can tailor the route more to your liking.

Expansion Questions

- Set up a shortest route between two locations familiar to you (such as home and work, home and school, or two well-known destinations in your local area), then position the cursor at junctions along the path. You can drag and drop the nodes to new locations and thus change the route to your liking. By altering the route, how specifically did it change in terms of time and distance?

- Each of the five mapping sites will give you information back about the source of the maps being generated (usually in small text at the bottom of the map). What is the copyrighted source of the maps for each of the four mapping sites?

last. In this case, you'd want to find the shortest paths from your house to the shoe store, from the shoe store to the music store, and finally from the music store to the bookstore.

2.   The second scenario involves finding the shortest path when you can arrange the order of visiting the stops: For instance, if the bookstore and the shoe store are near each other, it makes sense to rearrange your travels, so you visit them one after the other and then drive to the music store.

Geospatial technology applications can help determine some solutions to both of these scenarios. To visit stops in a pre-determined order, the system will find the shortest route (of all possible routes) from the origin to the first stop, then choose the shortest route to travel from the first stop to the second stop, and so on. An example of this in vehicle navigation systems is being able to set up a "Via Point" (another stop) between your origin and final destination—the device will then calculate the route from the origin (or current location) to the Via Point, then from the Via Point to the final destination.

With the ability to rearrange the order of stops, a system will evaluate options by changing the order of visiting stops to produce the overall shortest route. See **Figure 8.8** for an example of how the "shortest path" changes between visiting stops in order as opposed to being able to rearrange the order of visiting the stops. Additional constraints can be placed on the problem—for instance, you may have to return home (the starting point) after visiting all of the stops. In that scenario, when you rearrange the stops, the starting point also becomes the last stop, which may affect how the reordering is done. Conversely, you may want to have your starting point and ending point different from one another, and these types of parameters will have to be placed on the problem. Keep in mind that if you can arrange the order of stops, the program will likely give a decent solution to the problem; however, the only way to truly identify which configuration is the best is for the program to work through every possible arrangement—something that

FIGURE 8.8 Two different shortest paths through Youngstown, Ohio, both involving multiple stops at public libraries. The stops on the first path are made in a predetermined order (from #1 through #5 on the map). In the second map, the library visits are rearranged to find the shortest route (without a fixed starting or ending point). *[Source: Esri]*

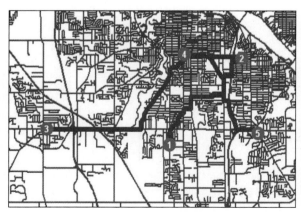
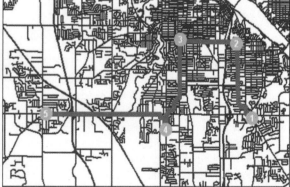

HANDS-ON APPLICATION 8.5

Finding the Best Route For Multiple Stops

A Web application like Google Maps or Bing Maps allows you to compute the shortest path between two stops. However, when you have multiple stops, a different tool can be used to find the shortest path among them. For an example of this, go to **www.mapquest.com /routeplanner**. This is the Website for the Map-Quest Route Planner, which will allow you to begin with two stops, add others, and then allow Map-Quest to reorder your stops to find the shortest route among all of them.

We'll find the shortest route between a set of addresses in Washington, D.C., from Table 8.2. Use the address for the White House (1600 Pennsylvania Avenue NW, Washington, D.C.) as the Start address and the Office of the Federal Register (800 North Capitol Street NW, Washington, D.C.) as the End address. Click on Get Directions and you'll see a map appear with the route between the two places. Click on the button marked Add Stop under the two addresses to add a third address, that of the U.S. Capitol (1 1st Street NE, Washington, D.C.) and then add a fourth address, that of the National Gallery of Art (401 Constitution Avenue NW,

Washington, D.C.). Drag the icons so that your starting point is the White House, second is the U.S. Capitol, third is National Gallery of Art, and then the final destination is the Office of the Federal Register. Click on Get Directions again and MapQuest will compute the shortest path of visiting these stops in this particular order.

However, MapQuest can also rearrange the stops—put a checkmark in the box that says "Allow MapQuest to re-order stops" and then click Get Directions one more time. You'll see that the route has changed and the stops have been rearranged for an overall shorter path.

Expansion Questions

- How did the shortest paths change between the stops in order and the rearranged order of stops?

- Add in the other two addresses from Table 8.2 as new fourth and fifth stops in between the White House and the Office of the Federal Register. How does the shortest path change the stops in order? How does it change after allowing MapQuest to rearrange the stops?

would be impossible with a larger number of stops. See *Hands-On Application 8.5: Finding the Best Route for Multiple Stops* for an online tool that works with these types of conditions.

How Are Networks Used in Geospatial Technology?

All of the things discussed in this chapter are used for a variety of applications with geospatial technology. Through GIS, these types of network data can be created and utilized in various ways. The vehicle navigation systems integrate many of these concepts—using GPS to pinpoint the device's location on a map, then using network base data, geocoding, and shortest paths to navigate through locales worldwide. Other options on these systems involve utilizing real-time data broadcast into the device to

FIGURE 8.9 A smartphone running a mapping application.
[Source: © iPhone/Alamy]

determine areas of congestion, high traffic, construction, or road accidents. This data can then be used in the shortest-path process to route you around these impediments. These same maps and technological resources have been integrated into smartphones and mobile devices to put GPS locations, mapping, geocoding, shortest-paths, and real-time routing information capabilities into the palm of your hand (**Figure 8.9**). As we have noted, however, these devices are only as accurate as the base maps they're utilizing, so regular updates are usually made available, or users are given the option to make their own updates.

The same types of network base maps are used online for services like MapQuest, Bing Maps, Yahoo! Maps, and Google Maps (see *Hands-On Application 8.4: Online Mapping and Routing Applications and Shortest Paths*), which allow you to perform geocoding and obtain directions for the shortest path between points. **Street View** allows Google Maps (or Google Earth) users to examine 360 degrees of location photography—on a street, say, as if you'd just stopped your car right there to take a look around. With this function, you can examine canned photography of an address or destination when you're planning a stop (**Figure 8.10**).

Cars equipped with special cameras capable of capturing a 360-degree view travel the nation's roads, taking in imagery along the way (see **Figure 8.11** and *Hands-On Application 8.6: Examining Google Street View*). Google is even extending their Street View into places inaccessible by motor vehicles

Street View a component of Google Maps and Google Earth that allows the viewer to see 360-degree imagery around an area on a road

FIGURE 8.10 The U.S. Capitol Building as seen from Google Maps Street View. *[Source: Google]*

by attaching the same type of camera equipment to bicycles and checking out the hiking and biking trails.

With this level of data availability, and geospatial networks being integrated into so many different applications, mapping (and providing accurate and up-to-date base maps) has become big business. Companies such as Tele Atlas and NAVTEQ produce these maps, and their products are often what you're accessing on an online service or through a vehicle navigation system. Next time you access one of these kinds of services, look for the copyright data somewhere in the map to see which company is producing the map data you're using. NAVTEQ, for instance, sends teams out in high-tech cars to collect data on the location and attributes of new roads, housing developments, and other items for updates of maps (as well as photography equipment to produce items like the Street View scenes).

FIGURE 8.11 An example of a car capturing Google Street View images. *[Source: AP Photo/dda, Michael Kappeler]*

HANDS-ON APPLICATION 8.6

Examining Google Street View

Google Street View is a very useful application that gives you a look at what the street network would look like if you were driving or riding past. Go to Google Maps at **www.google.com/maps**, and then enter a particular address—try typing in the address of the Rock and Roll Hall of Fame in Cleveland, Ohio (1100 Rock and Roll Blvd., Cleveland, OH). Zoom in until you can see the distinctive building of the Rock Hall and its surrounding streets. Below the zoom controls on the lower right-hand side of the map you'll see an orange icon that looks like a person (called the "pegman"). Grab that with the mouse and place it on top of the street in front of the Rock Hall (E. 9th Street). Streets that have been covered by Google Street View will be highlighted in dark blue. Place the icon on one of the available streets and the view will shift to photography of that area. Use the mouse to move the view about for 360-degree imagery of the area. You'll also see some white lines and arrows superimposed on the road—clicking on them will move you along the street and you'll see the imagery change. Move around until you can get a good street-level view of the Rock and Roll Hall of Fame.

Expansion Questions

* How would the imagery from Street View be helpful if you were traveling to the Rock Hall as a destination (and had never been there before)? There are no parking areas available immediately adjacent to the Rock Hall—how could Google Street View be more helpful to you than a road map or an overhead view in finding somewhere to park?

* Examine some local areas near you using Street View. Which roads are available with Street View imagery and which are not (for instance, is Street View imagery of residential areas available or is it available only on main roads)?

THINKING CRITICALLY WITH GEOSPATIAL TECHNOLOGY 8.2

What Kind of Issues Come with Google Street View?

Go to Google Maps at **www.google.com/maps**, and then enter the address of the White House (1600 Pennsylvania Avenue, Washington, D.C. 20500). When you try to move the Street View pegman to view the roads, you'll see that the streets immediately surrounding the White House are unavailable to view in Street View, presumably for security purposes. However, most of the surrounding streets can be viewed using Street View, allowing you to look at the exteriors of shops, residences, and other government buildings. Can the images being collected by Street View pose a security risk, not just for government areas, but for the private security of homes or businesses?

An article in the May 31, 2008, issue of the *Star Tribune* (available online at **www.startribune.com/lifestyle/19416279.html**) describes how the community of North Oaks, Minnesota, demanded that Google remove its Street View images of the privately owned roads in the area, citing laws against trespassing. Is Street View encroaching too far on people's private lives? Keep in mind that you can view or record the same things quite legally by driving or walking down the same streets.

Chapter Wrapup

Networks, geocoding, and routing are all powerful tools in everyday use in geospatial technology. With GIS, these concepts are used in a variety of applications and businesses today. Today 911 operators can geocode the address that a call is coming from, and emergency services can determine the shortest route to a destination. Delivery services can use geocoding and routing applications to determine locations quickly and reduce travel time by using the shortest paths to get to them. Also, the use and development of these types of techniques in satellite navigation systems and smartphone apps represent rapidly changing technology that keeps getting better. For current examples of geospatial technologies used for networks and routing, check out *Geocoding and Shortest Paths Apps, as well as Geocoding and Shortest Paths in Social Media.*

Geospatial Lab Application 8.1 uses Google Maps as well as Google Earth Pro for investigating geocoding and shortest path uses. In the next chapter, we're going to start looking at a whole different aspect of geospatial technology—remote sensing. All of these overhead images that you can see on applications like Google Maps or MapQuest have to come from somewhere to get incorporated into the program, and we'll start looking at the methods behind remote sensing in the next chapter.

Important note: The references for this chapter are part of the online companion for this book and can be found at **www.macmillanhighered .com/shellito/catalog**.

Geocoding and Shortest Path Apps

Here's a sampling of available representative apps using geocoding and shortest paths for your phone or tablet. Note that some apps are for Android, some are for Apple iOS, and some may be available for both.

- **Footpath Route Planner:** An app that allows you to draw a circle around an area and convert the drawing to a street route for use in walking, running, or cycling
- **Google Maps:** An app for a mobile version of Google's online maps
- **HERE:** Nokia's app for a mobile version of Here (for maps and routing)
- **MapMyRide:** An app that lets you map and track your biking route
- **MapQuest GPS Navigation & Maps:** An app for a mobile version of MapQuest
- **Maps (from Apple for iOS):** Apple's mapping app for mobile devices
- **MAPS.ME:** An app that allows you to browse and use maps and routing while offline

- **Moovit:** An app that provides information about the best route to a destination via buses or trains
- **Street View on Google Maps:** An app for a mobile version of Google Street View
- **Trucker Path Pro:** An app designed for navigation for professional truck drivers to locate truck stops, weigh stations, nearby fuel prices, and other essentials for truckers
- **Waze:** An app used for obtaining and sharing real-time local traffic and road conditions
- **Where Am I? Geocode:** An app that will determine coordinates for an address and plot it on a map and also allow you to access Street View imagery of that location, if available

Geocoding and Shortest Paths in Social Media

Here's a sampling of some representative Facebook, Twitter, and Instagram accounts, along with some blogs related to this chapter's concepts.

 On Facebook, check out the following:

- **Google Maps:**
  www.facebook.com/GoogleMaps
- **HERE:**
  www.facebook.com/here
- **MapQuest:**
  www.facebook.com/MapQuest

Become a Twitter follower of:

- **BatchGeo:** @batchgeo
- **Bing Maps:** @bingmaps
- **Google Directions:** @GDirections
- **Google Maps:** @googlemaps
- **HERE:** @here
- **MapQuest:** @MapQuest

Become an Instagram follower of:

- **Google Maps:** @googlemaps

For further up-to-date info, read up on these blogs:

Bing Maps Blog (the blog for Microsoft's Bing Maps):
http://blogs.bing.com/maps

Google Directions (a blog not affiliated with Google but dedicated to all kinds of online and mobile mapping technologies):
www.google-directions.com

Google Lat-Long Blog (a blog about Google Maps):
http://google-latlong.blogspot.com

HERE 360 (the official blog for HERE):
http://360.here.com

Key Terms

address matching (p. 281)
address standardization (p. 282)
algorithm (p. 287)
batch geocoding (p. 284)
connectivity (p. 277)
Dijkstra's Algorithm (p. 287)
edge (p. 277)
geocoding (p. 281)
junction (p. 277)
line segment (p. 278)
linear interpolation (p. 283)

network (p. 277)
parsing (p. 282)
reference database (p. 282)
satellite navigation system (p. 275)
shortest path (p. 287)
stops (p. 288)
street centerline (p. 278)
Street View (p. 292)
TIGER/Line (p. 279)
transit cost (p. 287)

Geocoding and Shortest Path Analysis

This chapter's lab will introduce you to the concepts of calculating a shortest path between stops along a network, as well as generating directions for the path using Google Maps. You'll also be performing geocoding using Google Earth Pro and examining the geocoding results (as well as performing shortest path analysis using Google Earth Pro).

Objectives

The goals for you to take away from this lab are:

▶ To use Google Maps to create and alter shortest paths to account for route changes

▶ To utilize Google Earth Pro to geocode a series of addresses and then examine the results

▶ To familiarize yourself with the shortest path and directions functions of Google Earth Pro

Using Geospatial Technologies

The concepts you'll be working with in this lab are used in a variety of real-world applications, including:

▶ Emergency services, which utilize geocoding to determine the location of 911 callers, so that the nearest available aid can be quickly dispatched to the right place

▶ Commercial promotion and postal services, where traveling sales representatives and delivery drivers make use of routing information to take the shortest and most obstacle-free paths between destinations

[Source: © Stocktrek Images, Inc./Terry Moore/Alamy]

Obtaining Software

The current version of Google Earth Pro (7.1) is available for free download at **www.google.com/earth/explore/products/desktop.html**.

Important note: Software and online resources can change fast. This lab was designed with the most recently available version of the software at the time of writing. However, if the software or Websites have significantly changed between then and now, an updated version of this lab (using the newest versions) will be available online at **www.macmillanhighered.com /shellito/catalog**.

Lab Data

Copy the folder Chapter8—it contains a spreadsheet (in CSV format) called VirginiaBeachLibraries.csv featuring a series of addresses to be geocoded.

Localizing This Lab

The datasets in this lab focus on the locations of public libraries and the road network in Virginia Beach, Virginia. However, this lab can be modified to examine your local area. Use your local county library's Website (or the phone book) as a source of names and addresses of local libraries and use Microsoft Excel to create a file like the one in the Chapter8 folder (where each column has a different attribute of data). When you've entered the information into Excel, save the file in CSV format (it will be one of the file formats available to save to within Excel). If there are not enough local libraries around, use the addresses of other local venues, like coffee shops, pizza shops, or drugstores.

8.1 Google Maps' Shortest-Path Functions

1. To begin, we'll look at a simple shortest route calculation using Google Maps. Navigate to the Google Maps Website at **www.google.com /maps**.

2. Type Virginia Beach, VA and click the Search button.

3. When the map of Virginia Beach appears, select Directions from the pop-up menu that appears under where you searched.

[Source: Google Maps]

4. Let's start with a typical summertime scenario. You're on vacation in Virginia Beach, visiting the Old Coast Guard Museum (which faces the oceanfront boardwalk). You have tickets to see a show at Norfolk Scope Arena, located in nearby downtown Norfolk. You want to take the shortest route to get from your beachfront location to the arena. Use the following addresses for directions:

   a. Starting point: 2400 Atlantic Avenue, Virginia Beach, VA 23451 (this is the museum you're driving from)

   b. Ending point: 201 East Brambleton Avenue, Norfolk, VA 23510 (this is Norfolk Scope Arena)

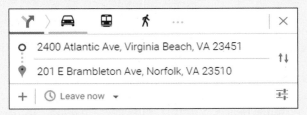

[Source: Google Maps]

5. By default, Google Maps will use driving a car as the way to compute the shortest route (above the directions you'll see icons for other options such as traveling by bus or walking). On the map, the shortest driving route between the museum and the arena will appear, highlighted in blue.

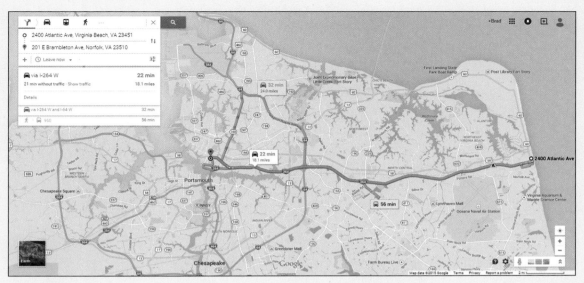

[Source: Google Maps]

> **Question 8.1** What is the approximate driving time between the two locations? What (in general) is the route that Google Maps suggests to take?

6. Select the option for walking between the museum and the arena (the icon is of a person on foot). A second path, shown with blue dotted lines, will appear.

> **Question 8.2** What is the new distance (and estimated walking time) of the route using this new path? Why is the walking route different than the driving route?

7. Return to the shortest-path driving distance. Zoom in closer to the starting point (the museum at 2400 Atlantic Avenue). Scroll the mouse over the blue shortest path line and you'll see a white circle appear along the path. This lets you change the calculated route to account for any variety of factors (such as your own travel preference, known congestion or construction areas, rerouting to avoid likely trouble spots or known delays, and so on).

8. Start by placing the mouse at the starting point and drag the circle up to state route 58 (also known as Laskin Road or Virginia Beach Boulevard—the other major east–west road just above and parallel to I-264). The route will change by first having you drive north on Atlantic Avenue, turning west on 58, then merging back onto I-264 again a little later on the route. Even though it's a small change, you'll see that the driving distance and estimated time have changed. Answer Question 8.3. When you're done, click the back arrow on your Web

browser to return the path to how it originally was (if you made multiple changes to the route, you may have to click the back arrow a couple times to return the map to its original path).

[Source: Google Maps]

> **Question 8.3** What is the new distance (and estimated driving time) of the route using this new path? What would account for this?

9. Scroll across the map until you see where the route crossed I-64 (about two-thirds of the way between the two points). I-64 is the major highway into the Hampton Roads area. Let's throw another change into the scenario—say, for instance, that there's heavy construction on I-264 west of the I-64 intersection. At the intersection, move the circle off the route and north onto I-64 far enough that the remainder of the route to Norfolk Scope diverts your shortest path headed north onto I-64, and not so that it returns you to I-264 west (you may have to move other circles as well if Google Maps diverts your path back onto I-264).

> **Question 8.4** What is the new route (in terms of the general roads now taken)? What is the new distance (and estimated driving time) of the route using this new path?

10. Reset the route back to its original state. Now, change the route so you would have to travel south on I-64 instead of north (still avoiding I-264 for the whole route).

> **Question 8.5** What is the new route (in terms of the general roads now taken)? What is the new distance (and estimated driving time) of the route using this new path?

11. You can now close your Web browser.

8.2 Geocoding and Google Earth Pro

Before any type of shortest path could be calculated, the system first had to match the addresses of the starting and ending points to their proper location. The next section of this lab will introduce a method of doing this address matching (geocoding) to create a new point layer for use in Google Earth Pro.

1. Open the VirginiaBeachLibraries.csv file using Microsoft Excel (or a similar program). You'll see the 10 Virginia Beach libraries with their address information split into multiple fields (columns): Name, Address, City, State, and Zip. These last four fields will be used to geocode the libraries and then the Name field will be used to reference them.

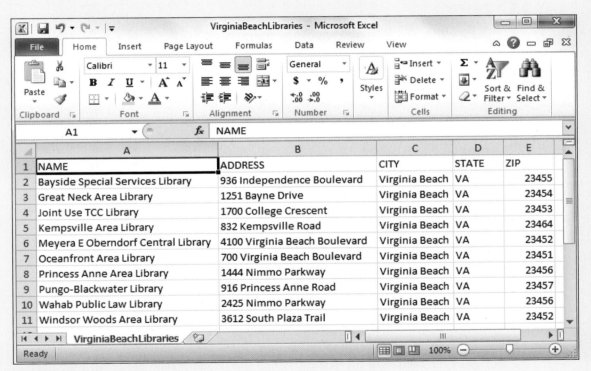

[Source: Microsoft Excel]

2. Close the VirginiaBeachLibraries.csv file (don't save it if prompted).

3. Start Google Earth Pro (GE), type Virginia Beach, VA, in the Search box, and then click on Search. Google Earth will switch to the view of Virginia Beach.

4. From the File pull-down menu, select Import. Navigate to the location of VirginiaBeachLibraries.csv and select that file to import into GE.

Data Import Wizard

Specify Delimiter
This step allows you to specify the field delimiter in your text file

Field Type

◉ Delimited   ◯ Fixed width

Delimited

Select the delimiter that separates each field. If there can be more than one delimiter between two fields (such as spaces), check the "treat consecutive delimiters as one" option. You can also provide your own custom delimiter by checking the "other" option

◯ Space   ☑ Treat consecutive delimiters as one
◯ Tab
◉ Comma
◯ Other   [   ]

Fixed Width

Column width 8 ⬍

Text Encoding

Supported encodings [System   ▾]

This is a preview of the data in your dataset.

	NAME	ADDRESS	CITY	STATE	ZIP	
1	Bayside Special ...	936 Independe...	Virginia Beach	VA	23455	
2	Great Neck Are...	1251 Bayne Drive	Virginia Beach	VA	23454	
3	Joint Use TCC L...	1700 College Cr...	Virginia Beach	VA	23453	
4	Kempsville Are...	832 Kempsville ...	Virginia Beach	VA	23464	
5	Meyera E Ober...	4100 Virginia Be...	Virginia Beach	VA	23452	

[Help]   [Next >] [Finish] [Cancel]

[Source: Google Earth]

5. In the Data Import Wizard that will open, select the radio button for Delimited for the Field Type, then select the radio button for Comma under the Delimited options. You'll see a preview of the data at the bottom of the Wizard, which should show each of the attribute fields split into their own column (like how they were in the initial VirginiaBeachLibraries.csv file). Click Next.

6. In the second screen of the Data Import Wizard, make sure the box for This dataset does not contain latitude/longitude information, but street addresses has a checkmark next to it (since you will be geocoding addresses rather than plotting points based on their latitude and longitude coordinates). Click Next.

7. In the third screen of the Data Import Wizard, select the radio button for Addresses are broken into multiple fields.

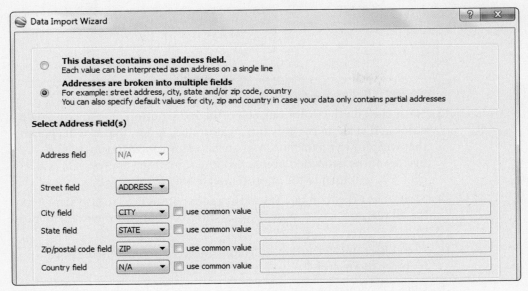

[Source: Google Earth]

8. In the options under Select Address Field(s), be sure the following options are chosen (if they are not, you can select them from the pull-down menu next to each):

   a. Street field: ADDRESS

   b. City field: CITY

   c. State field: STATE

   d. Zip/postal code field: ZIP

   e. Country field: N/A

9. Click Next.

10. In the next screen of the Data Import Wizard, make sure that string is selected for the Type for each of the following fields: NAME, ADDRESS, CITY, and STATE. Make sure that integer is selected for ZIP.

11. Click Finish. When prompted by the question: Do you want to apply a style template to the features you ingested? click Yes. The Style Template Settings dialog box will appear. This will give you several options as to how you want your geocoded points displayed in GE.

12. In the Style Template Settings dialog, select the Name tab. From the pull-down menu next to Set name field, choose NAME (this will tell GE to use the attribute listed in the NAME field when labelling the geocoded points).

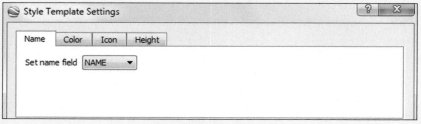

[Source: Google Earth]

13. Next, click on the Color tab. Select the radio button next to Use single color and click on the grey box next to that. A new dialog will appear that will allow you to choose whatever color you wish you for the display of the geocoded points. Choose something appropriate and click OK to return to the Style Template Settings dialog box.

14. Next, select the Icon tab. Select the radio button next to Use same icon for all features and choose the pushpin from the pull-down menu next to that.

15. Last, select the Height tab. Select the radio button next to Clamp features to ground.

16. Click OK. You'll be prompted to save your style template—save it with the name Chapter8style. This will allow you to load these same options later if you do more geocoding on your own.

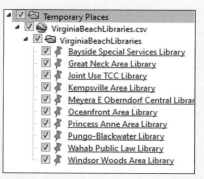

[Source: Google Earth]

17. In the GE Places box, you'll see an icon has been added called VirginiaBeachLibraries.csv. Expand this to see a folder under that called VirginiaBeachLibraries. Place a checkmark in both of these boxes and expand the folder option to see the labels for each of the libraries. In the main view, you will see the library addresses have been geocoded into points, which are shown with a pushpin in the color you selected.

18. To examine the results of the geocoding, locate the geocoded point for the Bayside Special Services Library and zoom in very closely so you can see where it was placed in relation to GE's crisp overhead imagery.

> **Question 8.6**   Where is the geocoded point placed in relation to the actual library building?

19. Next, examine each of the other nine geocoded points and verify that they've been geocoded into the correct location. Zoom in closely to each one (a quick way to do this is to double-click on the library's name in the Places box to find the location and then zoom in closer from there) and answer Question 8.7. Return to the full view where you can see Virginia Beach and all 10 geocoded library points when you're done.

> **Question 8.7**   For the other nine libraries in Virginia Beach, where were their geocoded points placed in relation to the library buildings themselves? Be specific in your answers.

8.3 Using Google Earth to Compute the Shortest Route Between Geocoded Points

1. The final step of the lab will investigate how to calculate shortest paths from the geocoded points in Google Earth. For instance, if you were driving between two of the geocoded libraries (such as the Great Neck Area Library and the Princess Anne Area Library), GE can find you the shortest route. To start, zoom in on the Great Neck Area Library point.

2. Right-click on the pushpin representing the Great Neck Area Library, and select the Directions from here option. In the Search box, you'll see the Search option switch to say "Great Neck Area Library" with its latitude and longitude.

3. Next, locate the Princess Anne Area Library, right-click on its pushpin, and select the Directions to here option.

4. In the Search box, the information about the new Princess Anne destination will be added. A new shortest path from the Great Neck Area Library (its address will be listed as A) to the Princess Anne Area Library (its address will be listed as B) will be calculated and displayed in purple on the view. Different routes between the destinations will also be shown in the Search box. By clicking on one of them, the corresponding route will be displayed in the view.

> **Question 8.8**   What are the suggested routes between Great Neck Area Library and the Princess Anne Area Library, and what is the distance (and estimated driving time) of each?

5. Locate both the Pungo-Blackwater Library and the Oceanfront Area Library.

> **Question 8.9** What are the suggested routes between the Pungo-Blackwater Library and the Oceanfront Area Library, and what is the distance (and estimated driving time) of each?

6. The geocoded points can also be treated by GE just like any other destination. For instance, you can search for a location using GE and find the shortest path to one of your geocoded points. For instance, search for Old Dominion University, 5115 Hampton Blvd, Norfolk, VA, 23529. GE will put a placemark at this location.

7. If you wanted to find the shortest distance from Old Dominion University to the Joint Library operated by the City of Virginia Beach and Tidewater Community College (TCC), GE can easily do this. To start, right-click on the newly added placemark for Old Dominion University and choose Directions from here.

8. Next, locate the Joint Use TCC Library geocoded point, right-click on it, and choose Directions to here.

> **Question 8.10** What are the suggested routes between Old Dominion University and the Joint Use TCC Library, and what is distance (and estimated driving time) of each?

Closing Time

This lab demonstrated several types of features associated with geospatial network data, including calculating different shortest paths and address matching with geocoding. Chapter 9 will switch gears and introduce some new concepts dealing with remote sensing of Earth (and the roads built on top of it). There's no need to save any data in this lab. Exit Google Earth Pro by selecting Exit from the File pull-down menu.

9

Remotely Sensed Images from Above

Where Aerial Photography Came From, UAS, Color Infrared Photos, Orthophotos, Oblique Photos, Visual Image Interpretation, and Photogrammetric Measurements

Whenever you take a picture with a digital camera, the image is captured and stored as a file on a memory card. Taking a picture at a birthday party or on vacation is actually a form of remote sensing—acquiring data without being in direct contact with the subject. For instance, you don't shove the camera right on top of the birthday cake to take a picture of it, you stand back a distance to do it. In essence, the camera is "sensing" (taking the picture) while you're some distance ("remotely") away from the target. In the same way, your eyes function as "remote sensing" devices—when you're looking at this book, you don't place your eyeball directly on the page. Instead, your eyes are "sensing" the data, and they're "remotely" located a few feet away from the page. Both a camera and your eyes are performing "remote sensing" as they're acquiring information or data (in this case, visual information) from a distance away without actually making contact with the item.

However, when it comes to **remote sensing** in geospatial technology, things are a little more specific. The data being acquired is information about the light energy being reflected off of a target. In this case, your eyes can still function like remote sensing devices, as what you're actually seeing is light being reflected off objects around you processed by your eyes. A digital camera does the same thing by capturing the reflection of light from whatever you're taking a picture of. In 1826, a French inventor named Joseph Niepce took the first photograph, capturing an image of the courtyard of his home. Photography has come a long way since those days (when it took 8 hours to

remote sensing the process of collecting information related to the reflected or emitted electromagnetic energy from a target by a device a considerable distance away from that target from an aircraft or spacecraft

expose the film), to the point where digital cameras are now commonplace, and it's difficult to find a cell phone without a built-in camera. In many ways, you can think of a camera as the original device for remote sensing—it captures data from the reflection of visible light nearly instantaneously from a distance away from the target.

aerial photography
taking photographs of objects on the ground from an airborne platform

People's fascination with the possibilities of remote observation led quickly to the development of **aerial photography**. Photographers have been capturing images of the ground from the sky for over 150 years. Initially they used balloons and kites to achieve the necessary height, but developing technologies of flight created new opportunities—airplanes, rockets, and, most recently, satellites, have enabled us to capture images of Earth's surface from greater and greater distances. When we're dealing with remote sensing in geospatial technologies, we're concerned with acquiring information from a platform on some type of aircraft or spacecraft. Currently, these types of imagery are obtained in both the public and private sector.

Whether for military reconnaissance, surveillance, studies of the landscape, planning purposes, or just for the good old purpose of being able to say, "Hey, I can see my house," people have been acquiring images of Earth from above for a long time. Today's aerial photographs are used for all manner of applications—for instance, aerial imagery serves as a base source for the creation and continuous updating of geospatial data and maps, such as the US Topo series of maps made by the USGS (see Chapter 13) and maps of road network data (see Chapter 8). Some of the extremely crisp high-resolution imagery you see on Google Earth is taken from aerial photography sources. In this chapter, we'll focus on how aerial images are captured and analyzed, and we'll learn to get used to looking at Earth from above.

How Did Aircraft Photography Develop?

A French photographer named Gaspar Felix Tournachon (known as "Nadar") took the first aerial photograph. In 1858, Nadar strapped himself into the basket of a hot-air balloon and captured an image of a landscape outside Paris. This first aerial photograph no longer exists, but this was clearly the beginning of something big. In 1860, aerial photographs were taken of Boston by James W. Black from a balloon above the city (**Figure 9.1**). However, development times, problems with landing at a prearranged location, and interference with the photographic plates from the balloon's gas all contributed to the view that balloons were not a practical platform for capturing aerial photographs.

By the end of the nineteenth century, photographic equipment was attached to kites (or a series of kites) and used to photograph the landscape remotely—while the photographer was safely on the ground. A famous example of kite photography comes from 1906, when a photographer named George Lawrence used a set of kites 1000 feet above San Francisco Bay to

FIGURE 9.1 An aerial photograph of Boston in 1860, taken by James W. Black. This is the earliest surviving aerial photograph taken from a balloon in the United States. *[Source: © Corbis]*

capture an aerial image of the city in the wake of the 1906 earthquake; he called it "San Francisco in Ruins" (**Figure 9.2**). These early images taken from the air set the stage for the use of aircraft as platforms for aerial photography.

The history of modern aviation started in 1903, with the first flight of the Wright Brothers at Kitty Hawk, North Carolina. In 1908, a cameraman took the first photograph from an airplane as he flew with Wilbur Wright near Le Mans, France. During World War I, aerial photography from an airplane was commonly used for mapmaking and the planning of military tactics. Aerial photography of the battlefields allowed for the patterns of trenches to be

FIGURE 9.2 "San Francisco in Ruins," an aerial photo taken in 1906 by George Lawrence, who suspended a remotely operated camera from a train of 17 kites, tethered 1000 feet over San Francisco Bay. *[Source: Library of Congress, Prints and Photographs Division (LC-USZC4-3870 DLC)]*

FIGURE 9.3 World War I
aerial photography
showing trench
formations. *[Source: IWM
via Getty Images]*

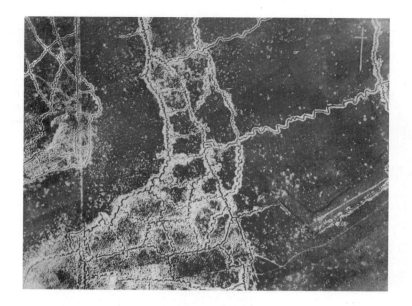

determined, enemy positions to be observed, troop movements and encampments to be plotted, and the location of artillery or supplies to be found. Countless aerial photos were shot and developed over the course of the war, and aerial photography opened up new areas in military intelligence. See **Figure 9.3** for an example of trench formations captured by an aerial camera.

Aerial photography continued to be extremely important for tactical planning and reconnaissance during World War II. Both the Allies and the Axis forces depended heavily on aerial photography for obtaining vital information about locations prior to military action, and it became a key reconnaissance technique. For instance, a V-2 rocket base in Peenemunde, Germany, was the subject of British aerial photography in June of 1943, and bombing strikes destroyed the base less than two months later. Similarly, aerial reconnaissance was used in obtaining photographs of German dams and the subsequent "dam-busting" bombing missions to destroy them. A huge amount of aerial photography was performed during the war, and a good source of historical imagery is The Aerial Reconnaissance Archives (TARA), currently held at the Royal Commission on the Ancient and Historical Monuments of Scotland (RCAHMS), which hosts a large archive of European World War II aerial photos (see *Hands-On Application 9.1: World War II Aerial Photography Online* for a glimpse of these historic photos).

Post–World War II aerial photography remained critical for military applications, especially for spy planes. The 1950s saw the development of the U-2 spy plane, capable of flying at an altitude of 70,000 feet, traveling at over 400 mph, and photographing the ground below. The U-2 was used for taking aerial photos over the Soviet Union, but also enabled the acquisition of imagery showing the placement of missile launchers at key points in Cuba during the Cuban Missile Crisis of October 1962 (**Figure 9.4**). Spy plane

HANDS-ON APPLICATION 9.1

World War II Aerial Photography Online

The National Collection of Aerial Photography is a great online resource for viewing aerial photography of Europe during World War II. Visit **http://ncap.org.uk/galleries**, which contains many different galleries of aerial imagery from several countries. Check out some of the countless historic aerial photos that have been archived online.

Expansion Questions

- A highlighted section of the National Collection of Aerial Photography showcases the "detective work" that has uncovered new information about historic events from the aerial archive. Look at **http://ncap.org.uk /news/detective-work-uncovers-hidden -secrets-aerial-reconnaissance-archive -tara**. What kinds of images linked to historic events have been found, and what kind of information has been gathered from them?

- Click on the Features section of the Website for access to several showcased historic aerial photos (and lessons on working with aerial photography). Select the option for The D-Day Landings and examine the aerial photos in this section. What aspects of the actions of the D-Day invasion have been captured through aerial photography? What particular features of the invasion can you interpret from the images?

- Also in the Features section, select the option for Camouflage, Concealment, and Deception. It's hard to hide something as big as an airfield or a warship from overhead cameras— but look closely at the images in this section. How were airfields, warships, oil refineries, and rocket launching sites concealed from aerial photos?

technology continued to advance with the development of the Lockheed SR-71 Blackbird, which could fly at an altitude of over 80,000 feet and travel at more than three times the speed of sound (see **Figure 9.5** on page 314), thus enabling it to go higher and faster than the U-2. The Blackbird was an

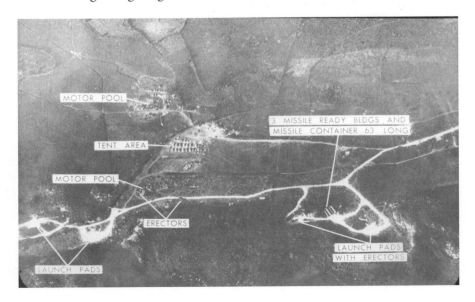

FIGURE 9.4 Aerial imagery of Cuba showing missile facilities, shot during the Cuban Missile Crisis of 1962. *[Source: U.S. Air Force Photo]*

FIGURE 9.5 The SR-71 Blackbird in action. *[Source: USAF/Judson Brohmer]*

essential aerial photography plane through the Cold War and beyond, until it was retired in the late 1990s. Today, digital aerial photography (among many other types of remotely sensed data) is obtained by unmanned aircraft systems (UAS).

What Are Unmanned Aircraft Systems?

UAS unmanned aircraft systems—a reconnaissance aircraft that is piloted from the ground via remote control

UAS (unmanned aircraft systems) are aircraft that are piloted from the ground via remote control. Sometimes these types of aircraft are also referred to as drones or unmanned aerial vehicles (UAVs). Like the name implies, UAS are not flown with a pilot in the cockpit—instead, the pilot remains behind on the ground and flies the UAS from there. UAS have been used for military operations to collect aerial imagery for reconnaissance as well as being weaponized for both imaging and combat strikes. Military UAS range from smaller drones such as the Predator to larger ones such as the RQ-4 Global Hawk (which is planned to eventually replace the U-2 for aerial imagery reconnaissance—see **Figure 9.6**).

UAS have increasingly branched out from military uses to civilian use as well. Digital cameras and instruments for remote sensing data collection to UAS are very useful for a variety of civilian applications. Police use UAS for collecting imagery to aid in fighting crime, border patrols and port security use UAS for monitoring and security, and farmers utilize UAS for crop and irrigation monitoring. The equipment onboard UAS collect remotely sensed aerial imagery and video, and UAS can be launched and used on demand as the user needs it. Today, UAS are becoming more affordable and can be controlled via an app on your phone. Recreational use of aerial imagery from UAS are becoming

FIGURE 9.6 An RQ-4 Global Hawk Unmanned Aircraft System. *[Source: U.S. Air Force photo/Stacey Knott]*

common; researchers and universities own UAS and fly them for imagery data collection to aid in projects (see **Figure 9.7**). You can even get a bachelor's degree in aeronautics with a major in UAS Operations from the University of North Dakota.

With this growing proliferation of UAS, there are a number of rules governing who can fly UAS, what kind of permits are required, and where they

FIGURE 9.7 A civilian drone used for collecting imagery for research. *[Source: Zack Borst]*

can be flown. The U.S. Federal Aviation Administration (FAA) regulates the use of UAS for civil use, and there are a variety of places where you're not permitted to fly UAS (see *Hands-On Application 9.2: No-Fly Zones for UAS* for more information, as well as some UAS videos). Whether from a UAS or a regular piloted aircraft, photography is used extensively for many different kinds of imagery data collection.

HANDS-ON APPLICATION 9.2

No-Fly Zones for UAS

There are several locations around the United States where it is illegal to fly a UAS, such as national parks, military reservations, and within five miles of an airport. To spatially examine what areas are off-limits to flying UAS, go to **http:// arcg.is/1EQu5XH**. This is an Esri story map (see Chapter 15) that has these restricted areas for UAS flights mapped out. The large panels on the left side of the story map display various features on the map (such as the boundaries of national parks or military areas, as well as buffer zones around airports). The

fourth panel will allow you to view a variety of different geotagged videos captured from UAS.

Expansion Questions

* Use the search feature (the magnifying glass) to look at the no-fly zones for your local area. What specific places in your own area are restricted from flying UAS?

* From viewing some of the drone videos, what kinds of applications can videos captured from UAS be used for?

THINKING CRITICALLY
WITH GEOSPATIAL TECHNOLOGY 9.1

How Can UAS Be Used for Security Purposes?

UAS are not limited to military operations. Aircraft that can be piloted via remote control and equipped with multiple types of aerial surveillance equipment have a variety of other uses. For example, a May 2013 article (available online at **www.newswise .com/articles/university-police-to-develop-uavs -for-campus-security**) indicates that the University of Alabama Huntsville is looking into using UAS drones to aid in campus security, while an October 2014 article in the Guardian (available online at **www.theguardian.com/world/2014/oct/01/drones -police-force-crime-uavs-north-dakota**) noted the

advances in crime fighting that have been made in the city of Grand Forks, North Dakota, through their fleet of police UAS. How else can UAS be utilized for functions such as border security or law enforcement? How could police or federal agents utilize UAS capabilities for gathering intelligence about dangerous situations or acquiring reconnaissance before entering an area? Also, if UAS are being used to remotely monitor civilian conditions, what potential is there for abuse of these methods of data collection? Is UAS surveillance in a domestic urban environment an invasion of privacy?

What Are the Different Types of Aerial Photos?

When you're flying in a plane and you look out the window, you're looking down on the landscape below. If you had a camera strapped on the bottom of the plane and took a picture straight down from under the plane, you'd be capturing a **vertical photo**. Take a look back at Figure 9.4—the Cuban landscape as seen by the U-2 is photographed as if you were looking down from the plane. Many aerial photos are vertical (although there may be some minor tilting from the camera). The spot on the ground directly beneath the camera is referred to as the **nadir**.

Aerial photos can be **panchromatic** or color. Panchromatic imagery only captures the visible portion of light in its entirety. As a result, panchromatic aerial photos will be grayscale (that is, composed from a range of gray color values that exist between black and white—like the Cuban aerial photo). Color imagery captures the three main bands of visible light—red, green, and blue—and the colors are composited together in the digital imagery (or film) as a true color composite (we'll discuss much more about light and composites in Chapter 10).

A color infrared or **CIR photo** is a distinctive type of aerial color photo. A CIR image is captured with a special type of film that's sensitive to infrared light. Infrared energy is invisible to the naked eye, but special film processes can capture it. To enable us to see infrared energy reflection, a special filter is applied to the film—the end result is a CIR photo, wherein near-infrared (NIR) reflection is displayed in the color red, red reflection is displayed in the color green, and green reflection is displayed in the color blue. Blue reflection is not shown in the image, as it is blocked by the filter and displayed as black (**Figure 9.8**).

vertical photo an aerial photo in which the camera is looking directly down at a landscape

nadir the location on the ground that lies directly below the camera in aerial photography

panchromatic black-and-white aerial imagery

CIR photo color infrared photo—a photo where infrared reflection is shown in shades of red, red reflection is shown in shades of green, and green reflection is shown in shades of blue

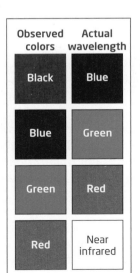

FIGURE 9.8 How the actual reflection of blue, green, red, and NIR light appears in its corresponding colors in a CIR image.

Observed colors	Actual wavelength
Black	Blue
Blue	Green
Green	Red
Red	Near infrared

FIGURE 9.9 A CIR aerial image of Burley, Idaho.
[Source: NASA Airborne Science Program image acquired by NASA ER-2 aircraft 26 August 1983]

CIR imagery can be very useful in environmental studies. For instance, the relative health of areas of vegetation can be observed very clearly in CIR imagery. Examine the CIR aerial photo of Burley, Idaho, in **Figure 9.9**. The bright red sections of the image are indicative of areas with a high amount of NIR reflection (since NIR reflection is being shown with the color red on the image)—probably grass, trees, or healthy agricultural fields (all elements that would reflect large amounts of NIR light).

The river appears black since water absorbs all types of light, except the color blue (which is then blocked in the CIR photo and shows up as black). Developed areas appear as a white or cyan color, as do fields of bare earth that have either been harvested and plowed over, or have been prepared for crops but not yet sown with seed. For more about examining different types of aerial photos, see *Hands-On Application 9.3: Examining CIR Photos* and *Hands-On Application 9.4: The National Aerial Photography Program.*

No matter what type of aerial photo is being used, the big problem is that the photo cannot be used as a map. Even though it shows all the features you might need (like roads or footprints of buildings) you can't rely on the photo to be a map for one reason: the photo doesn't have the same scale at every spot on the image. As a plane flies over the surface, some terrain

HANDS-ON APPLICATION 9.3

Examining CIR Photos

To look at the differences between color and colorinfrared images, check out the Map West Virginia viewer (also used back in *Hands-on Application 1.3*) at: **www.mapwv.gov/viewer/index .html**. Select the Layers tab and you'll see there are several basemaps that use aerial photos, including recent NAIP color imagery, and an option for color infrared (CIR) photos from 1996. Under the Search tab, type in the address of West Virginia University's downtown campus: 1550 University Avenue, Morgantown, WV 26506, and then search for it. The Viewer will zoom you to the WVU campus area. Back under Layers, select the option for 1996 CIR Imagery and the basemap will change to color infrared imagery. Examine the campus and then select the option for 2009 NAIP imagery (this is color aerial imagery). Zoom around the campus and switch back and forth between the CIR and regular color aerial imagery to see the differences between the two types of aerial photos.

Expansion Questions

- From examining the downtown campus area and its surroundings, how are the following items represented in CIR imagery: buildings, rivers, grass, and fields? Why do areas in each of these categories appear in these colors in CIR imagery, as opposed to regular color imagery? You will also see some areas in green—what do these areas represent?

- Move north from the downtown area to the Evansdale and Health Sciences section of the campus, where you'll see the football stadium (Milan Pusker Stadium) and other outdoor sports areas. Examine the football stadium and its surrounding areas. Why do the grass fields around the stadium and the football field itself appear so differently in the CIR imagery?

features will be closer to the camera and some will be further away—and thus, some areas will have a larger or smaller scale than others. The center of the photo is referred to as its **principal point**. In aerial photos, you'll see tall objects (such as terrain relief, towers, or buildings) have a tendency to "lean" away from this center point toward the edges of the photo—this effect is called **relief displacement**. See **Figure 9.10** on page 321 for an example of this—if you look at the area labeled "Principal Point," it's like looking straight down at that area. However, if you look outward from the principal point, you'll see that the tall buildings seem to be "leaning" away from the center of the photo (rather than looking straight down on all the buildings in the image).

Regular aerial photos do not have uniform scale. **Orthophotos**, however, do. To create an orthophoto, a process called **orthorectification** is performed on a regular aerial photo. This process removes the effects of terrain and relief displacement and enforces the same scale across the whole photo. A regular aerial photo can't be used as a map because it doesn't have uniform scale everywhere, but an orthophoto can. Even with uniform scale, some of the side appearances of objects can still be seen because when the photo was

principal point the center point of an aerial photo

relief displacement the effect seen in aerial imagery where tall items appear to "bend" outward from the photo's center toward the edges

orthophoto an aerial photo with uniform scale

orthorectification a process used on aerial photos to remove the effects of relief displacement and give the image uniform scale

HANDS-ON APPLICATION 9.4

The National Aerial Photography Program

The National Aerial Photography Program (NAPP) was a federal program that operated between 1987 and 2007. It was sponsored by several agencies, including the U.S. Department of Agriculture and the U.S. Department of the Interior. NAPP's goal was to provide regular aerial photography of the United States. NAPP imagery consisted of 1-meter-resolution photography and the program produced over 1.3 million aerial photos in twenty years. NAPP photos are now available online via the USGS's Earth Explorer utility. Open your Web browser and go to **http://earthexplorer.usgs.gov** to begin (make sure your computer can accept cookies for the Website to load properly).

Using Earth Explorer is a four-step process. First, enter your search criteria (such as a place name—for example, try Las Vegas, NV). Next, select the Data Sets tab and choose the datasets you want to search (in this case, expand the Aerial Photography option and select NAPP). In the third option, Additional Criteria, you can add other information to your search request to help narrow it down if you need to. Lastly, click on Results to see what NAPP imagery is available for the place you've searched for.

Click on the "footprint" symbol for each photo to see the highlighted outline of the section of Earth's surface the photo represents. Also, click on the Show Browse Overlay option (the button next to the footprint) to see the NAPP image overlaid on the map, so that you can zoom in on the image. A thumbnail preview of the images will be available—click on an image to view it full size. It should be noted that the USGS requires users to create an account and log in before they can access download options for the photos.

Expansion Questions

- Use EarthExplorer to examine NAPP imagery of the Empire State Building by searching for this address: 350 5th Avenue, New York, NY 10118. Examine the NAPP images associated with this location. What kinds of images are available (panchromatic, color, or CIR), and from what image dates?

- Use EarthExplorer to see what NAPP imagery is available for your local area or home address, and then examine the results. What kinds of aerial photos are available, and from what image dates?

---

true orthophoto an orthophoto where all objects look as if they're being seen from directly above

DOQ a Digital Orthophoto Quad— orthophotos that cover an area of 3.75 minutes of latitude by 3.75 minutes of longitude, or one-fourth of a 7.5 minute USGS quad

taken, some of the sides of the building may have been hidden due to relief displacement and not visible in the photo. This effect is removed in a **true orthophoto**, which gives the appearance of looking directly straight down on all objects in the image. This is achieved by using several images of the area for filling in the missing information while keeping constant scale in the final photo.

Orthophotos are used to create special products called Digital Orthophoto Quads, or DOQs. A **DOQ** is orthophotography that covers 3.75 minutes of latitude and 3.75 minutes of longitude of the ground, the same area as one-fourth of a 7.5 minute USGS topographic map (see Chapter 13 for more about topographic maps). DOQs (sometimes also referred to as DOQQs, or Digital Orthophoto Quarter Quads, as they cover one quarter of the size of a topographic map) have been georeferenced so that they can

Principal Point

FIGURE 9.10 The effect of relief displacement in an aerial photo. *[Source: © Johannes Mann/Corbis]*

be matched up with other geospatial datasets (see Chapter 3 for more information on georeferencing). Also note that the USGS produces certain other DOQ products that cover the area of the entire quad.

A similar resource is produced by the National Agriculture Imagery Program (**NAIP**), which is administered through the United States Department of Agriculture's Farm Service Agency. NAIP images are aerial photos of the United States that are taken on a regular basis, processed into DOQs (and also as mosaicked images of whole counties), and made publicly available. NAIP began in 2003 and continues to produce orthophotographic images today (see *Hands-On Application 9.5: NAIP Imagery Online* for a USDA FSA utility for viewing NAIP imagery and their DOQ boundaries).

In an **oblique photo**, the camera is tilted so that it's not positioned directly at the nadir, but rather at an angle. Take another look at Figure 9.2— the image of San Francisco taken from the kites—for an example of an oblique photo. Unlike a vertical photo, you can see the sides of some of the buildings, and you're really seeing the city from a perspective view instead of an overhead one. Oblique images allow for very different views of a target than those taken from straight overhead and can capture other types of information (such as the sides of buildings or mountains). By examining different oblique photos of a single building, law enforcement agencies can obtain detailed information (the locations of entrances, exits, windows,

NAIP the National Agriculture Imagery Program, which produces orthophotos of the United States

oblique photo an aerial photo taken at an angle

HANDS-ON APPLICATION 9.5

NAIP Imagery Online

The USDA maintains a GIS data viewer online for viewing NAIP imagery at **http://gis.apfo.usda.gov/gisviewer**. Available layers are on the right side of the map; these include DOQQ boundaries (for available states, covering multiple years), the film centers of the DOQQs (for some states), and other geographic data. By default, the NAIP imagery itself will be turned on. Under the Find a Place option, you can search for a state and city (such as San Francisco, CA). The results of the search will be shown in a separate pane. Note that you can also search by county or by DOQQ number. By right-clicking on the name of place you can select an option to Zoom To that location. You can

zoom in closer to view the NAIP imagery, and by turning on the NAIP QQ Coverage layers, you can see the boundaries of each DOQQ covered by NAIP imagery.

Expansion Questions

- In San Francisco, which DOQQ numbers would you have to obtain to have full 2012 NAIP coverage of the entire city?

- Which DOQQ numbers would you have to obtain to have full 2012 NAIP coverage of the entire city of East Lansing, MI?

- DOQQ numbers 3609463NE and 3609463SE contain NAIP imagery of what city?

and so on), which often prove vital information in deciding whether to send officers into a potentially dangerous situation. Companies such as Pictometry make extensive collections of high-resolution oblique aerial imagery to sell commercially. You can view oblique images of several locations online using Microsoft's free Bing Maps utility, if its "Bird's Eye" imagery is available (see **Figure 9.11** and *Hands-On Application 9.6: Oblique Imagery on Bing Maps* for examples of oblique images created on Bing Maps).

FIGURE 9.11 Oblique imagery of the Lincoln Memorial, Washington, D.C., as shown in Bing Maps' "Bird's Eye view" feature. *[Source: Pictometry Bird's Eye/Pictometry International Corp/NAVTEQ/Microsoft Corporation]*

HANDS-ON APPLICATION 9.6

Oblique Imagery on Bing Maps

Bing Maps is a great source of online oblique imagery supplied via Pictometry. Like Google Earth and Google Maps, Bing Maps allows you to view aerial or satellite images of areas, but it also gives you the option to view high-resolution oblique images of areas (where oblique imagery is available). To start, go to **www.bing.com/maps** and type the name of a location (try Atlantic City, NJ). In the Atlantic City map, zoom in until you can see the ocean and boardwalk. Select the option for "Bird's Eye." The view will switch to oblique photography of Atlantic City. You can zoom in and out and pan the imagery (when you move past the edge of a photo, a new tile of oblique imagery should load).

Reposition the images so that you can see the boardwalk and oceanfront at an oblique angle, and then move down the boardwalk to see the casinos, piers, and developments. In addition, Bing Maps provides oblique imagery from multiple angles—click on one of the arrows pointing downward to the left and right of the compass rose (around the letter "N") to rotate the direction and load oblique imagery of the same location but shot from a different angle. You can use this function to examine a spot from multiple viewpoints. Once you get used to navigating and examining oblique images of the New Jersey shore, search for your local area with Bing Maps to see if Bird's Eye imagery is available to view your home, school, or workplace.

Expansion Questions

- What advantages are there for examining the Atlantic City boardwalk from oblique views rather than the usual overhead view? What specific features of the boardwalk can be seen in the oblique images that cannot be seen in from a vertical perspective?

- If Bird's Eye imagery of your local area is available, examine it using Bing Maps. What specific features of your local area can be seen in the oblique images that cannot be seen from an overhead view?

How Can You Interpret Objects in an Aerial Image?

Understanding what you're looking at when you're viewing Earth from the sky instead of from ground level takes a completely new skill set. Objects that you're used to looking at head-on have a whole different appearance from directly above. For example, **Figure 9.12** on page 324 shows two different views of the Stratosphere Hotel and Casino in Las Vegas, Nevada—one showing an overhead aerial view and the other an oblique photograph showing details of the tower and its buildings. Obviously, the complex looks completely different from the oblique angle than from directly above. If you look closely, you can see where areas in the two photos match up, such as the tower on the right-hand side of the image and the remainder of the hotel and casino on the left. However, if you're only given the aerial image and asked what building this is (of all possible structures in the world), it'll be a whole different ballgame.

(a)   (b)

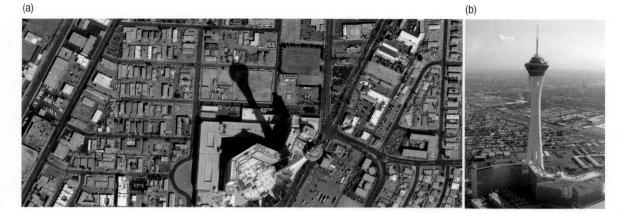

FIGURE 9.12 The Stratosphere Hotel and Casino in Las Vegas, Nevada: (a) an overhead aerial view of the building and (b) an oblique image showing the tower and the hotel layout. *[Source: (a) DigitalGlobe/Getty Images (b) Ethan Miller/AFP/Getty Images/Newscom]*

From the aerial image alone, you'll have to search for clues to determine what's really being shown. For instance, based on the size of the complex (roughly a city block) you'll probably guess that it's some sort of hotel, resort, entertainment attraction, or museum. By looking at its location amidst other large buildings and multiple-lane roads, you may guess that the building's in a big city. On closer examination, you might make out a different colored object (the light blue) among the main complex and determine that it's a swimming pool, which narrows the chances further on the building being some sort of hotel or resort.

Although the height of the tower on the right-hand side of the image isn't really discernible (since the image was taken looking down onto the tower itself), the huge shadow cast by the tower, stretching upward, is visible. Based on the length and shape of the shadow being cast, you may determine that the object casting it is a tower (or a similar very large, very tall object). Putting all of these clues together (city-block-sized hotel/resort in a big city with a massive tower in front of it), with the help of some outside information, like a few travel books and maybe a quick Web search, you should be able to determine pretty quickly that you're looking at the Stratosphere Hotel in Las Vegas.

When you try to interpret features in an aerial photo or a satellite image (for instance, objects in developed areas or physical features in a natural landscape), you act like a detective searching for clues in the image to figure out what you're really looking at. As in the Stratosphere Hotel example, clues like the size and shape of objects or the shadows they cast help to determine what you're really looking at. **Visual image interpretation** is the process of identifying objects and features in an aerial (or other remotely sensed) image based on a number of distinct elements:

▶ **Pattern:** This is the physical arrangement of objects in an image. How objects are ordered (or disordered) will often help to interpret an

visual image interpretation the process of examining information to identify objects in an aerial (or other remotely sensed) image

pattern the arrangement of objects in an image, used as an element of image interpretation

image. A large array of cars set up as if for inspection in a parking lot around a relatively small building will probably be a car dealership rather than some other type of shopping area. Evenly spaced rows of jet airplanes suggest a military base, while a haphazard arrangement of aircraft could be a group of statics on display at an aircraft or military museum.

▶ **Site and association:** Site deals with the location characteristics of an item, while association relates an object in an image to other nearby features in the image. For example, a football field itself has enough distinctive features to identify it, but the related phenomena you could see in the image (the number of bleachers, the number and distribution of the seats, and perhaps the amount of nearby parking) would help in determining if you're looking at a field used by a high school team, a 1-AA college team, or a professional NFL team.

site and association the information referring the location of objects and their related attributes in an image, used as elements of image interpretation

▶ **Size:** This describes the length, width, and area on the ground of objects in the image. The relative size of objects in an image can offer good clues in visual image interpretation. For instance, the average length of a car is about 15 feet. If a car is present in an image, you can gain information about other objects by comparing their length with that of the car—you can tell, for instance, if a structure near the car is the size of a house or a shopping center. The sizes of some features remain the same, whatever the image. If baseball diamonds or football fields are present in an image, elements in them (the 90 feet between bases, or the 100 yards between goal lines) can always be compared with other objects in the image to calculate their relative sizes.

size the physical dimensions (length, width, and area on the ground) of objects, used as an element of image interpretation

▶ **Shadow:** This is the dark shape cast by an object with a source of light shining on it. Shadows help to provide information about the height or depth of the objects that are casting them. For instance, in the Stratosphere Hotel example, the height of the tower itself was partially hidden due to the nature of being photographed from directly above, but we can figure out that it must be very tall because of the shadow it casts. Shadows can help us to identify objects that are virtually unidentifiable from a viewpoint directly above them. From directly above, a set of railings looks like a thin, almost invisible thread around a park or in front of a building, and telephone poles look like dotted lines. However, if you look down at them from directly above late on a sunny afternoon, railings and telephone poles can be identified immediately by their shadows.

shadow the dark shapes in an image caused by a light source shining on an object, used as an element of image interpretation

▶ **Shape:** This is the particular form of an object in an image. The distinctive shapes of objects in an aerial photo can help us to identify them. A baseball field has a traditional diamond shape, a race track has a distinctive oval shape (and even an abandoned horse race track may still show evidence of the oval shape on the landscape), and the circular shape of crops may indicate the presence of center-pivot irrigation in fields.

shape the distinctive form of an object, used as an element of image interpretation

texture repeated shadings or colors in an image, used as an element of image interpretation

▶ **Texture:** This refers to the differences of a certain shading or color throughout parts of the image. The texture of objects can be identified as coarse or smooth. For instance, different types of greenery can be quickly distinguished by their texture—a forest of trees and a field of grass may have the same tone, but the trees appear very rough in an image, while grass will look very smooth. The texture of a calm lake will look very different from the rocky beach surrounding it.

tone the grayscale levels (from black to white), or range of intensity of a particular color discerned as a characteristic of particular features present in an image, used as an element of image interpretation

▶ **Tone:** This is the particular grayscale (black to white) or intensity of color in an image. The tone of an object can convey important information about its identity. In the Stratosphere example, the light-blue color of the swimming pool made it easy to identify. Similarly, a wooden walkway extending into a sandy beach will have different tones that help in distinguishing them from one another in a photo.

These same elements can be used to identify items in satellite images as well. **Figure 9.13** is an overhead view of an area containing several objects that all add up to represent one thing. We'll apply these elements to this image to determine what is being shown.

▶ Pattern: The three main objects are arranged in a definite diagonal line.

▶ Texture: The area around the three objects is of very fine texture, compared with some of the rougher textures nearby, especially in the east.

FIGURE 9.13 An overhead view of an area containing several different objects. *[Source: Space Imaging Europe/Science Source]*

▶ Tone: The area around the three objects is much lighter than the area of rougher texture (and that area's probably a town or city of some sort, given the pattern of the objects in this range). Given the tone and texture of the ground around the three objects, it's likely not smooth water or grass or concrete, so sand is a likely choice.

▶ Site and association: The three objects are located in a large sandy area, adjacent to a city, with what resembles roads leading to the objects.

▶ Size: Comparing the size of the objects to the size of the buildings indicates that the objects are very large—even the smallest of the three is larger than many buildings, while the largest is greater in size than blocks of the city.

▶ Shadow: The shadows cast by the objects indicate that not only are they tall, but they have a distinctive pointed shape at their top.

▶ Shape: The objects are square at the base, but have pointed, triangular tops, making them pyramid shaped.

Bringing all of these elements of interpretation together, we can find three pyramids of different sizes (although still very large), in a definite fixed arrangement, in sand near a busy city. All of these clues combine to identify the Great Pyramids at Giza, just outside Cairo, Egypt. Aerial images often contain clues that help to identify the objects they contain, or at least narrow down the options to a handful. Some brief research among collateral material (like groupings of large desert pyramids in that particular arrangement) should nail the identification.

Visual image interpretation skills enable the viewer to discern vital information about the objects in a remotely sensed image. The elements of interpretation can be used to identify, for instance, the specific type of fighter jet on a runway based on the size and shape of its features, wingspan, engines, and armaments. A forestry expert examining a stand of trees in a remotely sensed image might use visual image interpretation to identify the species of trees based on their features.

How Can You Make Measurements from an Aerial Photo?

Once identification of objects in an image has been positively made, these objects can be used for making various types of measurements. **Photogrammetry** is the process of obtaining measurements from aerial photos. Photogrammetric techniques can be used for determining things like the height and depth of objects in an aerial photo. For instance, the heights of buildings and the lengths of visible features in a photo can be calculated. There are a lot of possible photogrammetric measurements—the following are two simple examples of how photogrammetric measurements can be made.

> **photogrammetry**
> the process of making measurements using aerial photos

photo scale the representation used to determine how many units of measurement in the real world are equivalent to one unit of measurement on an aerial photo

As every map has a scale (discussed in Chapter 7), so every photo has a **photo scale,** which is listed as a representative fraction. For example, in a 1:8000 scale aerial photo, one unit of measurement in the photo is equivalent to 8000 of those units in the real world. Using the photo scale, you can determine the real-world size of features. Say you measure a section of railroad track in an aerial photo as a half-inch, and the photo scale is 1:8000. Thus, 1 inch in the photo is 8000 inches in the real world, and so a measurement of 0.5 inches in the photo will be 4000 inches (0.5 times 8000), or 333.33 feet.

The scale of a photo relies on the focal length of the camera's lens and the altitude of the plane when the image is captured. The problems start when you try to make measurements from an orthophoto if you don't know the photo's scale. If you don't know how many real-world units are equal to one aerial photo unit, you can't make accurate measurements. The photo scale of a vertical photo taken over level terrain can be determined by using a secondary source that has a known scale, as long as an item is visible in both the photo and the secondary source and as long as you can measure that item in both. A topographic map (see Chapter 13) with a known scale will make a good secondary source because it will contain many of the features that will also be clearly seen in an aerial photo (like a road, or a section of road, where the beginning and end are visible). When you make these types of measurements, it's important to remember that a regular aerial photo will not have the same scale everywhere in the photo, whereas an orthophoto will have uniform scale throughout the image.

By being able to make the same measurement on the map (where the scale is known) and on the aerial photo (where the scale is unknown), you can determine the photo scale. This is the way it works: Your photo could be 1:6000 or 1:12000 or 1:"some number." Just like the representative fraction (RF) discussed in Chapter 7, the photo scale can be written as a fraction—1:24000 can be written as 1/24000. So, assume your unknown photo scale is the RF—this is equal to the distance measured on the photo, known as the photo distance (PD), divided by the real-world distance measured on the ground, known as the ground distance (GD), or:

$$RF = \frac{PD}{GD}$$

Say you have a 1:12000 scale map showing an oceanfront boardwalk, and you also have an orthophoto of unknown scale of the same area, showing the same features. For the photo to be useful, you have to determine its scale. You can find the same section of boardwalk in both the map and the photo. By measuring the boardwalk section on the map, you find it is 0.59 inches. However, that's not the ground distance, or how long that section of the boardwalk is in the real world—because of the map scale, every one inch measured on the map translates into 12,000 inches in the real world. So, 0.59 inches on the map is actually 7080 inches in the real world. This measure is the GD variable in the equation above (the ground distance):

$$RF = \frac{PD}{7080 \text{ in}}$$

You can measure the same section of boardwalk on the photo and find that it's 1.77 inches. This measurement is the PD variable in the equation (the photo distance):

$$RF = \frac{1.77 \text{ in}}{7080 \text{ in}}$$

Doing some quick division, you find that RF is equal to 1/4000.

$$RF = \frac{1}{4000}$$

So the photo scale is 1:4000. One unit measured on the photo is equal to 4000 units in the real world.

A second type of measurement that you can make from the elements found in an aerial photo is to calculate the height of an object in the photo simply by examining its shadow. At first blush, you might think you'll need to know all sorts of other information—where the photo was taken, what time of day it was taken, the date on which it was taken—all variables related to the relative location of the Sun and how the shadows would be cast. Chances are you won't be able to get your hands on a lot of this type of information easily, so you're probably thinking there must be a better way to do this.

You'd be right—photogrammetric measurements give you a simple way of determining the heights of objects in a photo from their shadows, without needing all that other data. They rely on three things: (1) knowing the scale of the photo (which we just figured out); (2) being able to see, fully and on level ground, the shadows of all objects (from the top of each object) whose height you want to measure; and (3) already knowing the height of one object with a shadow you can measure. If you have all three, measuring heights is a snap.

Let's take that hypothetical 1:4000 photo from the last example and assume that it's got a number of large hotels casting shadows on the boardwalk. You know the height of one building (115 feet), and its shadow in the photo (from the base to the top) is 0.10 inches. You can use this information to calculate the angle of the Sun, which is casting the shadows in the photo, as follows:

$$\tan a = \frac{h}{L}$$

In this equation, a is the angle of the Sun, h is the real-world height of the object, and L is the real-world length of the shadow. From basic trigonometry, the tangent of a right angle ($\tan a$) is equal to its opposite value (h)

FIGURE 9.14 The relationship between the height of an object (*h*), the length of its shadow (*L*), and the angle of the Sun's rays.

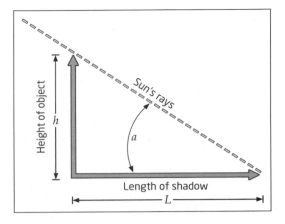

divided by its adjacent value (*L*). See **Figure 9.14** for a diagram that shows how this works. You already know the height, *h* (115 feet). The length, *L*, can be found by taking the length of the shadow measured in the photo (0.10 inches) and multiplying it by the photo scale (1:4000). 0.10 inches on the photo is 400 inches in the real world, or 33.33 feet. This means that if you want to check your calculations by getting into a time machine and traveling back to the boardwalk on the day the aerial photo was taken, you'll find the shadow of the building will be 33.33 feet long. Plugging these numbers into the formula, we find that:

$$\tan a = \frac{115 \text{ ft}}{33.33 \text{ ft}}$$

and that the tangent of angle *a* (or "tan *a*") is equal to 3.45.

Now, since an aerial photo represents a single snapshot in time, we can assume that the angle of the Sun is going to remain the same across the photo for all of the buildings casting shadows. So, the measure for "tan *a*" will be the same value when applied to all heights we're trying to determine. We can use this information to calculate the height of any object in the photo casting a shadow that we can measure. Say you measure the shadow cast by a second building in the photo on the boardwalk and find it to be 0.14 inches. The 1:4000 photo scale tells us that the length of this shadow in the real world will be 560 inches, or 46.67 feet. The only thing that's unknown now is the actual height of the building. Using the equation:

$$3.45 = \frac{h}{46.67 \text{ ft}}$$

and solving for *h*, we can find that the height of the new building is about 161 feet. This type of calculation can be applied to other objects casting shadows, meaning we can quickly determine the height of multiple objects in a single photo. Be cautious, however, when you employ these techniques—to determine height from shadows, the object must be straight up and down, and it must cast a full shadow on level ground.

Chapter Wrapup

Aerial photography has a wide variety of uses and is an integral part of geospatial technology. Aerial photos are used for interpretation, photogrammetry, and as additional data sources for other aspects of geospatial technology, such as GIS. Despite its myriad uses and applications, aerial photography is just one part of remote sensing. The next chapter will delve into how the whole remote sensing process actually works. First, however, see *UAS Apps* for some mobile apps related to this chapter's concepts, as well as *Aerial Imagery in Social Media* for several blogs, Facebook pages, Twitter accounts, and YouTube videos related to aerial photography.

This chapter's lab will put you to work with several remotely sensed images and help you to apply elements of visual image interpretation to items in the images.

Important note: The references for this chapter are part of the online companion for this book and can be found at **www.macmillanhighered .com/shellito/catalog**.

UAS Apps

Here's a sampling of available representative UAS apps for your phone or tablet. Note that some apps are for Android, some are for Apple iOS, and some may be available for both.

- **FreeFlight:** An app that allows you to control a special type of UAS from your phone
- **iDrone Control:** An app that allows your mobile device to interact with specific types of drones
- **UAS RAVEN Mobile:** An app that allows you to control a specific type of UAS from your mobile device

Aerial Imagery in Social Media

Here's a sampling of some representative Facebook, Twitter, and Instagram accounts, along with some YouTube videos and blogs related to this chapter's concepts.

On Facebook, check out the following:

- **Bluesky** (aerial remote sensing company): **www.facebook.com/BlueskyWorld**

- **Old Aerial Photos:**
  www.facebook.com/OldAerialPhotos
- **Pictometry** (bird's eye imagery):
  www.facebook.com/PictometryInternational

Become a Twitter follower of:

- **Bluesky** (aerial remote sensing company): @Bluesky_int
- **Flightglobal UAVs** (news about UAVs): @FG_UAVs
- **FLYTMedia** (a company using UAS for video production): @FLYTmedia
- **Pictometry** (bird's eye imagery): @Pictometry
- **senseFly** (a UAS company): @sensefly
- **The Best Drone Info** (news about UAS): @BestDroneInfo
- **UAS.Technology** (news about UAS): @UAS_Technology

Become an Instagram follower of:

- **FLYT Media:** @_flyt_

On YouTube, watch the following videos:

- **Pictometry** (this is the YouTube channel for Pictometry, showcasing high resolution aerial photography applications):
  www.youtube.com/user/Pictometry
- **UAS—Transforming the Horizon** (a NASA video about UAS):
  www.youtube.com/watch?v=XMr9W6WYWcc

For further up-to-date info, read up on these blogs:

- **EagleView Blog** (a blog from EagleView technologies and Pictometry):
  http://blog.eagleview.com
- **Plane-ly Spoken** (a blog about aviation regulations, including UAS):
  www.planelyspokenblog.com

Key Terms

aerial photography (p. 310)
CIR photo (p. 317)
DOQ (p. 320)
nadir (p. 317)
NAIP (p. 321)
oblique photo (p. 321)
orthophoto (p. 319)
orthorectification (p. 319)
panchromatic (p. 317)
pattern (p. 324)
photo scale (p. 328)
photogrammetry (p. 327)
principal point (p. 319)

relief displacement (p. 319)
remote sensing (p. 309)
shadow (p. 325)
shape (p. 325)
site and association (p. 325)
size (p. 325)
texture (p. 326)
tone (p. 326)
true orthophoto (p. 320)
UAS (p. 314)
vertical photo (p. 317)
visual image interpretation
   (p. 324)

Visual Imagery Interpretation

This chapter's lab will get you started thinking about what objects look like from the sky rather than the ground. You'll be examining a series of images and applying the elements of visual image interpretation we discussed in the chapter—think of it like being a detective and searching the image for clues to help you figure out just what it is you're looking at. You should be able to figure out exactly what you're looking at, or at least narrow down the possibilities to a short list before you turn to support data for help.

Although the items you'll be examining will all be large and prominent—buildings or other very visible features—the application of visual image interpretation elements for these simple (and hopefully, fun) examples will help get you started with looking at the world from above.

Objectives

The goals for you to take away from this exercise are:

▶ To think of how objects look from an aerial perspective

▶ To apply the elements of visual image interpretation (as described in the chapter) in order to identify objects in the images

Using Geospatial Technologies

The concepts you'll be working with in this lab are used in a variety of real-world applications, including:

▶ Military intelligence, which relies on knowing the locations of items on the ground—interpretation of objects in aerial reconnaissance photographs is often a key component in planning military operations

▶ Forest resource management, which utilizes aerial photo interpretation to monitor the health and development of different types of trees by careful observation of foliage and the height and density of the tree canopy

[Source: © Agencja Fotograficzna Caro/Alamy]

Obtaining Software

There is no special software used in this lab, aside from whatever program your computer uses to view graphics and images. You may find Google Earth useful during the lab, however.

Lab Data

Copy the folder Chapter9—it contains a series of JPEG images (.jpg files) in which you'll find items that you'll be attempting to interpret. These are numbered, and tend to increase in difficulty as the numbers go up.

Localizing This Lab

The images used in this lab were taken from a variety of locations around the United States. You can use Google Earth, Google Maps, or Bing Maps to locate good overhead views of nearby areas (showing prominent developed or physical features) to build a dataset of local imagery that you can use for visual image interpretation.

9.1 Applying Elements of Visual Image Interpretation

Take a look at each of the images (there are questions related to each further down) and try to determine just what it is you're looking at. Although everyone's application of the elements may vary, there are some guidelines that you may find useful.

Several images may contain multiple items, but all of them work together to help define one key solution.

1.   We'll start with a sample image (a digital copy of this sample image is also in the Chapter9 folder labeled "sample"):

[Source: Bing Map/ Microsoft Corporation/Esri]

2.   First, look at the image as a whole for items to identify:

   a.   The central object is obviously some sort of structure—and a large one at that, judging from the shadow being cast. The relative size of the structure (when compared with the size of several of the cars visible throughout the image) suggests that it's probably a building and not a monument of some kind.

   b.   It's set on a body of water (the tone of the water is different from the tone of the nearby concrete or greenery). Thus, the site can be fixed.

   c.   One of the most notable features is the large triangular shape of the main portion of the building—note too that its texture is very different from the rest of the building (the three white sections).

   d.   The shape of the entranceway is also very distinctive, being large and round, with a couple of concentric circles.

3.   Second, take the interpretation of the initial items and start looking for specifics, or relationships between the items in the image:

   a.   There's something about the pattern of those concentric circles at the entrance plaza that's striking—two outer rings and a third inner ring with some sort of a design placed in it. Judging from the relative size of a person (pick out shadows near the plaza or on the walkway along the water), the plaza is fairly large. The pattern sort of resembles a giant vinyl record album (and the concept solidifies further with the shape of the large curved feature that follows around the plaza on its right—that looks like the arm of an old record player).

   b.   The texture of the triangular portions of the building suggests that they could be transparent, as if that whole portion of the structure were made of glass.

   c.   There's no parking lot associated with the building, so you'd have to park somewhere else and walk there. Further, there are no major roads nearby, again indicating that this is not some sort of destination that you just drive up to and park in front. The large body of water gives us a clue that this building is in a large city.

4.   Third, put the clues together and come up with an idea of what the object or scene could be:

   a.   The building's distinctive enough to be a monument, but probably too large to be one. It could be some sort of museum, casino, shopping center, or other attraction. However, the building's too small to be a casino or resort center, and lacks the necessary parking to be a casino, office, or shopping center.

   b.   The very distinct "record" motif of the front plaza seems to indicate that music (and more specifically older music, given the whole "vinyl record" styling) plays a large part in whatever the building is.

   c.   From the clues in the image, this could be some sort of music museum, like the Rock and Roll Hall of Fame, which is located in Cleveland, Ohio, right on the shore of Lake Erie.

5.   Fourth and final, use a source to verify your deduction, or to eliminate some potential choices.

   a.   By using Google Earth to search for the Rock and Roll Hall of Fame, the identification will be confirmed. An image search on Google will also turn up some non-aerial pictures of the site to verify your conclusion.

Not all images will be this extensive, and not all may make use of all elements—for instance, there may be one or two items in the image that will help you make a quick identification. However, there are enough clues in each image to figure out what you're looking at—or at least narrow the choices down far enough that some research using other sources may be able to lead to a positive identification.

For instance, by putting all of the clues together, you could start searching for large museums dedicated to music—add the triangular glass structures, and the Rock and Roll Hall of Fame will certainly come up.

9.2 Visual Image Interpretation

1.   To examine an image, open the Chapter9 folder and double-click on the appropriate image (.jpg) file. Whatever program you have on your computer for viewing images should open with the image displayed. You may find it helpful to zoom in on parts of an image as well.

2.   For each image, answer the questions presented below, then explain which of the elements of visual image interpretation led you to this

conclusion and how they helped you to decide. Writing "shadow and shape helped identify traits of the building" will be unacceptable. However, writing something that answers the question, "What was so special about the shadows in the scene or the shape of items that helped?" will be much better.

3. When you identify the items, be very specific (for instance, not just "a baseball stadium," but "Progressive Field" in Cleveland, Ohio). *Important note:* All images used in this exercise are from areas within the United States.

4. You may want to consult some outside sources for extra information, such as Websites, search engines, books, or maps. Lastly, when you finally think you have the image properly identified, you may want to use something like Google Earth or Bing Maps to obtain a view of the object in question to help verify your answer.

Question 9.1 Examine image1. There are several items in this image, but there's one that's more prominent than the others. What, specifically, is this item? What elements of visual image interpretation lead you to this conclusion?

Question 9.2 Examine image2. What (specifically) is being displayed in this image? What elements of visual image interpretation lead you to draw this conclusion?

Question 9.3 Examine image3. What (specifically) is being displayed in this image? What elements of visual image interpretation lead you to draw this conclusion?

Question 9.4 Examine image4. There are several items in this image, but there's one that's more prominent than the others. What, specifically, is this item? What elements of visual image interpretation lead you to draw this conclusion?

Question 9.5 Examine image5. There are many similar objects in this image, but you'll notice some differences too. What (specifically) is being displayed in this image, and what is its location? What elements of visual image interpretation lead you to draw this conclusion?

Question 9.6 Examine image6. There are several items in this image, but they all add up to one specific thing. What, specifically, is this image showing, and what is its location? What elements of visual image interpretation lead you to draw this conclusion?

Question 9.7 Examine image7. What (specifically) is being displayed in this image? What elements of visual image interpretation lead you to draw this conclusion?

Question 9.8 Examine image8. There are several items in this image, but they all add up to one specific thing. What, specifically, is this image showing, and what is its location? What elements of visual image interpretation lead you to draw this conclusion?

Question 9.9 Examine image9. What feature is prominent in the image? Be specific as to what the area is and what geographic location is being shown here. What elements of visual image interpretation lead you to this conclusion? What are all those white objects throughout the image (and how do they help identify the area)?

Question 9.10 Examine image10. What feature is prominent in the image? Be specific as to what the object is and what geographic location is being shown here. What elements of visual image interpretation lead you to this conclusion? Several airplanes are very clearly visible—how do they help identify the feature?

Closing Time

Now that you've gotten your feet wet with some visual image interpretation and have hopefully gotten used to viewing Earth from above, Chapter 10 will delve into satellite imagery, and its exercise will have you doing (among other things) some more interpretation tasks based on satellite capabilities.

Once you've identified all of the images and explained your application of the elements of image interpretation, you can close the Chapter9 folder and any other programs or resources you have open.

Electromagnetic Energy, the Remote Sensing Process, Spectral Reflectance, NDVI, Digital Imagery, and Color Composites

When you use Google Earth to get an overhead view of your house, Google had to get those images from somewhere, but a number of steps must be taken before that image of your house can be acquired and finally turned into the picture you can view on the computer screen. In Chapter 9, we discussed how what's really being captured via **remote sensing** is reflected light—this is a form of electromagnetic energy, and a sensor is measuring the amount of energy that reflects off a target on the ground. In addition, in remote sensing the sensor will be on an airborne or spaceborne platform that will be a considerable distance away from a target on the ground below. Some remote sensing devices measure other forms of energy emitted by objects on the ground (heat, or the thermal properties of items, for instance). This chapter, however, focuses on measuring the reflection of electromagnetic energy.

If remote sensing measures the reflectance of electromagnetic energy, there needs to be a source for that energy. Luckily, we have one nearby—the Sun. The Sun is almost 93 million miles away from Earth, and sunlight takes about 8.3 minutes to reach Earth. This solar energy radiates through space at the speed of light (almost 300 million meters per second), and passes through and interacts with Earth's atmosphere before it reaches Earth's surface. The energy then interacts with objects on Earth's surface, and some of the energy is reflected back into space. That reflected energy passes back through the atmosphere, and is picked up and recorded by a sensor, either on an airborne or spaceborne platform (**Figure 10.1**).

These measurements made at the sensor are being captured as remotely sensed data. Thus, the remotely sensed information you see on

remote sensing the process of collecting information related to the electromagnetic energy reflected or emitted by a target on the ground, using a device a considerable distance away from the target on board an aircraft or spacecraft

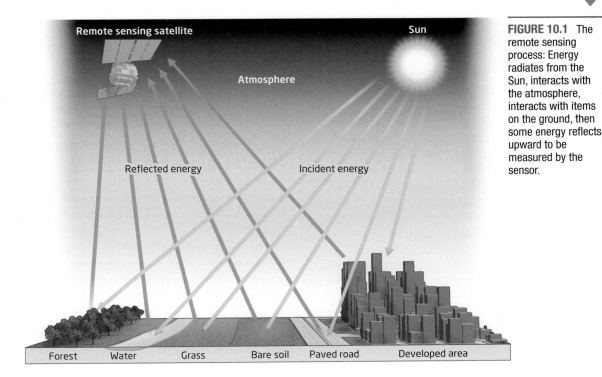

FIGURE 10.1 The remote sensing process: Energy radiates from the Sun, interacts with the atmosphere, interacts with items on the ground, then some energy reflects upward to be measured by the sensor.

things like Google Earth is actually the reflection of energy off targets on the ground that has been measured, processed, and turned into imagery to view. We'll go through each of these stages in more detail to see how you can start with energy from the Sun and end up with a crisp satellite image of a target on the ground (also see *Hands-On Application 10.1: Viewing Remotely Sensed Imagery Online* for some finished products of remote sensing).

Note that this process (and this chapter) describes passive remote sensing—where the sensor doesn't do anything but simply measure and record reflected or emitted energy. In active remote sensing, the sensor generates its own energy, casts it at a target, and then measures the re-bound, echo, or reflection of that energy. Radar is a good example of active remote sensing—a device on a plane throws radar waves (microwaves) at a target, those waves interact with the target, and the device measures the backscatter of the returning radar waves. Lidar (see Chapter 13) is a similar kind of active remote sensing technique that shoots a laser beam at a target and uses the beam's reflection to measure the heights of objects on the ground.

HANDS-ON APPLICATION 10.1

Viewing Remotely Sensed Imagery Online

An excellent tool for viewing different types of remotely sensed data in your Web browser (without using a separate program like Google Earth) is Flash Earth. As the name implies, make sure you have Flash installed on your computer, then go to **www.flashearth.com** to get started. When Flash Earth starts, you can pan around Earth and zoom in on an area (or search by location). To get started, look at the United States and search for New York City. Zoom in on Manhattan and examine what types of imagery become available from the Map source options—you can select from options such as Bing Maps, Bing Maps with Labels, HERE Maps, ArcGIS, and more. As you zoom in, new imagery will load for different scales. Conversely, by selecting some of the broader-scale remotely sensed options (such as NASA Terra or NASA Aqua) you'll be able to view daily imagery of whole swaths of a continent, rather than fine details.

Expansion Questions

- Examine some of the areas in Manhattan— what types of remotely sensed images are available at each option as you continue to zoom in on the city?

- When you're done looking around Manhattan, type your home location (or nearby local areas) into the Search box. What kinds of remotely sensed imagery can you view of your current location (and what kinds of details can be seen as you zoom in)?

What Is Remote Sensing Actually Sensing?

The information that is being sensed is the reflection of electromagnetic (EM) energy off a target. There are two ways of thinking of light energy: as a particle or as a wave. For our purposes, we'll stick to the concept of light energy as a wave. When the Sun radiates light energy, it radiates in the form of waves. Think of it like sitting on the beach and watching the ocean waves come in and crash on the shore. The distance between the crests of two waves is called the **wavelength** (λ)—some wavelengths are very long, and some are very short. If it's a calm day at the ocean, you see waves less frequently than on a rough day— there will be a longer distance between waves (and therefore a longer wavelength). If it's a rough day on the ocean, you'll see waves a lot more frequently, and the distance between waves will be shorter (that is, a shorter wavelength).

As we can observe, there's a definite connection between wavelengths and the frequency of the waves themselves. With a longer wavelength, you'll see waves far less frequently, and with a shorter wavelength, you'll see the waves a lot more frequently (**Figure 10.2**). If you sit on the beach and count the waves, you'll count a higher number of waves when the wavelength between them is small, and you'll count fewer waves when the wavelength between them is longer.

Waves of energy work on a similar principle. Energy travels at a constant speed—that is, the speed of light (c), which is about 300 million meters per

> **wavelength** the distance between the crests of two waves

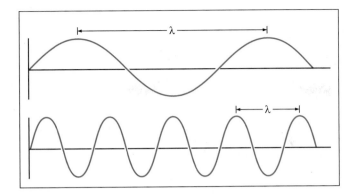

FIGURE 10.2 The relationship between wavelength and frequency: Long wavelengths equate to low frequencies, whereas short wavelengths equate to high frequencies.

second. If you multiply the frequency by the wavelength, the result is always equal to the speed of light. When the value for the wavelength of light energy is high, the frequency is low, and if the value for the frequency is high, the value for the wavelength has to be lower to make the product of the two values always come out to the same number.

Electromagnetic energy at different wavelengths also has different properties. Light energy with very long wavelengths will have different characteristics from light that has very short wavelengths. The **electromagnetic spectrum** is used to examine the properties of light energy in relation to wavelength. The values of the wavelengths are usually measured in extremely small units of measurement called **micrometers** (μm): A micrometer is one-millionth of a meter (roughly the thickness of a single bacterium). Sometimes an even smaller unit of measurement is used, called a **nanometer** (nm), which is one-billionth of a meter. Forms of electromagnetic energy with very short wavelengths are cosmic rays, gamma rays, and x-rays, while forms of electromagnetic energy with very long wavelengths are microwaves and radio waves (**Figure 10.3**).

It's important to remember that the Sun radiates many types of energy, but they're invisible to the human eye. When you put a slice of cold pizza in the microwave oven to heat it, you don't actually see the microwaves

electromagnetic spectrum the light energy wavelengths and the properties associated with them

micrometer a unit of measurement equal to one-millionth of a meter, abbreviated μm

nanometer a unit of measurement equal to one-billionth of a meter, abbreviated nm

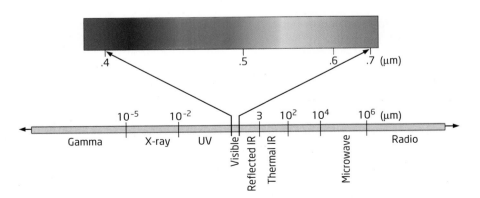

FIGURE 10.3 The electromagnetic spectrum, showing the wavelengths of energy and the properties they correspond with. The visible light portion of the spectrum occurs at wavelengths between 0.4 and 0.7 micrometers.

bombarding the pizza. In the same way, you can't see the x-rays that a doctor uses in an exam or the radio waves that are picked up by your car's antenna. All this energy is out there, but you just can't see it. This is because the composition of your eyes (which are structured to respond to reflected energy) are only sensitive to the reflected light that has a wavelength between 0.4 and 0.7 micrometers. This portion of the electromagnetic spectrum is called the **visible light spectrum** or visible spectrum (see Figure 10.3 to see what portion of the whole electromagnetic spectrum is occupied by the visible light spectrum). If your eyes were sensitive to the reflection of slightly shorter wavelengths of energy, you'd be able to see ultraviolet (**UV**) light (as bees can), and if you could see longer wavelengths of energy, you'd be able to see infrared light. However, the human eye is only sensitive to electromagnetic energy in this very small range of wavelengths.

The visible spectrum can be broken down further to the colors that it comprises. Shorter visible light wavelengths (between 0.4 and 0.5 micrometers) represent the blue portion of the spectrum. Medium visible light wavelengths (between 0.5 and 0.6 micrometers) represent the green portion of the spectrum. Longer visible light wavelengths (between 0.6 and 0.7 micrometers) represent the red portion of the spectrum. When you see a bright red shirt, your eyes are actually sensing the reflection of light with a wavelength measurement somewhere between 0.6 and 0.7 micrometers.

Each of these narrow portions of the electromagnetic spectrum that represent a particular form of light is referred to as a **band** of energy in remote sensing. For example, the shorter wavelengths of 0.4 to 0.5 micrometers are referred to as the **blue band** of the electromagnetic spectrum, because the properties of those wavelengths of light correspond to the characteristics of blue energy. Wavelengths between 0.5 and 0.6 micrometers make up the **green band** of the electromagnetic spectrum and wavelengths between 0.6 to 0.7 micrometers make up the **red band**.

Keep in mind that while the human eye is restricted to seeing only these narrow bands of the electromagnetic spectrum, remote sensing devices have the capability to sense energy from other sections of the spectrum. For instance, the infrared (**IR**) light portion of the electromagnetic spectrum covers wavelengths between 0.7 and 100 micrometers. A sensor could be tuned to measure only the reflection of infrared energy between 0.7 and 0.9 micrometers, and even though this energy is not visible to the human eye, the sensor would have no difficulty measuring this narrow band of reflected energy. As we'll see, an enormous amount of information can be gained from examining these other forms of energy as they are reflected off objects. In remote sensing of infrared light, there are a few wavelength ranges that are utilized much more than the others:

▶ 0.7 to 1.3 micrometers: Bands in this portion, referred to as "near infrared" (**NIR**), are frequently used in satellite remote sensing (see Chapter 11) and in color infrared photography (see Chapter 9).

▶ 1.3 to 3.0 micrometers: Bands of this portion, referred to as "shortwave infrared" (**SWIR**) or "middle infrared" (MIR), are also utilized by various

---

visible light spectrum the portion of the electromagnetic spectrum with wavelengths between 0.4 and 0.7 micrometers

UV ultraviolet; the portion of the electromagnetic spectrum with wavelengths between 0.01 and 0.4 micrometers

band a narrow range of wavelengths that may be measured by a remote sensing device

blue band the range of wavelengths between 0.4 and 0.5 micrometers

green band the range of wavelengths between 0.5 and 0.6 micrometers

red band the range of wavelengths between 0.6 and 0.7 micrometers

IR infrared; the portion of the electromagnetic spectrum with wavelengths between 0.7 and 100 micrometers

NIR near infrared; the portion of the electromagnetic spectrum with wavelengths between 0.7 and 1.3 micrometers

SWIR shortwave infrared; the portion of the electromagnetic spectrum with wavelengths between 1.3 and 3.0 micrometers

HANDS-ON APPLICATION 10.2

Wavelengths and the Scale of the Universe

The Scale of the Universe is an online tool for comparing items of different sizes—everything from the size of the known universe all the way down to sub-atomic particles. It's a fun visualization tool for measuring sizes, including making comparisons of the wavelengths of energy used in remote sensing with other items. To get started, go to **http://htwins.net/scale2**. Click on the Start button to begin. You'll start with the size of a human relative to other objects. As you move the scroll bar at the bottom of the screen to the right you'll see the scale increase and show larger objects, while moving the scroll bar to the left will decrease the scale and allow you to view smaller items. By moving the bar slightly to the right you'll see the wavelengths for FM radio waves. By clicking on the FM radio waves icon, you'll receive information about its measurement (these radio wavelengths are about one meter). Move the scroll bar to the left to start examining items at a smaller scale, including the wavelengths used in remote sensing.

Expansion Questions

- What is the wavelength of infrared energy measured as and what is it compared to in size?

- What is the wavelength of red light measured as and what is it compared to in size?

- What is the wavelength of ultraviolet measured as and what is it compared to in size?

other satellite sensors. For instance, measurements of bands of SWIR energy are used regularly in studies related to the water content of plants.

▶ 3.0 to 14.0 micrometers: Bands of this portion are referred to as "thermal infrared" (**TIR**), and are widely used for measuring heat sources and radiant heat energy.

> **TIR** thermal infrared; the portion of the electromagnetic spectrum with wavelengths between 3.0 and 14.0 micrometers

For a quick example of examining the size of these wavelengths that we deal with in remote sensing, check out *Hands-On Application 10.2: Wavelengths and the Scale of the Universe* for a very neat online tool that will help demonstrate just how small the wavelengths we'll be working with really are.

What Is the Role of the Atmosphere in Remote Sensing?

Earth's atmosphere acts like a shield around the planet, and the energy from the Sun has to pass through it before it reaches Earth's surface. As a result, a significant amount of the Sun's electromagnetic energy never actually makes it to the ground—and if it never makes it to the ground, it's never going to be reflected back to the remote sensing device (and will never be part of the whole remote sensing process). Earth's atmosphere contains a variety of gases (including carbon dioxide and ozone) that serve to absorb electromagnetic energy of numerous wavelengths. For instance, ozone absorbs wavelengths of energy that correspond to ultraviolet light; these are "UV rays" that can be harmful to humans. Thus, a considerable amount of ultraviolet radiation gets trapped in the atmosphere and doesn't pass through to Earth. Very short wavelengths and some very long wavelengths are also absorbed by the atmosphere.

Those wavelengths of energy that pass through the atmosphere (rather than being absorbed) are referred to as **atmospheric windows**. Remote sensing of the wavelengths that make up these "windows" is possible since they are the ones that transmit most of their energy through the atmosphere to reach Earth. The visible light wavelengths of the electromagnetic spectrum (wavelengths between 0.4 and 0.7 micrometers) are examples of atmospheric windows. Most of the energy at these wavelengths is not absorbed by the atmosphere and instead transmits to Earth, where it reflects off objects and surfaces to be viewed by our eyes as color (and viewed also by cameras and satellite sensors). Other windows used in remote sensing include portions of the infrared and thermal infrared sections of the spectrum (see **Figure 10.4** for a diagram of the atmospheric windows). Sensors are set up to measure the energy at the wavelengths of the windows. It won't do any good whatsoever to have a sensor measuring wavelengths (such as the entire ultraviolet portion of the spectrum) that are completely absorbed by the atmosphere, as there will be no reflectant energy to measure.

While some light energy is absorbed by the atmosphere, other light energy is scattered by it. Scattering is caused by particles, such as dust and water, in the atmosphere—it is always present, but its effects are always unpredictable. Light hitting these particles will be absorbed and redirected back out in random directions, causing scattering. There are three types of scattering that can occur: The first of these is **Rayleigh scattering**, which occurs when the particles causing the scattering are significantly smaller than the wavelengths being affected. Keeping in mind that the visible-light wavelengths are smaller than one-millionth of a meter, Rayleigh scattering is caused by molecules in the atmosphere. In addition, shorter wavelengths are scattered to a much greater degree than longer wavelengths. This explains why the sky is blue during the day: The shorter blue wavelengths are

atmospheric windows those wavelengths of electromagnetic energy in which most of the energy passes through Earth's atmosphere

Rayleigh scattering scattering of light caused by atmospheric particles smaller than the wavelength being scattered

FIGURE 10.4 The absorption of various wavelengths by different gases in Earth's atmosphere and the atmospheric windows used in remote sensing.

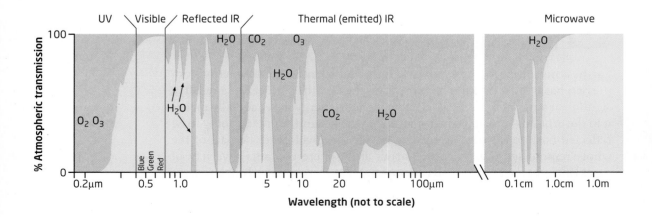

scattered far more than the longer red wavelengths. However, at sunset or sunrise, sunlight has a longer way to travel to reach Earth, and thus with all of the blue wavelengths already scattered, we see the reds and oranges of longer wavelengths as the Sun goes out of view.

The second is **Mie scattering**, which occurs when the particles causing the scattering are roughly the same diameter as the wavelengths they're scattering (this would include things like dust or smoke). The third is **non-selective scattering**, where what's causing the scattering—say, water droplets or clouds—is larger than the wavelengths being scattered and scatters all wavelengths equally. Non-selective scattering also explains why we see clouds as white—the visible primary colors of blue, green, and red are all scattered the same, and equal levels of red, green, and blue will produce the color white.

Mie scattering
scattering of light caused by atmospheric particles the same size as the wavelength being scattered

non-selective scattering scattering of light caused by atmospheric particles larger than the wavelength being scattered

What Happens to Energy When It Hits a Target on the Ground?

After so much energy has been absorbed and scattered by the atmosphere, many of the wavelengths that correspond to the windows finally make it to Earth's surface and interact with targets there. One of three things can happen to that energy (on a wavelength-by-wavelength basis)—either it can be transmitted through the target, absorbed by the target, or reflected off the target. **Transmission** occurs when a wavelength of energy simply passes through a surface to interact with something else later. Think of light passing through a windshield of a car—most of the light will pass through the glass to affect things in the car rather than the windshield itself.

transmission when light passes through a target

Absorption occurs when the energy is trapped and held by a surface rather than passing through or reflecting off it. Think of walking across a blacktop parking lot during a hot summer day—if you don't have something on your feet, they're going to hurt from the heat in the pavement. Because the blacktop has absorbed energy (and converted it into heat), it's storing that heat during the day. Absorption also explains why we see different colors. For instance, if someone is wearing a bright green shirt, there's something in the dyes of the shirt's material that is absorbing all other colors (that is, the portions of the visible light spectrum) aside from that shade of green, which is then being reflected to your eyes.

absorption when light is trapped and held by a target

Different surfaces across Earth have different properties that cause them to absorb energy wavelengths in various ways. For instance, we see a clear lake as a blue color—it could be dark blue, lighter blue, maybe even some greenish blue, but it's nonetheless a shade of blue. We see it this way because water strongly absorbs all electromagnetic wavelengths except for the range of 0.4 to 0.5 micrometers (and some lesser absorption past the edges of that range), which corresponds to the blue portion of the spectrum. Thus, other wavelengths of energy (such as the red and infrared portions) get absorbed and the blue portion gets reflected.

What remote sensing devices are really measuring is the reflectance of energy from a surface—the energy that rebounds from a target to be collected by a sensor. Just as different objects and surfaces absorb energy in different ways, so different objects and surfaces reflect energy differently. The total amount of energy that strikes a surface (the **incident energy**) can be calculated by adding up the amounts of energy that were transmitted, absorbed, and reflected per wavelength by a particular surface as follows:

> **incident energy** the total amount of energy (per wavelength) that interacts with an object

$$I = R + A + T$$

where I is the incident energy (the total amount of energy of a particular wavelength) that strikes a surface and is made up of R (reflection) plus A (absorption) plus T (transmission) of that particular wavelength. Since remote sensing is focused on the reflection of energy per wavelength, let's change that equation to focus on calculating what fraction of the total incident energy was reflected (and to express this value as a percentage instead of a fraction, we'll multiply by 100):

$$\rho = (R/I) \times 100$$

> **spectral reflectance** the percentage of the total incident energy that was reflected from that surface

where ρ is the portion of the total amount of energy composed by reflection (rather than absorption or transmission) per wavelength. This final value is the **spectral reflectance**—the percentage of total energy per wavelength that is reflected off a target, makes its way toward the sensor, and is utilized in remote sensing.

THINKING CRITICALLY WITH GEOSPATIAL TECHNOLOGY 10.1

How Does Remote Sensing Affect Your Privacy?

The remote sensing processes described in this chapter are systems that measure energy reflection and turn it into images without interfering with what's happening below them. For instance, when an aircraft flies overhead to take pictures or a satellite crosses more than 500 miles above your house, it collects its imagery and moves on. This data collection process doesn't physically affect you or your property in the slightest, and all of this is done without your explicit permission and usually without your knowledge. If you can see your car in your driveway on Google Earth, then it just happened to be parked there when the airplane or satellite collected that image. Does this unobtrusive method of data collection affect your privacy? Is the acquisition of images via remote sensing an invasion of your personal space? In what ways could this use of geospatial technology intrude on someone's life? Also, a lot of this kind of imagery is being acquired by private companies—is this more or less intrusive to your privacy than a government agency doing the same thing?

How Can Spectral Reflectance Be Used in Remote Sensing?

Everything on Earth's surface reflects energy wavelengths differently from everything else. Green grass, bare soil, a parking lot, a sandy beach, and a large lake—all reflect portions of the visible, near-infrared, and middle-infrared energy differently (these items emit thermal energy differently as well). In addition, some things reflect energy differently depending on different conditions. As an example of examining these remote sensing concepts, let's look at something simple—measuring reflectance of energy from the leaves on a tree. During late spring and summer, a tree canopy will be dense and full, as the leaves increase in number, stay on the trees, and are a healthy green color. The human eye sees healthy leaves as green because the chlorophyll in the leaves is reflecting the green portion of the spectrum and absorbing the red and blue portions. What can't be seen by the human eye (but can be seen by remote sensing instruments) is that, due to their structure, healthy leaves also very strongly reflect near-infrared energy. Thus, measurement of near-infrared energy is often used (in part) as an indicator of the relative health of leaves and vegetation.

However, leaves aren't always going to stay green. As autumn progresses, leaves go through a senescence process in which they lose their chlorophyll. As this happens, the leaves reflect less green energy and begin to absorb less (and thus, reflect more) red and blue energy, causing the leaves to appear in other colors, such as yellow, orange, and red. As the leaves turn brown and fall off the trees, they have a lot more red reflection, causing their brownish appearance (and there will also be less reflectance of near-infrared energy from the tree).

If a sensor were able to measure all of these energy wavelengths simultaneously for different objects, it would give the person examining the data the ability to tell objects apart by examining their reflectance values in all of these wavelengths. If you were to chart the spectral reflectance values against the wavelengths being measured for each item, you would find that each of these things would have a different line on the chart. This is referred to as an item's **spectral signature** (also called a spectral reflectance curve), as each set of charted measurements will be unique to that item, just like your handwritten signature on a piece of paper is different from everyone else's. Remote sensing analysts can use these spectral signatures to distinguish items in an image and to tell the difference between different types of plants or minerals (see **Figure 10.5** on page 350 for comparisons of various spectral signatures).

One useful application of remotely sensed imagery is its use in assessing the health of green vegetation (such as fields, grass, and foliage). When vegetation is very healthy, it will have a strong reflection of near-infrared energy and absorption of red energy. As the vegetation becomes more stressed, less near-infrared energy will be reflected and more red energy will be reflected rather than absorbed. An image can be processed to measure the health of

> **spectral signature**
> a unique identifier for a particular item, generated by charting the percentage of reflected energy per wavelength against a value for that wavelength

FIGURE 10.5 Four spectral signatures, generated for different items by charting the percentage of reflection on the *y*-axis and wavelengths on the *x*-axis.

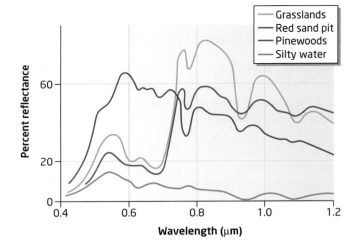

NDVI Normalized Difference Vegetation Index; a method of measuring the health of vegetation using near-infrared and red energy measurements

various kinds of vegetation. The Normalized Difference Vegetation Index (**NDVI**) is a means of doing so—the process takes a remotely sensed image and creates a new NDVI image from it, containing values related to the relative health of the vegetation in the image. As long as the sensor can measure the red and near infrared portions of the electromagnetic spectrum, NDVI can be calculated.

NDVI is computed using the measurements for the red and near-infrared bands. The formula for NDVI is:

$$\frac{(\text{NIR} - \text{Red})}{(\text{NIR} + \text{Red})}$$

NDVI returns a value between -1 and $+1$—the higher the value (closer to 1), the healthier the vegetation is at the area being measured. Low values indicate unhealthy vegetation or a lack of biomass. Very low or negative values indicate areas where nothing is growing (like pavement, clouds, or ice). **Figure 10.6** shows an example of NDVI being calculated for both a healthy vegetation source (one that reflects a high amount of near-infrared energy and less red energy), resulting in a high NDVI value (0.72) and an unhealthy source (which reflects less near-infrared energy and more red energy), which results in a lower NDVI value (0.14).

NDVI can be calculated by a variety of remote sensing devices, as long as their sensors can measure the red and near-infrared bands (which several of the specific sensors discussed in the next two chapters can do). NDVI gives a quick means of assessing the health of various forms of vegetation and is used in a variety of environmental applications, both local and global—from crop monitoring to surveying the planet's natural resources (see **Figure 10.7** for an example and *Hands-On Application 10.3: Examining NDVI with NASA ICE* to use some remotely sensed imagery to examine the health of a rainforest).

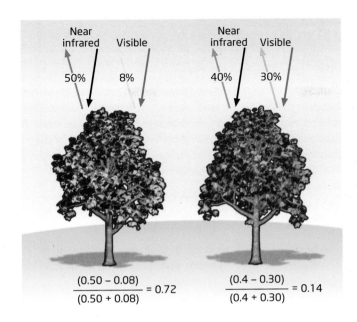

FIGURE 10.6 An example of NDVI calculation for healthy (left) and unhealthy (right) vegetation.

[Source: Adapted from NASA/ Robert Simmon]

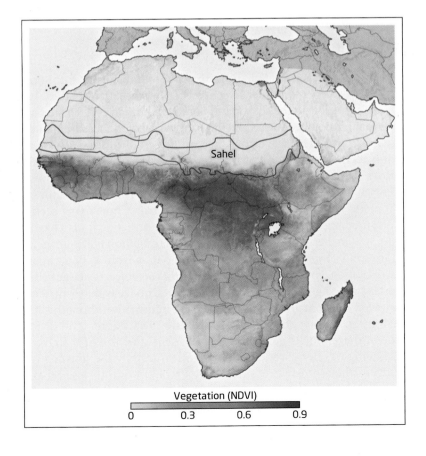

FIGURE 10.7 An NDVI image of Africa, showing the health (or presence) of vegetation on the continent.

[Source: Adapted from NASA/ University of Maryland]

HANDS-ON APPLICATION 10.3

Examining NDVI with NASA ICE

NASA (the National Aeronautics and Space Administration) has an online tool called ICE (Image Composite Explorer) that allows users to analyze different applications of remotely sensed data, and one component of ICE features an interactive NDVI tool. To use ICE's NDVI exercise, go to **http://earthobservatory.nasa.gov /Experiments/ICE/panama/panama_ex2.php**. This section of ICE is part of a larger exercise designed to examine the health of tropical rainforests. The tools will allow you to build NDVI images using the near-infrared and red bands of a Landsat satellite image. To see the NDVI imagery fully, select channel 4 (this is

the near-infrared band) in the pull-down menu next to the red Channel 1 text and then select Channel 3 (this is the red band) in the pull-down menu next to the green Channel 2 text. Click Build and you can view the NDVI image of the area. Change the color table and zoom around the imagery.

Expansion Questions

- How can NDVI be used to measure the health of the rainforests?
- In the NDVI image, what do the very dark areas represent? What do the very light areas represent?

How Do You Display a Digital Remotely Sensed Image?

Once the digital data is collected, the next step is to transform it into something that can be viewed on a computer screen, especially since many of these bands of energy are only visible to the sensor and not to the human eye. Sensors can only "see" a certain area of the ground at a time and will record measurements within that particular area. The size of this area on the ground is the sensor's **spatial resolution**: it's the smallest unit of area the sensor can collect information about. If a sensor's spatial resolution is 30 meters, each measurement will consist of a 30 meter × 30 meter area of ground, and the sensor will measure thousands of these 30 meter × 30 meter sections at once. No matter how many sections are being measured, the smallest area the sensor has information for is a block that is 30 meters × 30 meters. Keep in mind, the spatial resolution of the image is going to limit the kind of visual information you can obtain from it. In an image with 4-meter resolution, you may be able to discern that an object is a large vehicle, but you won't be able to go into any details about it. An image with 0.5-meter resolution, however, will contain much more detailed information about the vehicle.

We'll discuss more about spatial resolution in Chapter 11, but for now, each one of these blocks being sensed represents the size of the area on the ground having its energy reflection being measured by the sensor. Owing to the effects of the atmosphere, there's not a straight one-to-one relationship between the reflectance of energy and the radiance being measured at the satellite's sensor. The energy measurements at the sensor are converted to a series of pixels, with each pixel receiving a **brightness value**, or BV (also

---

spatial resolution the size of the area on the ground represented by one pixel's worth of energy measurement

brightness values the energy measured at a single pixel according to a predetermined scale; also referred to as digital numbers (DNs)

referred to as a *digital number*, or *DN*), for the amount of radiance being measured at the sensor for that section of the ground. This measurement is made for each wavelength band being sensed.

Brightness values are scaled to fit a range specified by the sensor. We'll discuss these ranges in more detail in Chapter 11 (with regard to a sensor's radiometric resolution), but for now, the scale of values in a range represents the energy being recorded—from a low number (indicating very little radiance) to a high number (indicating a lot of radiance)—and measured as a number of bits. For instance, an 8-bit sensor will scale measurement values between 0 (the lowest value) and 255 (the highest value). These values represent the energy being recorded in a range from 0 to 255 (these values are related to the amount of energy being measured). For instance, all values in a band of an image could possibly only be in a range such as 50 to 150, with nothing reaching the maximum value of 255. Other types of sensors can have larger ranges such as 0–2047 or 0–65535.

Each wavelength band that is being sensed is assigned a BV in this range for each pixel in the image. Brightness values (BVs) can be translated to a grayscale to be viewed on a computer screen. With the range of values associated with **8-bit imagery**, values of 0 represent the color black and values of 255 represent the color white. All other integer values between 0 and 255 are displayed in shades of gray, with lower numbers being darker shades and higher numbers being lighter shades. In this way, any of the wavelength band measurements (including the wavelengths normally invisible to the human eye) can be displayed in grayscale on a computer screen (see **Figure 10.8**). For instance, a near-infrared image will have the brightness value for each pixel representing the near-infrared measurements at the sensor for that area.

If a sensor is measuring the visible portion of the spectrum and treating the entire 0.4- to 0.7-micrometer range as if it were one band, the result will be black-and-white **panchromatic imagery**. However, remote sensing devices are capable of sensing several wavelengths of the electromagnetic spectrum simultaneously. Some widely used remote sensing devices can sense several different bands at once—U.S. government satellite sensors are often capable of

8-bit imagery a digital image that carries a range of brightness values from 0 to 255

panchromatic imagery black and white imagery formed by viewing the entire visible portion of the electromagnetic spectrum

FIGURE 10.8 Pixels of a sample 8-bit image, with their corresponding brightness values and grayscale shading.

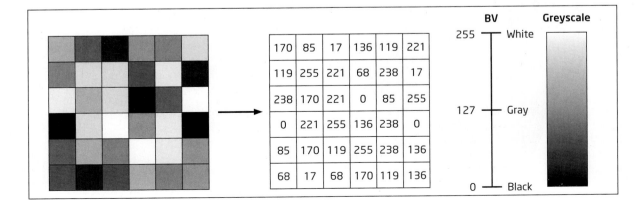

multispectral imagery remotely sensed imagery created from the bands collected by a sensor capable of sensing several bands of energy at once

hyperspectral imagery remotely sensed imagery created from the bands collected by a sensor capable of sensing hundreds of bands of energy at once

color gun equipment used to display a color pixel on a screen through the use of the colors red, green, and blue

color composite an image formed by placing a band of imagery into each of the three color guns (red, green, and blue) to view an image in color rather than grayscale

FIGURE 10.9 Colors that can be formed by adding different colors together (left) and subtracting different colors from each other (right).

sensing 7, 10, or even 36 bands at the same time. Imagery created by sensing multiple bands together is referred to as **multispectral imagery**.

Some technology can simultaneously sense over 200 bands of the electromagnetic spectrum, producing **hyperspectral imagery**.

When multiple bands of imagery are available, they can be combined so that the image may be viewed in color, rather than grayscale. Displaying imagery in color requires three bands to be displayed simultaneously, each in a separate color other than gray. This is simplifying the technical aspects a bit, but a display monitor (like a computer screen) is equipped with devices that can draw a pixel on the screen in varying degrees of red, green, and blue. These are often referred to in remote sensing as **color guns** (a name that refers back to equipment in old CRT monitors and tube televisions for displaying colors in red, green, and blue). If an 8-bit image is drawn using the red color gun, the values of 0 to 255 will not be in a range of black, gray, and white, but instead will be in a range of the darkest red, medium red, and very bright red. The same goes for the blue and green color guns—each gun is able to display an image in its respective color on a range of 0 to 255.

A **color composite** can be generated by displaying the brightness values of one band of imagery in the red gun, a second band of imagery in the green gun, and a third band of imagery in the blue gun. Each pixel will then be drawn using brightness values as the intensity of the displayed color. For instance, a pixel can be displayed using a value of 220 in the red gun, 128 in the green gun, and 75 in the blue gun, and the resulting color shown on the screen will be a combination of these three color intensities. By using the three primary colors of red, green, and blue and combining them to varying degrees using the range of values from 0 to 255, many other colors can be created—using the 256 numbers in the 0–255 range, with three colors, there are 16,777,216 possible color combinations (see **Figure 10.9** for examples of color formation).

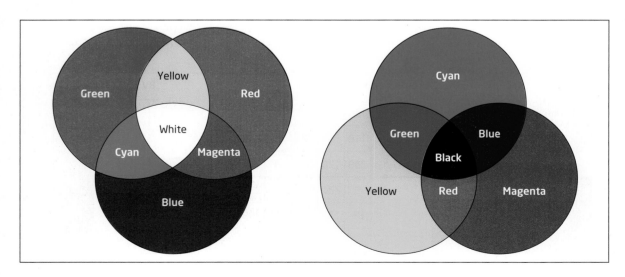

Table 10.1 Examples of colors generated through 8-bit combinations of red, green, and blue.

Red Value	Green Value	Blue Value	Color Formed
255	255	0	Yellow
255	0	255	Magenta
0	255	255	Cyan
255	128	0	Orange
0	64	128	Dark Blue
255	190	190	Pink
255	255	115	Bright Yellow
115	76	0	Brown
0	197	255	Sky Blue
128	128	128	Gray
255	255	255	White
0	0	0	Black

To create colors, different values can be assigned to the red, green, and blue guns. **Table 10.1** shows some examples of ways to combine values together to form other colors (keeping in mind that there are millions of possible combinations of these values). See also *Hands-On Application 10.4: Color Tools Online: Color Mixing* for an online tool that can be used to create colors from various 0 to 255 values.

The color for each pixel in an image corresponds to particular values of red, green, and blue being shown on the screen. For a remotely sensed image, those numbers are the brightness values of the band being displayed with that color gun. Any of the bands of an image may be displayed with any

HANDS-ON APPLICATION 10.4

Color Tools Online: Color Mixing

The Color Tools Website provides several great resources related to color. The Color Mixer tool available at **www.colortools.net/color_mixer.html** allows you to input values (between 0–255) for red, green, and blue and see the results. On the Website, type a value into the box in front of the "/255" for each of the three colors and check out the result. Try the values listed in Table 10.1 as well as some of your own to see how different numbers create different colors.

Expansion Questions

- What colors do the following combinations generate? (Try to figure out the result first, then use the Color Mixer to verify your results.)
  - R: 0, G: 128, B: 255
  - R: 175, G: 175, B: 175
  - R: 85, G: 85, B: 128
  - R: 20, G: 100, B: 50
  - R: 120, G: 90, B: 10

FIGURE 10.10
Combining the display of
three 8-bit imagery
bands, each being
displayed in a different
color gun, to form a color
composite image.
[Source: NASA Images acquired
by Landsat 5]

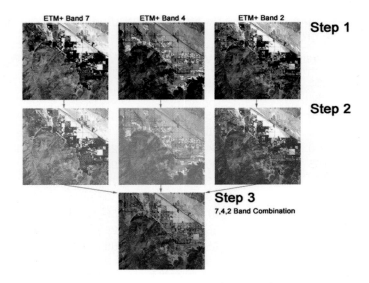

ETM+ Band 7   ETM+ Band 4   ETM+ Band 2   Step 1

Step 2

Step 3
7,4,2 Band Combination

of the guns, but as there are only three color guns, only three bands can be displayed as a color composite at any time (**Figure 10.10**).

For example, the near-infrared band values could be displayed using the red color gun, the green band values could be displayed using the green color gun, and the red band values could be displayed using the blue color gun. Using this arrangement, a brown pixel can be seen on the screen. On closer examination, it's discovered that the red gun (which has the near-infrared band) has a value of 150, the green gun (which has the green band) has a value of 80, and the blue gun (which has the red band) has a value of 20. These values correspond to the brightness values of their respective bands and combine together to form that pixel color. If you change the arrangement of bands and guns, you'll see different colors displayed on the screen, but the actual brightness values will not change. **Table 10.2** shows how the display color will change depending on which band is being displayed with which gun.

Note that while the pixel being displayed on the screen will change color depending on what band is being displayed with a different gun, none of the brightness values of that pixel will be altered. Different band combinations

Table 10.2   **Different combinations of brightness values for various bands displayed with different color guns from different colors on the screen.**

Red Value	Green Value	Blue Value	Color Formed
NIR band (150)	Green band (80)	Red band (20)	Brown
Red band (20)	Green band (80)	NIR band (150)	Deep blue
Green band (80)	NIR band (150)	Red band (20)	Green
Red band (20)	NIR band (150)	Green band (80)	Paler green

can reveal different information about objects in an image due to the varying brightness values of the bands for those items.

Several different types of composites can be formed using the various bands and guns. Whenever a composite has the red band displayed in the red gun, the green band displayed in the green gun, and the blue band displayed in the blue gun (the way natural colors would be shown), a **true color composite** is formed. A true color composite appears as you would expect a remotely sensed image to appear if you were looking down at the ground from an airplane (that is, the trees would be green and the water would be blue). Whenever the distribution of bands to guns deviates from this arrangement, a **false color composite** is generated. For instance, all four of the combinations described in Table 10.2 would result in false color composites as none of them display the red band in the red gun, the green band in the green gun, and the blue band in the blue gun. See **Figure 10.11** for a comparison of true color and false color composites.

With all of these different possible combinations, there is a **standard false color composite** that is used in many remote sensing studies. In this standard false color arrangement, the near-infrared band is displayed in the red gun, the red band is displayed in the green gun, and the green band is displayed in the blue gun. The false color composite image is similar to the CIR photos discussed back in Chapter 9. Objects that have a strong reflection of near-infrared energy (like healthy plants, grass, or other green vegetation) appear in a variety of bright reds, while water (which has very little reflection of the near-infrared, red, and green energy) appears as black. With composites, multispectral imagery can be displayed in color, regardless of what bands are being used, and these composites can be used for many different types of image analysis (see *Hands-On Application 10.5: Comparing True Color and False Color Composites* for online methods of comparing how objects on the ground are shown in true color and false color composites).

true color composite an image arranged by placing the red band in the red color gun, the green band in the green color gun, and the blue band in the blue color gun

false color composite an image arranged when the distribution of bands differs from placing the red band in the red color gun, the green band in the green color gun, and the blue band in the blue color gun

standard false color composite an image arranged by placing the near-infrared band in the red color gun, the red band in the green color gun, and the green band in the blue color gun

FIGURE 10.11 Two views of the Cleveland area: (a) a true color composite, and (b) a false color composite.
[Source: NASA/Purdue Research Foundation]

(a)

(b)

HANDS-ON APPLICATION 10.5

Comparing True Color and False Color Composites

An excellent online resource for examining Landsat imagery in different band combinations from different dates is the Esri ChangeMatters tool. Open your Web browser and go to **http://changematters.esri.com/compare** to get started. ChangeMatters will show you three remotely sensed images—two Landsat images from different dates and then an NDVI comparison of the two. By default, the initial images should be those of Mount St. Helens (from 1975, 2000, and an NDVI change between the two dates).

Start by selecting the dates 2005–2010 from the Select Dates pull-down menu (if a warning message appears, just close it and continue). This will change the first image to an image from a 2005 date and the second image to that taken from a 2010 date. Next, choose the option for Natural Color from the Select Image Map pull-down menu. This will change both the 2005 and 2010 images to true color composites (note that the colors may be a little different in each one). You can zoom in on one of the two images to examine some of the landscape features (such as Spirit Lake, the volcano itself, or the vegetated areas surrounding it) and all three images will zoom to the same extent. To examine the same area in a false color composite, choose the option for Infrared from the Select Image Map pull-down menu. Check out the same features in a false color arrangement to see how they appear differently.

In the Search box, you can type the name of a location (such as Cleveland, Ohio) and the images will switch to Landsat images from 2005 and 2010 (and the NDVI change comparison) of Cleveland. Take a close look at the urban areas of Cleveland and the surrounding non-urban areas, as well as Lake Erie in both the Natural Color and Infrared settings. In addition, use the Search box to view your own local area and examine it on both Natural Color and Infrared imagery from the two dates.

A second Esri resource for examining remotely sensed imagery in different band combinations is available online here: **www.esri.com/landing-pages /software/landsat/unlock-earths-secrets**. Click on Try it Live or scroll down to the You're in Command section. With this application, you can view Landsat imagery in different band combinations that are commonly used for analysis of different features on the landscape. For instance, clicking on the "eye" icon (labeled Natural Color) will allow you to see Earth in a true color composite, while clicking on the "red grass" icon (labeled Color Infrared) will allow you to see Earth in a standard false color composite. Check out the different band combination options (including Agriculture, SWIR, Geology, and Bathymetric) as well as the Panchromatic option for a sharper 15-meter resolution image and the Vegetation Index and Moisture Index options for other images derived from Landsat imagery.

Expansion Questions

- How did all the features change in appearance (from Mount St. Helens and your own local area) when you switched from the true color composite (Natural Color) to the false color composite (Infrared), and why?

- Check out the parts of Cleveland, Ohio, around Lake Erie, both rural and urban. How do the lake and areas around the lake appear in true color and false color composites, and why?

- In the second application, why do features like vegetation appear bright red in the Color Infrared option, bright green in the Agriculture option, and dark green in the Natural Color option?

Chapter Wrapup

The basics described in this chapter cover the process of remote sensing and how a digital image can be formed from the measurement of electromagnetic energy reflection. Remote sensing provides a valuable technique for a variety of applications, including monitoring the health of vegetation and mapping land cover. See *Remote Sensing Apps* for some representative apps for your phone or tablet and *Remote Sensing in Social Media* for some Facebook pages, Twitter accounts, blogs, and YouTube videos related to remote sensing and this chapter's concepts. In the next chapter you'll be looking at how these concepts can be applied to perform remote sensing from a satellite platform in space. Chapters 11 and 12 will examine different types of remote sensing platforms and look at some of the uses to which their imagery is put, including assessment of air pollution, wildfire monitoring, and measurements of water quality.

The lab for this chapter uses a free program called MultiSpec to examine a digital remotely sensed image as well as different band combinations and color composites.

Important note: The references for this chapter are part of the online companion for this book and can be found at **www.macmillanhighered .com/shellito/catalog**.

Remote Sensing Apps

Here's a sampling of available representative remote sensing apps for your phone or tablet. Note that some apps are for Android, some are for Apple iOS, and some may be available for both.

- **Flat Earth HD:** An app for viewing global remotely sensed imagery of Earth
- **Images of Change:** A NASA app that compares satellite imagery (and sometimes regular photography) of environmental features and changes from two time periods
- **iNSIGHT Color Mixing:** An app that allows you to mix the primary colors of red, green, and blue to different degrees and generate new colors

Remote Sensing in Social Media

Here's a sampling of some representative Twitter accounts, along with some YouTube videos and blogs related to this chapter's concepts.

 On Facebook, check out the following:

- **Google Earth Outreach:**
  www.facebook.com/earthoutreach

Become a Twitter follower of:

- **Google Earth Outreach:** @earthoutreach
- **USGS Land Cover** (the USGS Land Cover Institute): @USGSLandCover

You Tube On YouTube, watch the following videos:

- **Google Earth Outreach** (a Google channel on YouTube showcasing how geospatial technology, including remotely sensed imagery, is applied in numerous areas of study):
  www.youtube.com/user/EarthOutreach
- **Monitoring Forests from the Ground to the Cloud** (a Google Earth Outreach video showing how remotely sensed imagery is used for monitoring deforestation):
  www.youtube.com/watch?v=ymKHb3WJLz4
- **San Francisco-True&FalseColorImagery-Landsat7.mov** (a video showing the differences between true color and false color composites using Landsat 7 bands):
  www.youtube.com/watch?v=9pVF8LK_KMs

For further up-to-date info, read up on this blog:

- **ArcGIS Resources—Imagery Category** (blog concerning topics related to working with remotely sensed imagery):
  http://blogs.esri.com/esri/arcgis/category/subject-imagery

Key Terms

8-bit imagery (p. 353)
absorption (p. 347)
atmospheric windows (p. 346)
band (p. 344)
blue band (p. 344)
brightness values (p. 352)
color composite (p. 354)
color gun (p. 354)
electromagnetic spectrum (p. 343)
false color composite (p. 357)
green band (p. 344)
hyperspectral imagery (p. 354)
incident energy (p. 348)
IR (p. 344)
micrometer (p. 343)
Mie scattering (p. 347)
multispectral imagery (p. 354)
nanometer (p. 343)
NDVI (p. 350)

NIR (p. 344)
non-selective scattering (p. 347)
panchromatic imagery (p. 353)
Rayleigh scattering (p. 346)
red band (p. 344)
remote sensing (p. 340)
spatial resolution (p. 352)
spectral reflectance (p. 348)
spectral signature (p. 349)
standard false color
  composite (p. 357)
SWIR (p. 344)
TIR (p. 345)
transmission (p. 347)
true color composite (p. 357)
UV (p. 344)
visible light spectrum (p. 344)
wavelength (p. 342)

The large mountain icon will zoom in and the small mountain icon will zoom out.

10. Zoom around the image, paying attention to some of the city, landscape, and water areas.

> **Question 10.1** What wavelength bands were placed into which color guns?

> **Question 10.2** Why are the colors in the image so strange compared to what we're normally used to seeing in other imagery (such as Google Earth or Google Maps)? For example, why is most of the landscape red?

> **Question 10.3** In this color composite, what colors are the water, vegetated areas, and urban areas displayed as in the image?

11. Reopen the image, this time using band 4 in the red gun, band 3 in the green gun, and band 2 in the blue gun (referred to as a 4-3-2 combination). Once the image reloads, pan and zoom around the image, examining the same areas you just looked at.

> **Question 10.4** What kind of composite did we create in this step? How are the bands being displayed in this color composite in relation to their guns?

> **Question 10.5** Why can we not always use this kind of composite (from Question 10.4) when analyzing satellite imagery?

12. Reopen the image yet again, this time using band 7 in the red gun, band 5 in the green gun, and band 3 in the blue gun (referred to as a 7-5-3 combination). Once it reloads, pan and zoom around the image, examining the same areas you just looked at.

> **Question 10.6** Once again, what kind of composite was created in this step?

> **Question 10.7** How are vegetated areas being displayed in this color composite (compared with the arrangement in Question 10.4)? Why are they displayed in this color?

case of Sputnik II, animals) could be placed into orbit and that the equipment could then be put to use. As a result, **NASA** (the National Aeronautics and Space Administration) was established in 1958 to head up the U.S. space and aeronautics program. Satellites were first used for remote sensing reconnaissance and surveillance in 1960, when the **Corona** program was initiated.

A Corona satellite had a camera onboard that took pictures of areas on the ground. When the roll of film was used up, it would be ejected from the satellite and collected by a U.S. plane. After the film was developed, it would then be interpreted by agents to provide intelligence information about locations around the world (see **Figure 11.1** for an example of Corona imagery). Corona and its counterpart—the USSR's Zenit satellite—provided numerous images from orbit, with Corona itself providing an estimated 800,000 remotely sensed satellite images. Corona ceased operation in 1972, although similar satellite reconnaissance programs continued, such as the Gambit and Hexagon programs, which remained active until 1984. Corona's imagery was declassified in 1995 but satellite remote sensing has continued, and a series of satellites are now used by the U.S. government for acquiring digital imagery of areas around the world.

NASA the National Aeronautics and Space Administration, established in 1958; it is the U.S. government's space exploration and aerospace development branch

Corona a U.S. government remote sensing program, which utilized film-based camera equipment mounted in a satellite in Earth's orbit; Corona was in operation from 1960 to 1972

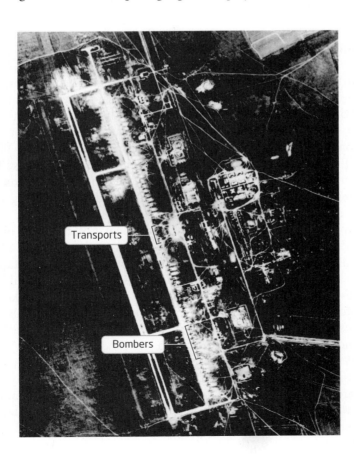

FIGURE 11.1 Corona imagery of Dolon Air Field in the former USSR (taken in 1966). *[Source: National Reconnaissance Office]*

Beyond government surveillance, satellites are used today for a wide variety of purposes, constantly collecting data and imagery of the world below. How does satellite imaging compare with the aerial photography we discussed in Chapter 9? Using a satellite instead of an aircraft has several apparent advantages. Satellites are constantly orbiting Earth and taking images—there's no need to wait for a plane to fly over a particular area of interest. Satellites can image a much larger area than a single aerial photograph can. Also, remote sensing satellites provide global coverage and are not restricted to geographic boundaries or constraints the way aircraft are, since satellites orbit hundreds of miles above Earth's surface.

How Do Remote Sensing Satellites Collect Data?

Remote sensing satellites are placed in a specific orbital path around Earth. One type of orbit used in remote sensing is **geostationary orbit,** in which satellites rotate at the same speed as Earth. Since a geostationary satellite takes 24 hours to make one orbit, it's always in the same place at the same time. For instance, although it's not a remote sensing system, satellite television utilizes a geostationary orbit—the satellite handling the broadcasting is in orbit over the same area continuously, which is why a satellite reception dish needs to be always pointed at the same part of the sky. Likewise, the WAAS and EGNOS satellites discussed in Chapter 4 are in geostationary orbit so that they can provide constant coverage to the same area on Earth. There are some remote sensing satellites that use geostationary orbit to always collect information about the same area on Earth's surface (such as the GOES series of satellites used for obtaining weather imagery; see Chapter 12).

Many remote sensing satellites operate in **near-polar orbit,** a north-to-south path wherein the satellite moves close to the North and South Poles while it makes several passes a day about Earth. For instance, the Landsat 8 satellite (a U.S. satellite that we'll discuss in more detail shortly) completes a little more than 14 orbits per day (each orbit takes about 99 minutes)—see **Figure 11.2** for Landsat 8's near-polar orbit. Also, check out *Hands-On Application 11.1: Examining Satellite Orbits in Real Time* for two methods of tracking satellite orbits in real time.

While a satellite in continuous orbit passes over Earth, it will be imaging the ground below it. However, during an orbit, a satellite can only image a certain area of ground at one time—it can't see everything at once. The **swath width** is the measurement of the width of ground the satellite can image during one pass. A Landsat satellite's sensor has a swath width of 185 kilometers, meaning that a 185-kilometer-wide area on the ground is

geostationary orbit an orbit in which an object follows precisely the direction and speed of Earth's rotation and is therefore always directly above the same point on Earth's surface

near-polar orbit an orbital path close to the North and South Poles that carries an object around Earth at an unvarying elevation

swath width the width of the ground area the satellite is imaging as it passes over Earth's surface

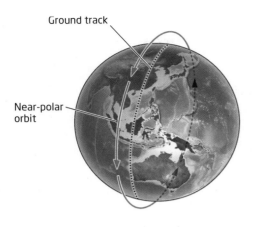

Ground track

Near-polar orbit

imaged as Landsat flies overhead. As satellites orbit, Earth is rotating beneath them, meaning that several days may elapse before an orbital path (and thus, the swath of ground being imaged) takes a satellite over nearby geographic areas. For example, a Landsat satellite may pass over the middle

HANDS-ON APPLICATION 11.1

Examining Satellite Orbits in Real Time

Think about all the jobs that satellites are doing—communication, satellite TV, weather monitoring, military applications, and GPS—and then add remote sensing to that list. When you consider how useful these tasks are, you can see the need for many, many satellites in orbit.

JSatTrak is a very cool program that you can use to view the locations of numerous satellites, including remote sensing ones, in real time. JSatTrak can be downloaded from **www.gano.name/shawn /JSatTrak.** When JSatTrak opens, you can access a list of satellites via the Windows pull-down menu and by choosing Satellite Browser. A list of available satellites will appear. (Many of the remote sensing satellites described in this chapter and in Chapter 12 can be found by expanding the Weather and Earth Resources option, then selecting Earth Resources.) When you drag the name of a satellite from the list to the Object List window on the right side of the screen, the orbit track and current location of the satellite will be shown in the display window. Check out Landsat 7 and Terra, and also look for others. JSatTrak has many features that you can explore

(such as the 3D Earth Window for looking at satellite tracking).

NASA also provides an online satellite tracking tool called J-Track 3D that allows users to track satellite positions in real time, using only a Web browser and Java. This utility is available at **http:// science.nasa.gov/realtime/jtrack/3d/JTrack3D .html.** It will plot the positions of all satellites currently being tracked as a series of dots around Earth. Selecting a dot will identify which satellite it is and show its orbital path.

Expansion Questions

- What is the current position of the following remote sensing satellites: EO-1, IKONOS 2, GeoEye-1, WorldView-2, and SPOT 5?

- The tools at the top of the JSatTrak interface allow you to play animations of the satellite paths, while a clock tells you the time a particular satellite will pass over a particular area. Use these tools to answer the question, "How many orbits around Earth will WorldView-2 make in one day?"

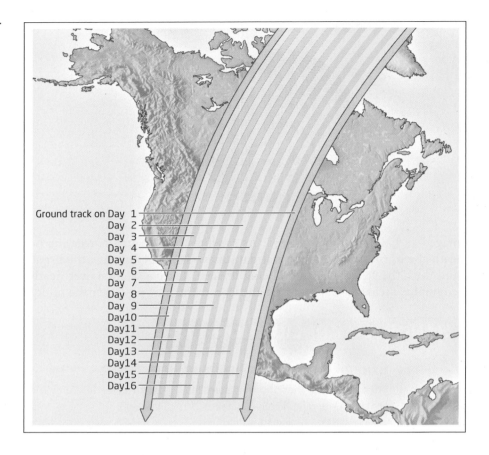

FIGURE 11.3 An example of the paths covered for one section of the United States during the 16-day revisit cycle of a Landsat satellite.

of the United States on Day 1 of its orbit, pass hundreds of miles to the west on Day 2, and not return to the path next to the first one until Day 8 (**Figure 11.3**).

After a certain amount of time, the satellite will completely pass over all areas on Earth and again resume its initial path. For instance, each one of the Landsat satellites is set up in an orbital path so that it will pass over the same swath on Earth's surface every 16 days. Thus, when Landsat's orbit carries it to the swath where it's collecting imagery of your current location, it will be another 16 days before it's able to image your location again. A **sun-synchronous orbit** occurs when a satellite's orbit always takes it over the same swath of Earth's surface at the same local time. As a result, images collected from the satellite will have similar sun-lighting conditions and can be used for comparing changes to an area over time. For instance, Landsat 8 has a sun-synchronous orbit, allowing it to cross the Equator at the same time every day.

The sensors onboard the satellites have different ways of actually scanning the ground and collecting information in the swath being imaged. In **along-track scanning,** a linear array scans the ground along

sun-synchronous orbit an orbital path set up so that the satellite always crosses the same areas at the same local time

along-track scanning a scanning method using a linear array to collect data directly on a satellite's path

the satellite's orbital path. As the satellite moves, the detectors in the array collect the information from the entire swath width below them. This type of sensor is referred to as a "pushbroom." Think of pushing a broom in a straight line across a floor—in this case, the broom itself is the sensor, and each bristle of the broom represents a part of the array that senses what is on the ground below it. A second method is **across-track scanning**; here, a rotating mirror moves back and forth over the swath width of the ground to collect information. This type of sensor is referred to as a "whiskbroom." Once again, think of sweeping a floor with a broom, swinging the broom back and forth across a path on the floor—the broom is the sensor, and the bristles represent the detectors that collect the data on the ground below them. Satellite systems usually use one of these types of sensors to obtain the energy reflectance measurements from the ground (see **Figure 11.4** for examples of the operation of both sensor types).

After a satellite's sensors have collected data, the data need to be off-loaded from the satellite and sent to Earth. Because data collection is continuous, this information needs to get to people on the surface so that it can be processed and utilized as it becomes available. The Earth Resources Observation Science (**EROS**) Center located outside Sioux Falls, South Dakota, is one of the downlink stations in the United States that receives data from numerous remote sensing satellites. Satellite data can be directly sent to a receiver at a station or transmitted to other tracking and data relay satellites, which then send the data to Earth. Once received on the ground, the data can be processed into imagery.

across-track scanning a scanning method using a rotating mirror to collect data by moving the device back and forth across the width of the satellite's swath

EROS the Earth Resources Observation Science Center, located outside Sioux Falls, South Dakota, which serves (among many other things) as a downlink station for satellite imagery

(a) Across-track scanner

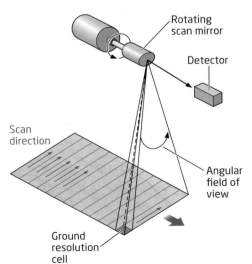

(b) Along-track scanner

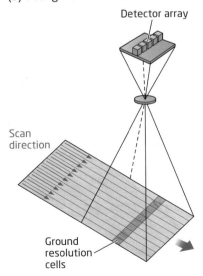

FIGURE 11.4 The operation of across-track and along-track scanners.

THINKING CRITICALLY WITH GEOSPATIAL TECHNOLOGY 11.1

What Effect Does Satellite Remote Sensing Have on Political Borders?

When imagery is acquired by satellite remote sensing for any part of the globe, the satellite sensors effectively ignore political boundaries. After the 2010 earthquake in Haiti, imagery of the country was quickly acquired by satellites and made available for rescue and recovery efforts. When satellite images of nuclear facilities in North Korea or Iran are shown on the news, the satellite's sensors were able to image these places from orbit. Satellites can acquire global imagery of events ranging from the 2016 Olympics in Rio de Janeiro to parades in Beijing; and in all cases, the political boundaries of these places are being ignored by the "eye in the sky" that is looking down from Earth's orbit. Satellites are not restricted by no-fly zones or other restrictions involving a country's airspace, and they are capable of acquiring imagery from nearly everywhere. Given these imaging capabilities, what effect does acquisition of images via satellites have on political borders? There are many areas that can't be approached from the ground without causing international tensions, but these same places can easily be viewed from space. How are political boundaries affected by these types of geospatial technologies?

What Are the Capabilities of a Satellite Sensor?

A satellite sensor has four characteristics that define its capabilities: spatial resolution, radiometric resolution, temporal resolution, and spectral resolution. In Chapter 10, we discussed the concept of resolution in the context of **spatial resolution**, with the area on the ground represented by one pixel in a satellite image. A sensor's spatial resolution will affect the amount of detail that can be determined from the imagery—a sensor with a 1-meter spatial resolution has much finer resolution than a sensor with spatial resolution of 30 meters (see **Figure 11.5** for examples of the same area viewed by different spatial resolutions). A satellite's sensor is fixed with one spatial resolution for the imagery it collects—a 30-meter resolution sensor can't be "adjusted" to collect 10-meter resolution images.

If two bands (one with a higher resolution than the other) are used by the sensor to image the same area, the higher-resolution band can be used to sharpen the resolution of the lower-resolution band. This technique is referred to as **pan-sharpening** because the higher-resolution band used is a panchromatic band. As we discussed in Chapter 10, in terms of imagery, a **panchromatic sensor** will measure only one large band of wavelengths at once (usually the entire visible portion of the spectrum or the entire visible and part of the near-infrared spectrum). For example, a satellite (like Landsat 8) that senses the blue, green, and

spatial resolution the ground area represented by one pixel of satellite imagery

pan-sharpening the technique of fusing a higher-resolution panchromatic band with lower-resolution multispectral bands to improve the clarity and detail seen in the image

panchromatic sensor a sensor that can measure one range of wavelengths

FIGURE 11.5 The same area on the landscape as viewed by three different sensors, each with a different spatial resolution. *[Source: NASA Marshall Space Flight Center]*

red portions of the electromagnetic spectrum at 30-meter resolution could also have a sensor equipped to view the entire visible portion of the spectrum and part of the near-infrared spectrum as a panchromatic band at 15-meter resolution. This panchromatic image can then be fused with the 30-meter resolution color imagery to create a higher-resolution color composite. Many satellites with high spatial resolution capabilities sense in a panchromatic band that allows for pan-sharpening of the imagery in the other bands.

As we mentioned in Chapter 10, a sensor scales the energy measurements into several different ranges (referred to as *quantization levels*) to assign brightness values to a pixel. The sensor's **radiometric resolution** refers to its ability to measure fine differences in energy. For instance, a sensor with 8-bit radiometric resolution places the energy radiance values on a scale of 0 (lowest radiance) to 255 (highest radiance). The wider the range of measurements that can be made, the finer the sensor's radiometric resolution is. The radiometric resolution (and the resulting number of levels) is a measure of the number of the sensor's bits of precision. To calculate the number of levels, take the number of bits and apply it as an exponent to the number 2. For instance, a 6-bit sensor would have 64 levels (2^6) ranging from a value of 0 to a value of 63. An 8-bit sensor would have 256 levels (2^8) ranging from 0 to 255. An 11-bit sensor would have 2048 levels (2^{11}) ranging from 0 to 2047.

The finer a sensor's radiometric resolution, the better it can discriminate between smaller differences in energy measurements. For example, a 2-bit sensor (2^2 or four values) would have to assign every pixel in an image a brightness value of 0, 1, 2, or 3, which would result in most pixels having the same value, and it would be difficult to distinguish between items or to create effective spectral signatures. However, an 11-bit sensor would assign a value anywhere from 0 to 2047, creating a much wider

radiometric resolution the degree of a sensor's ability to determine fine differences in a band of energy measurements

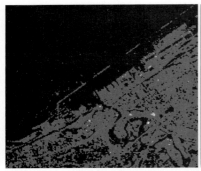

FIGURE 11.6 The effect of different levels of radiometric resolution on an image (simulated 2-bit, 4-bit, and actual 8-bit imagery of one band). [Source: NASA]

temporal resolution the length of time a sensor takes to come back and image the same location on the ground

off-nadir viewing the capability of a satellite to observe areas other than the ground directly below it

spectral resolution the bands and wavelengths measured by a sensor

range of value levels and allowing much finer distinctions to be made between items. See **Figure 11.6** for examples of differing effects of radiometric resolution on an image when performing remote sensing.

The sensor's **temporal resolution** refers to how often the sensor can return to image the same spot on Earth's surface. For instance, a sensor with 16-day temporal resolution will collect information about the swath containing your house and not return to image it again until 16 days have passed. The finer a sensor's temporal resolution, the fewer days it will take between return times. A way to improve a sensor's temporal resolution is to use **off-nadir viewing** capability, in which the sensor is not fixed to sense what's directly below it (the nadir point). A sensor that is capable of viewing places off-nadir can be pointed to image locations away from the orbital path. Using off-nadir viewing can greatly increase a sensor's temporal resolution, as the sensor can image a target several times during its orbits, even if the satellite isn't directly overhead. This capability is a great help in monitoring conditions that can change quickly or require rapid responses (such as mitigating disasters resulting from hurricanes, tornadoes, or widespread fires). See *Hands-on Application 11.2: Seeing What the Satellites Can See* for an online tool that shows how much area on the ground a satellite can image (and the extent of any off-nadir capabilities).

The sensor's last parameter is its **spectral resolution,** which refers to the bands and their wavelengths measured by the sensor. For instance, a sensor on the GeoEye-1 satellite can measure four bands of energy simultaneously (wavelengths in the blue, green, red, and near-infrared portions of the spectrum), and the sensor onboard the Landsat 8 satellite can sense in seven different wavelengths at the same time. A sensor with a higher spectral resolution can examine several finer intervals of energy wavelengths (for example, the MODIS sensor onboard the Terra satellite—see Chapter 12—can sense in 36 different bands at once). Being able to sense numerous smaller wavelengths in the near-infrared portion of the spectrum gives much finer spectral resolution than just sensing one wider wavelength of near-infrared energy.

HANDS-ON APPLICATION 11.2

Seeing What the Satellites Can See

Satellites can't see everything at once. Because their orbits carry them over certain places at certain times, they can only image what's in their swath, and off-nadir viewing can let the satellite see only so far away from its orbital path. Knowing when a satellite will be passing over an area (or if it can image areas outside its immediate swath) can be critical for a number of applications. For instance, emergency rescue and relief efforts rely on up-to-date imagery of what's happening at a disaster site, so it's extremely important to know when a particular satellite will be acquiring new imagery of that locale. If scientists are taking water quality samples of a lake in conjunction with multispectral imagery of the lake, they should time their data collection with a satellite overpass.

The Remote Sensing Planning Tool (RESPT) will show you, for a given day and time, what areas a satellite will pass over, how wide of a swath can be imaged, and what can be imaged through off-nadir capabilities. To get started, go to **ww2.rshgs.sc.edu.** Select the option for Satellite Modeling, then select Predict Collection (multiple satellites). On the next screen, first select a satellite sensor (to start, choose GeoEye-1). Next, select today's date in the calendar and a location—to start, choose Seattle, Washington, from the City Center list (note that you can enter a set of coordinates or another city if you want to predict satellite overpasses for a specific location); and for

number of days, select 3 (to see all possible opportunities over a 3-day stretch). Next, click on Get all Opportunities. A list of all opportunities for imaging from the satellite will appear. From the first opportunity, click on Map Field of View (for GeoEye-1, there will be multiple options, but some satellites may only have one opportunity during the specified time). On the map, an icon showing the satellite will appear, as well as info showing its swath (in blue or green) and the range of off-nadir capabilities in red.

Expansion Questions

- For today's date, how many opportunities (through off-nadir viewing at an angle) will GeoEye-1 have to image Seattle? Of these opportunities, which day and time has the smallest off-nadir angle for collecting an image?

- Over the next 8 days, how many opportunities will Landsat 7 have to image an area on the ground that contains Seattle's city center?

- For your own local area (specified as a city center; or use coordinates for your own location), over the next 3 days, how many opportunities will GeoEye-1, Landsat 7, and Terra–ASTER (all satellites we'll discuss in this chapter or in Chapter 12) have to image the area? Which of these have off-nadir capabilities?

What Is a Landsat Satellite, and What Does It Do?

This chapter and Chapter 10 have mentioned a satellite system called Landsat but, until now, haven't really explained what it is, how it operates, or why it's so important. What would eventually become the U.S. **Landsat** program was conceived in the late 1960s as a resource for doing remote sensing of Earth from space. The program's first satellite, Landsat 1, was launched in 1972 and the program continues today with the most recent launch,

> **Landsat** a long-running and ongoing U.S. remote sensing project that launched its first satellite in 1972

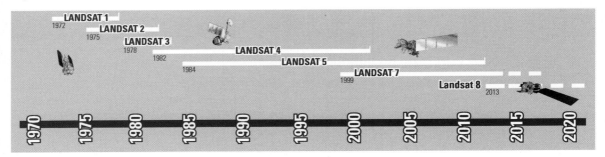

MSS the Multispectral Scanner aboard Landsat 1 through 5

multispectral sensor a sensor that can measure multiple wavelength bands simultaneously

TM the Thematic Mapper sensor onboard Landsat 4 and 5

Landsat 7 the seventh Landsat mission, launched in 1999, which carries the ETM+ sensor

ETM+ the Enhanced Thematic Mapper sensor onboard Landsat 7

Landsat 8 the latest Landsat mission, launched in 2013

LDCM the Landsat Data Continuity Mission—a previous name for the Landsat 8 mission

Landsat 8, in 2013 (see **Figure 11.7** for a graphical timeline of Landsat launches).

Landsat 1's sensor was the Landsat **MSS** (Multispectral Scanner), which had a temporal resolution of 18 days. The MSS would continue to be part of the Landsat missions through Landsat 5. The MSS was capable of sensing four bands (red and green, plus two bands in the near infrared) at 79-meter spatial resolution and 6-bit radiometric resolution. As the name implies, MSS is an example of a **multispectral sensor,** which can measure multiple wavelengths at once.

Landsats 4 and 5 were also equipped with a new remote sensing device called the **TM** (Thematic Mapper). The TM was also a multispectral sensor that allowed for sensing in seven bands in the electromagnetic spectrum simultaneously, including the three visible bands, a band of NIR, two bands of SWIR, and a thermal band (see **Table 11.1** for specifics about these bands and their wavelengths). The TM bands had improved radiometric resolution over MSS of 8 bits while the spatial resolution of the multispectral bands was improved to 30 meters (the thermal band had a spatial resolution of 120 meters, however) and had a swath width of 185 kilometers. TM also had an improved temporal resolution of 16 days.

Landsat 7, launched in 1999, carried neither the MSS nor the TM; instead, it carried a new sensor called the **ETM+** (Enhanced Thematic Mapper Plus). This new device senses the same seven bands and their wavelengths as TM at the same swath width as well as the same radiometric and spatial resolutions (although the thermal band's spatial resolution was improved to 60 meters). It also added a new band 8—this was a panchromatic band (covering wavelengths of 0.52–0.90 micrometers) with a spatial resolution of 15 meters. With this improved spatial resolution, the panchromatic band could be used to pan-sharpen the quality of the other imagery. Landsat 7's temporal resolution was still 16 days, but it could collect over 500 images per day.

The latest Landsat mission is **Landsat 8** (also previously known as the Landsat Data Continuity Mission, **LDCM**) launched on February 11, 2013

Table 11.1 The sensors from the various Landsat satellites and their remotely sensed bands and wavelengths (all wavelength values are listed in micrometers -μm).

Bands and Sensors	OLI / TIRS	ETM+	TM	MSS 1, 2, 3	MSS 4, 5
Band 1	Coastal (0.43–0.45)	Blue (0.45–0.52)	Blue (0.45–0.52)		Green (0.5–0.6)
Band 2	Blue (0.45–0.51)	Green (0.52–0.60)	Green (0.52–0.60)		Red (0.6–0.7)
Band 3	Green (0.53–0.59)	Red (0.63–0.69)	Red (0.63–0.69)		NIR (0.7–0.8)
Band 4	Red (0.64–0.67)	NIR (0.77–0.9)	NIR (0.77–0.9)	Green (0.5–0.6)	NIR (0.8–1.1)
Band 5	NIR (0.85–0.88)	SWIR (1.55–1.75)	SWIR (1.55–1.75)	Red (0.6–0.7)	
Band 6	SWIR (1.57–1.65)	Thermal (10.4–12.5)	Thermal (10.4–12.5)	NIR (0.7–0.8)	
Band 7	SWIR (2.11–2.29)	SWIR (2.09–2.35)	SWIR (2.09–2.35)	NIR (0.8–1.1)	
Band 8	Pan (0.5–0.68)	Pan (0.52–0.9)			
Band 9	Cirrus (1.36–1.38)				
Band 10	Thermal (10.6–11.19)				
Band 11	Thermal (11.5–12.51)				

(**Figure 11.8** on page 384). Landsat 8 currently collects over 700 images per day, with both instruments covering a 185-kilometer swath (see **Figure 11.9** on page 384 for an example of Landsat 8 imagery and *Hands-On Application 11.3: Live Landsat Imagery* on page 385 for some examples of recent imagery collected by Landsat).

Landsat 8's sensor is the next iteration of Landsat technology, called the **OLI** (Operational Land Imager), which has capabilities similar to ETM+. OLI senses visible and infrared bands at 30-meter resolution, a panchromatic band at 15-meter resolution, and two new 30-meter bands aimed at monitoring clouds and coastal zones. A thermal band is not part of OLI; instead, a separate instrument on the satellite, called **TIRS** (the Thermal

OLI the Operational Land Imager, the multispectral sensor onboard Landsat 8

TIRS the Thermal Infrared Sensor, the instrument that acquires thermal imagery onboard Landsat 8

FIGURE 11.8 Landsat 8 in action. *[Source: USGS/NASA Landsat]*

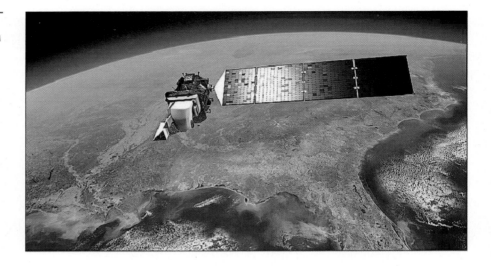

Infrared Sensor), collects two different bands of thermal data listed as bands 10 and 11 for Landsat 8. This thermal data is acquired at 100-meter resolution but is then resampled to 30-meter resolution when delivered to users. See Table 11.1 on page 383 for a comparison of the characteristics of OLI and TIRS with previous Landsat sensors.

These bands were chosen for Landsat sensors because they were (and still are) key to a variety of studies. The blue band is used for examining bathymetry due to its lack of absorption by water. The green, red, and NIR bands are used to monitor vegetation (and the red and NIR bands are also used to examine NDVI, as discussed in Chapter 10). The shorter of the SWIR bands is useful when examining the moisture content of plants and soils, while the wavelengths of the longer SWIR band are useful in distinguishing various rock and mineral types. The thermal bands enable the monitoring of heat energy in a variety of settings.

FIGURE 11.9 Landsat 8 imagery showing the Kenai National Wildlife Refuge in Alaska before, during, and after a massive wildfire in 2014. *[Source: USGS/NASA]*

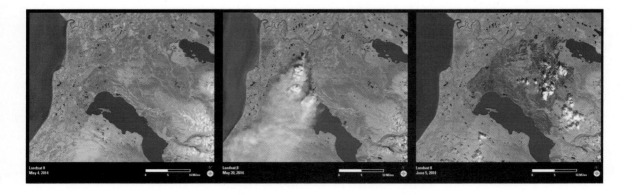

HANDS-ON APPLICATION 11.3

Live Landsat Imagery

Landsat Live is a Web application developed by Mapbox for viewing the most recently available Landsat imagery—check it out at **www.mapbox.com/bites/00113**. Every 16 days Landsat collects imagery of Earth, and this application shows you the latest available image from Landsat. As these are the most current Landsat images, some places around Earth may show up as completely clouded in—if so, check back in another 16 days to see what the view from space looks like. To get started with Landsat Live, click on the magnifying glass and type in the name of a location (such as San Diego, CA) and choose the corresponding location option (usually the first choice) to adjust the view to that location. The most recently available Landsat imagery (displayed as a true color composite) will be displayed. You can zoom in and out to get a good look at what the capabilities of Landsat imagery are. When you're done with San Diego, check out some other places around Earth (such as Melbourne, Australia, or Moscow, Russia) as well as your own local area to see the most recent imagery that Landsat acquired of those places.

Expansion Questions

- What kinds of features in the San Diego region can you easily identify even with the Landsat 30m resolution?

- In the most recent available Landsat imagery is your local area clear or clouded in? What types of local features can you identify with 30m imagery?

Each **Landsat scene** measures an area about 170 kilometers long by 183 kilometers wide and is considered a separate image. At this size, it takes several images just to cover one state. Thus, if you want access to imagery of a particular area, you need to know which scene that ground location is in. Landsat uses a **Worldwide Reference System** that divides the entire globe into a series of Paths (columns) and Rows. For instance, imagery of Columbus, Ohio, extends from the scene at Path 19, Row 32, while imagery of Toledo would be found in Path 20, Row 31 (**Figure 11.10** on page 386). At an area of 170 kilometers long by 183 kilometers wide, a scene will often be much larger than the area of interest—for example, Path 19, Row 32 contains much more than just Columbus; it encompasses the entire middle section of the state of Ohio, so to examine only Columbus, you would have to work with a subset of the image that matches the city's boundaries. Also, an area of interest may sometimes straddle two Landsat scenes—for instance, Mahoning County in Ohio is entirely located in Path 18, but part of the county is in Row 31 and part of it is in Row 32. In this case, the scenes would need to be merged together and then have the boundaries of the county used to define the subsetted area.

The Landsat program has not been without its issues and challenges throughout its lifetime. For instance, a Landsat satellite has a prime technology life of five years, but will often continue far beyond that timespan—Landsat 5 was launched in 1984 and its successor, Landsat 6, was not

> **Landsat scene** a single image obtained by a Landsat satellite sensor
>
> **Worldwide Reference System** the global system of Paths and Rows that is used to identify what area on Earth's surface is present in which Landsat scene

FIGURE 11.10 The total Landsat coverage for the state of Ohio. It requires nine Landsat scenes to cover the entire state.

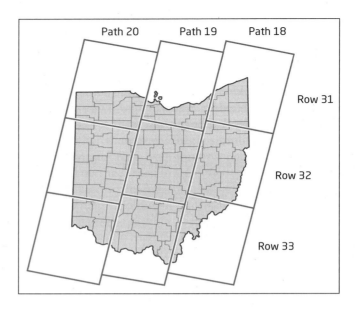

launched until 1993. However, Landsat 6 failed to achieve orbit at launch and the satellite crashed into the Indian Ocean and was lost. This forced the Landsat 5 mission to continue on until the launch of Landsat 7 in 1999.

However, in 2003, Landsat 7's **SLC** (Scan Line Corrector) failed and could not be repaired. Because of the malfunctioning SLC, the ETM+ produces imagery that's missing certain data. Without a working SLC, an ETM+ image contains only about 75% of the data that should be there (the imagery contains what looks like diagonal black stripes of missing pixels). These gaps in ETM+ imagery are filled by substituting pixels from other Landsat 7 scenes of the same area taken on a different date, as well as by other methods. With these types of issues emerging in Landsat 7, Landsat 5 continued to be used as a primary source of imagery. Landsat 5 continued collecting imagery (with many mechanical issues along the way) until it finally ended its mission in 2012 and was shut down in 2013. Landsat 5 was not likely intended to be the cornerstone of the program, but that's what it became, with its mission lasting for 28 years.

In 2008, the USGS announced that the entire Landsat archive would be made freely available. The USGS maintains an online utility called **GloVis** (Global Visualization Viewer), which is used to distribute Landsat (and other satellite) imagery across the Internet (see **Figure 11.11** on page 388 and *Hands-On Application 11.4: Viewing Landsat Imagery with GloVis and LandsatLook*). EarthExplorer (see Chapter 9) can also be used for downloading Landsat images. Landsat imagery is also available through another online resource called **LandsatLook.** This USGS decision has allowed free worldwide access to a vast library of remote sensing data for

SLC the Scan Line Corrector in the ETM+ sensor; its failure in 2003 caused Landsat 7 ETM+ imagery to not contain all data from a scene

GloVis the Global Visualization Viewer set up by the USGS for viewing and distribution of satellite imagery

LandsatLook an online utility used for the viewing and downloading of Landsat imagery

THINKING CRITICALLY
WITH GEOSPATIAL TECHNOLOGY 11.2

What If There Is No Landsat 9?

The Landsat program, which has been providing global coverage since 1972, is the longest running remote sensing program in existence. However, the U.S. remote sensing mission currently relies on the Landsat 7 satellite (launched in 1999 and now impaired since 2003 due to the SLC error) and also on Landsat 8, which was launched in 2013. Despite Landsat 5's successful 28-year tenure, the satellite missions are not intended to last for that long—a Landsat satellite has about five years of prime operating shelf life, though the satellites can last longer than that. At present, Landsat 7's mission is only expected to last until about 2019, at best, when it runs out of fuel.

With these issues, the future of the Landsat program must be in question, since there has been only one Landsat launch since 1999. It takes several years to develop, build, and launch a Landsat satellite, and as of the time of writing (summer 2015), Landsat 9 remains in the planning stages, and its current proposed launch date is not until 2023. A separate satellite with thermal sensors has been proposed to be launched before Landsat 9 to supplement the thermal sensors on Landsats 7 and 8, but, similarly, no information about the construction of this separate free-flying satellite is available.

Since there was a 14-year gap between the launches of Landsat 7 and Landsat 8, and, at present, a 10-year stretch between the launch of Landsat 8 and the proposed launch for Landsat 9, what might happen to data acquisition if there are gaps between successful Landsat satellites? This may sound like a doomsday scenario, but if Landsat 8 develops unfixable problems as Landsat 7 did or if the satellites somehow become unavailable for imagery acquisition, then what happens? Will the United States have to purchase imagery from other sources or countries to make up the data gap until another Landsat mission can be launched? Are there other sources that the U.S. government could use to make up for the lack of Landsat imagery? What impact would a lack of continuous Landsat imagery have on the numerous applications and studies previously described?

everyone, which can only help with furthering research and education in remote sensing.

There's a strong, ongoing need for Landsat imagery that monitors large-scale phenomena, such as environmental effects of land-use changes and urban expansion. Long-term gaps in available current data and between the dates of imagery would have a huge negative impact on these types of studies. Being able to monitor large-scale conditions on Earth on a regular basis at 30-meter resolution with a variety of multispectral bands (including near-infrared, middle-infrared, and thermal characteristics) gives users a rich dataset for numerous applications.

With Landsat imagery stretching back for more than 40 years and providing a constant source of observations, long-term tracking of environmental phenomena (such as ongoing land-use or land-cover changes over a particular area) can be performed. For instance, the growth of urban areas

FIGURE 11.11 The USGS Global Visualization Viewer utility used to examine and download a Landsat ETM+ scene (note the size of one scene in relation to the entire state of Michigan). *[Source: USGS]*

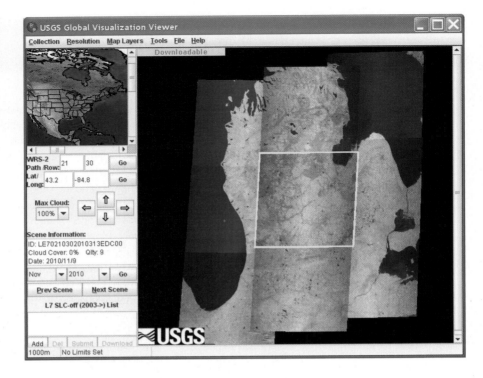

over time can be observed every 16 days (cloud cover permitting, of course), enabling planners to determine the spatial dimensions of new developments across a region. **Figure 11.12** on page 390 shows an example of utilizing Landsat imagery over time to track the patterns of deforestation in Brazil through examination of clear-cut areas and new urban areas built up in the forest. The color composite images show the rain forest in dark green shades, while the deep-cut patterns of forest clearing are clearly visible and the encroachment of development into the forest can be tracked and measured over time.

Beyond examining land-cover changes over time, the multispectral capabilities of Landsat imagery can be used for everything from measuring drought conditions and their effects on crops, to glacier features, to the aftermath of natural disasters. By examining the different band combinations provided by Landsat imagery, scientists can find clues to understanding the spread of phytoplankton in coastal waters, enabling environmentalists to try to head off potential algae blooms or red-tide conditions. Threats to animal habitats can be assessed through forest and wetland monitoring via Landsat imagery. Landsat images also allow scientists to track the movements of invasive species that threaten timber industries. *Hands-On Application 11.5: Applications of Landsat Imagery* on page 390 lets you dig further into the myriad uses of Landsat remote sensing.

Sentinel a satellite program operated by the European Space Agency

HANDS-ON APPLICATION 11.4

Viewing Landsat Imagery with GloVis and LandsatLook

With GloVis, Landsat imagery from multiple dates and sensors can be viewed, examined (for the amount of cloud cover and the available imagery dates), and downloaded for use by a GIS or an image processing or viewing software package. To use GloVis, go to **http://glovis.usgs .gov** (also make sure that you have Java installed on your machine and your browser will allow for a pop-up window from this Website). Either select a location from the map or enter latitude/longitude coordinates for a place. The Landsat scenes for the area will appear in a separate window, allowing you to select imagery by a specific path and row, obtain information about the scene (such as the percentage of cloud cover), and choose available imagery from different sensors. If the word "downloadable" appears in red in the upper-left-hand corner of the image, the raw band data for that Landsat scene can be obtained directly via GloVis. You'll also see that several other satellites and sensors have imagery available that can be viewed via GloVis.

The USGS has a second utility for viewing and downloading Landsat imagery called the Landsat-Look Viewer, available at **http://landsatlook.usgs .gov/viewer.html**. In the search box, type the name of the area you want to view Landsat imagery for and the Viewer will zoom to that location. Next, click on Select Scenes, and LandsatLook will assemble all available images for that location. Using the Time tools, you can slide the bar to view imagery of a specific date or use the arrows to advance between images. You can also adjust the transparency of the Landsat images to see how they match up with the base map. Under the Advanced Query options you can limit what dates you want to view and the maximum percentage of cloud cover that an image could have to be displayed, as well as which sensors the images are obtained from. By clicking on the Table button, you can obtain further information about each scene and also options for downloading the raw data for the scenes (by adding them to a download cart).

Expansion Questions

- What Path and Row is your local area located in? What Landsat imagery is available via GloVis for that Path/Row combination?

- How many Landsat images are available for your local area through LandsatLook? Are there any large-scale land-cover changes that you can see over time by viewing imagery from past and current years for your local area?

Other programs exist with similar capabilities to Landsat. For instance, the **Sentinel** program, operated by the European Space Agency, is a new series of satellites designed to acquire various types of remote sensing imagery. The Sentinel-2 mission will be a pair of satellites most similar to Landsat. The first **Sentinel-2** satellite, launched in 2015, will operate with a 10-day revisit cycle and collect multispectral imagery using its **MSI** (Multispectral Instrument) sensor in 13 bands similar to that of Landsat 8's OLI sensor. Sentinel-2 imagery is 10m, 20m, or 60m, depending on the bands being sensed. The USGS plans to archive and distribute Sentinel-2 imagery in conjunction with Landsat imagery.

NASA also operates a remote sensing project with capabilities related to the Landsat program. **EO-1** (Earth Observing 1) was launched in 2000 as part of NASA's New Millennium Program and placed in orbit so it was traveling right behind Landsat 7 (within 1 minute). EO-1 features a

Sentinel-2 a Sentinel mission, consisting of two satellites with characteristics similar to Landsat

MSI Multispectral Instrument, the multispectral sensor onboard the Sentinel-2 satellites

EO-1 a satellite launched in 2000 as part of NASA's New Millennium Program, which orbits 1 minute behind Landsat 7

FIGURE 11.12 Multiple dates of Landsat imagery of Rondonia, Brazil, showing the forested areas (in dark green) and the growing patterns of deforestation over time (in lighter shades). *[Source: NASA/Goddard Space Flight Center Scientific Visualization]*

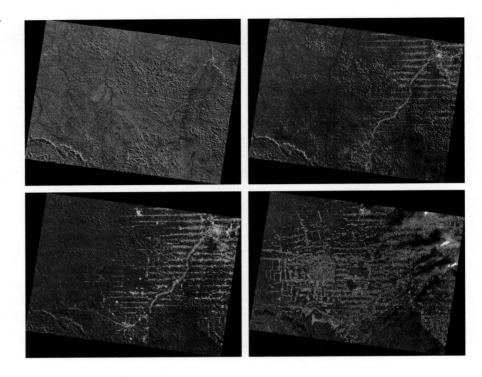

HANDS-ON APPLICATION 11.5

Applications of Landsat Imagery

Landsat imagery has been adopted for numerous environmental monitoring, measuring, and tracking applications. An excellent example of Landsat applications can be seen online at **http://world.time.com/timelapse**. The Website shows various places around Earth (such as Las Vegas, Dubai, or the Columbia Glacier) in a series of Landsat images between 1984 and 2012. You can also select the Explore the World option to type a location and see time-lapse Landsat imagery of that place from the last several years. As you scroll down on the Web page, you'll see further discussion of the applications of Landsat imagery (on topics such as climate change or urban growth) as well as several videos showcasing these topics as well.

To investigate these uses further, go to **http://landsat.gsfc.nasa.gov/?p=3501**. This Website, maintained by NASA, highlights several different categories of Landsat imagery applications, including Agricultural, Coastal, Geologic, Hydrologic, Mapping, and Environmental Uses. Select one of the categories of interest to you or related to your own field, and explore some of the case studies and applications presented in that column.

Expansion Questions

- How is Landsat imagery being applied in your field of choice?

- How does the long-term consistent use of multispectral imagery from Landsat lend itself to these kinds of applications?

multispectral sensor called **ALI** (Advanced Land Imager) that is similar to ETM+ but can image 10 different bands in a 37-kilometer swath width at 30-meter resolution (along with a 10-meter panchromatic band). ALI was also used to test and validate sensor technology planned for use in Landsat 8. EO-1 also carries **Hyperion,** a sensor that detects 220 bands at 30-meter resolution in a smaller swath width of 7.7 kilometers. Hyperion is an example of a **hyperspectral sensor,** which can measure hundreds of wavelengths simultaneously.

Although Landsat data has been widely used for numerous environmental studies, 30-meter resolution isn't nearly enough to do very detailed, high-resolution studies. At 30-meter resolution, broad land-use change modeling and studies can be performed, but a much finer spatial resolution would be required to pick out individual details in an image. High-resolution sensors have a different purpose than Landsat imagery—with a very fine spatial resolution, crisp details can emerge from a scene. Spy satellites or military surveillance satellites require this type of finer spatial resolution, but of course the imagery they acquire remains classified. Much of the non-classified and available high-resolution satellite imagery is obtained and distributed commercially rather than from government sources.

ALI the multispectral sensor onboard EO-1

Hyperion the hyperspectral sensor onboard EO-1

hyperspectral sensor a sensor that can measure hundreds of different wavelength bands simultaneously

 ## What Satellites Have High-Resolution Sensors?

There are several remote sensing satellites with high-resolution capabilities; this section describes a few of the more prominent ones. Much of the available high-resolution satellite imagery is distributed and sold by private companies from the satellites that they own and operate. For instance, the high-resolution satellite imagery seen in some images on Google Earth comes from a commercial provider. One thing to keep in mind is that prior to June 2014, U.S. government regulations did not permit the sale of imagery of less than 0.5-meter resolution to commercial customers (so imagery of finer resolution was resampled to this level before it was distributed). Commercial imagery after this date was permitted to be sold at its regular resolution (which is often much finer than 0.5 meters).

A common characteristic of satellites with high-resolution capabilities are a panchromatic band that allows pan-sharpening of other multispectral bands to provide very crisp and detailed image resolution—details of urban environments can be distinguished with little difficulty, although details of objects that are smaller than its resolution cannot be detected. Another common characteristic of these types of satellites is the ability to perform off-nadir viewing for satellite **tasking,** whereby the sensor can be directed on-demand to take imagery of specified locations up to several orbital paths away.

In the United States, the premiere commercial high-resolution satellite imagery provider is DigitalGlobe. In 2013, the two most prominent U.S.

tasking the ability to direct the sensors of a satellite on demand to collect imagery of a specific area

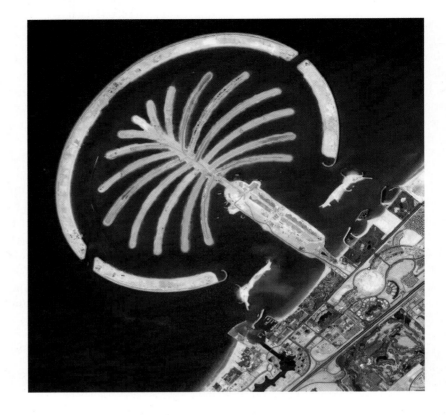

IKONOS a satellite launched in 1999 that features sensors with a multispectral spatial resolution of 3.2 meters and panchromatic resolution of 0.8 meters

QuickBird a satellite launched in 2001 by DigitalGlobe whose sensors have 2.4-meter multispectral resolution and 0.61-meter panchromatic resolution

WorldView-1 a satellite launched in 2007 by DigitalGlobe whose panchromatic sensor has 0.5-meter spatial resolution

GeoEye-1 a satellite launched in 2008 that features a panchromatic sensor with a spatial resolution of 0.46 meters

companies (DigitalGlobe and GeoEye) merged together to become one company simply called DigitalGlobe, which operates the set of satellites referenced here. Two early commercial launches for high-resolution satellite imagery were the IKONOS (launched in 1999) and Quickbird (launched in 2001) satellites. Both satellites featured off-nadir capabilities to reduce their revisit time to 1–3 days and their sensors had very high radiometric resolution (11 bits). **IKONOS** has two sensors, a four-band (blue, green, red, and near-infrared) multispectral sensor that has 3.2-meter spatial resolution and a panchromatic sensor with 0.8-meter spatial resolution. (See **Figure 11.13** for an example of IKONOS imagery that demonstrates the kind of spatial resolution this sensor can provide.) **Quickbird** was de-orbited and retired from remote sensing in early 2015. During its operational time, it collected multispectral imagery of 2.4-meter resolution and panchromatic imagery of 61 centimeters.

Improvements to the spatial resolution of commercial satellite imagery continued. In 2007, DigitalGlobe launched **WorldView-1,** which featured a single sensor, a panchromatic one, with 0.5-meter spatial resolution. The **GeoEye-1** satellite, launched in 2008, has two sensors—a four-band multispectral (blue, green, red, and near-infrared) sensor with 1.65-meter spatial

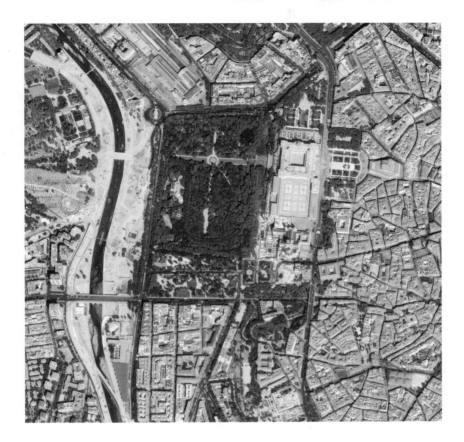

FIGURE 11.14 The Royal Palace and Campo del Moro gardens of Madrid, Spain, as seen by GeoEye-1. *[Source: GeoEye/ Science Source]*

resolution and also a panchromatic sensor with 0.46-meter spatial resolution. See **Figure 11.14** for an example of the high-resolution imagery acquired by the GeoEye-1 satellite. In October 2009, DigitalGlobe launched **World-View-2,** with 0.46-meter panchromatic and 1.84-meter 8-band multispectral spatial resolution.

The highest spatial resolution imagery currently commercially available is from the sensors onboard **WorldView-3,** launched in 2014, which provides 0.31-meter panchromatic and 1.24-meter 8-band multispectral resolution. WorldView-3 also contains advanced multispectral capabilities with 8 additional shortwave infrared bands and an additional 12 CAVIS bands for monitoring clouds, aerosols, and water. This spatial resolution represents the finest commercial resolution on the market today, currently making WorldView-3 the satellite with the sensors to beat when it comes to high-resolution imagery.

Another satellite, **WorldView-4** (previously named GeoEye-2), is, at the time of writing, being readied for launch in 2016. It is expected to feature both a multispectral sensor with an improved spatial resolution of 1.36 meters

WorldView-2 a satellite launched in 2009 by DigitalGlobe that features an 8-band multispectral sensor with 1.84-meter resolution and a panchromatic sensor of 0.46-meter resolution

WorldView-3 a satellite launched in 2014 by DigitalGlobe that features shortwave infrared and CAVIS bands, plus a panchromatic sensor of 0.31-meter resolution and 8 multispectral bands with 1.24-meter resolution

WorldView-4 a satellite planned for a future launch that is expected to feature a panchromatic sensor with a spatial resolution of 0.34 meters

and also a panchromatic sensor with an improved spatial resolution of 0.34 meters.

Another example of high-resolution commercially available imagery is the **Skysat** program, operated by Skybox Imaging, which in turn is owned by Google. At present, two Skysat satellites have been launched, with an additional 13 Skysats planned to be launched in the near future. Skysat-1 and Skysat-2 feature a panchromatic sensor with 0.9-meter resolution and a multispectral sensor with 2-meter resolution. The Skysat satellites also contain video cameras capable of recording up to 90 seconds of video as well. A Skysat satellite is an example of a **small satellite,** a type of remote sensing satellite that is simpler in design and smaller in size (it weighs between 220 and 1100 pounds) as well as cheaper to produce. With these kinds of factors, small satellites like Skysat are becoming more common and more commercially viable to put in orbit.

Even smaller remote sensing satellites, such as **nanosatellites** (which weigh between 2.2 and 22 pounds), are being designed and launched today. The **Cubesat** design is a version of these nanosatellites, where a simple box-like remote sensing satellite weighing less than three pounds and measuring about four inches long in size is placed into orbit and used for capturing high-resolution imagery. **Flock-1,** a constellation of 28 Cubesats (called Doves) designed by the PlanetLabs company and released into orbit from the International Space Station, beginning in 2014 is an example of series of Cubesats (see Figure 11.15). Each Dove satellite in Flock-1 can provide imagery with 3-5-meter resolution.

Skysat a high-resolution, small satellite program operated by Skybox Imaging and Google

small satellite a type of satellite that weighs between 220 and 1100 pounds

nanosatellite a type of satellite that weighs between 2.2 and 22 pounds

Cubesat a type of satellite with a weight of less than three pounds and a length of about four inches

Flock-1 a constellation of high-resolution remote sensing Cubesats operated by PlanetLabs

FIGURE 11.15 The Dove Cubesat satellites of Flock-1 being released from the International Space Station into Earth orbit. [Source: Image courtesy of NASA]

A different satellite program, **SPOT** (Satellite Pour l'Observation de la Terre), is currently owned and operated by Airbus Defence and Space. The SPOT series of satellites was initiated by CNES, the French space agency, and developed by Swedish and Belgian space agencies with the first satellite (SPOT 1) launched in 1986. SPOT 5 (launched in 2002) has a revisit cycle of 26 days but has off-nadir viewing capabilities that significantly improve some of its sensors' temporal resolution. SPOT 5 contains several sensors with differing resolutions, but the highest spatial resolution of a SPOT 5 image is 2.5 meters, allowing a significant amount of detail to be determined from a single image.

Airbus Defence and Space also developed SPOT 6 and SPOT 7, launched in 2012 and 2014, respectively. SPOT 6 and SPOT 7 feature improved resolution capabilities (including a 1.5-meter panchromatic spatial resolution and 6-meter resolution for four multispectral bands) as well as the stereo capability for production of digital elevation models (see Chapter 13). At the end of 2014, Airbus sold SPOT 7 to the Azerbaijan space agency, which renamed it Azersky. A related Airbus program is the **Pléiades** series of satellites (currently consisting of Pléiades-1A and Pléiades-1B), the first of which launched in 2011. Pléiades-1A and Pléiades-1B feature very high-resolution panchromatic (0.5-meter spatial resolution) and multispectral (2-meter spatial resolution of four bands, which can then be pan-sharpened to 0.5-meters) capabilities (see **Figure 11.16** for an example of Pléiades-1A's high-resolution imagery).

> **SPOT** a high-resolution imagery satellite program operated by Airbus Defence and Space

> **Pléiades** a series of satellites operated by Airbus Defence and Space, which feature sensors with a multispectral spatial resolution of 2 meters and panchromatic resolution of 0.50 meters

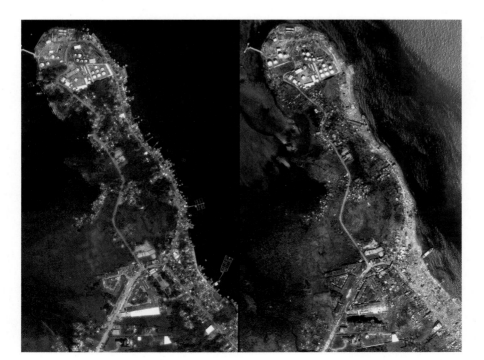

FIGURE 11.16 Pléiades-1A images of Tacloban, Philippines, showing the area before and after the Haiyan hurricane in 2013.
[Source: CNES 2013/dpa/Corbis]

How Can Satellites Be Used for Monitoring?

One of the advantages of the SPOT 6 and SPOT 7 (Azersky) satellites and the Pléiades series is that by having several satellites with similar capabilities in orbit, the amount of revisit time it takes for a location on the ground to be seen by one of them is greatly reduced. For instance, Pléiades-1A can view a location on day one, then Pléiades-1B can image the area on day two. Having multiple satellites (like the Pléiades twins) improves the overall temporal resolution of the satellites by being able to deliver imagery on a much faster basis.

This notion of using a constellation of satellites to be able to swiftly collect imagery of areas greatly enhances the ability to monitor areas on the ground. An example of this kind of high-resolution and quick-repeat capability is the **Rapideye** constellation operated and owned by Blackbridge. Rapideye is a series of five satellites with identical capabilities (5-meter resolution in five multispectral bands). With five satellites in orbit (see **Figure 11.17**), Rapideye can deliver daily imagery from anywhere on Earth. With such high-resolution sensors and a daily imagery capability, Rapideye is an excellent example of using remote sensing as a monitoring tool for environmental conditions such as land use or forestry. In 2014, Blackbridge announced plans to create a new Rapideye+ constellation with improved spatial resolution and additional multispectral bands.

The satellites operated by DigitalGlobe, Skybox, PlanetLabs, Airbus Defence and Space, and Blackbridge are only a few prominent examples of high-resolution imagery sources. Other companies and governments have launched or are planning to launch their own satellites with similar capabilities. See *Hands-On Application 11.6: Viewing High-Resolution Satellite Imagery* for examples of all of these high-resolution images online.

Rapideye a constellation of five satellites operated by Blackbridge that feature sensors with 5-meter multispectral sensors and a daily global revisit

FIGURE 11.17 The Rapideye constellation of satellites in orbit. *[Source: BlackBridge]*

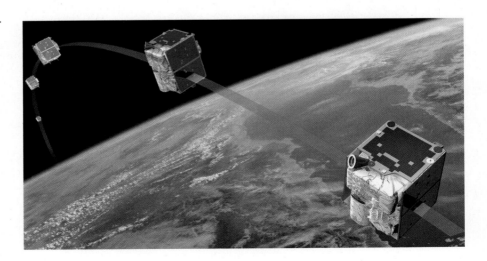

HANDS-ON APPLICATION 11.6

Viewing High-Resolution Satellite Imagery

While the data acquired by satellites like WorldView-3 and **Pléiades**-1A is sold commercially, you can view samples of these satellite images at several places online. The Satellite Imaging Corporation's Website provides a means for viewing high-resolution imagery. Open your Web browser and go to **www.satimagingcorp.com**. Select the option for Gallery. From the next Web page, your available options include GeoEye-1, IKONOS, Pléiades-1A, Pléiades-1B, Rapideye, QuickBird, Skysat-1, Skysat-2, SPOT 6, SPOT 7, WorldView-1, WorldView-2, and WorldView-3. Examine some examples of each sensor, keeping in mind the specifications of each. You can zoom in on the sample imagery from each satellite to get a real sense of what can be seen via high-spatial-resolution satellites. Note that the images available on the Website are just crisp graphics, not the actual satellite data itself.

Expansion Questions

- From viewing the gallery examples of high-resolution imagery, what kinds of analysis or interpretation are these images being used for?

- What current events have samples of high-resolution images available for viewing through the Website (and which satellite sensors were they acquired by)?

Using satellite imagery as a monitoring device has become more common now with the ever-increasing number of satellites and space agencies across the planet. Having high-resolution satellites that can be tasked to view specific off-nadir locations allows end users the capacity to rapidly capture imagery of events around the world, especially natural disasters such as floods, fires, tsunamis, or landslides. The **International Charter** "Space and Major Disasters" was put in place by the international community for just this purpose. The International Charter is an agreement between numerous government and commercial space agencies from around the world (including the United States, Canada, Russia, China, India, and many more) to provide satellite imagery and resources to aid in the event of a disaster emergency that activates the Charter.

International Charter an agreement between multiple international space agencies and companies to provide satellite imagery and resources to help in times of disasters

Utilizing satellite imagery for disaster management has extended throughout the international community. For instance, **UN-SPIDER** (United Nations Space-Based Information for Disaster Emergency and Response) is a program established in 2006 by the United Nations as a means to help get satellite imagery and other geospatial resources into the hands of disaster management personnel who need them to help aid in disaster mitigation and recovery.

UN-SPIDER the United Nations Space-Based Information for Disaster Emergency and Response program

An additional way of using satellite imagery resources in an emergency is to get them into the hands of everyone so that people can use their own geographic knowledge to help, regardless of where they are in the world. The notion of using satellite imagery to help crowdsource (see Chapter 1) disaster recovery is demonstrated by the Tomnod program set up by Digitalglobe (the word "tomnod" means "big eye" in Mongolian). Through **Tomnod**,

Tomnod an online crowdsourcing initiative that applies satellite imagery for a variety of real-world issues

HANDS-ON APPLICATION 11.7

Crowdsourcing Satellite Imagery

The search for the missing Malaysian Airlines flight MH370 is perhaps the most publicized usage of Tomnod, but it's far from the only thing that can be done with it. Digitalglobe uses Tomnod to involve people from all over the globe in applying their time and knowledge to a variety of different global events. To check some of these out (or to even volunteer your own geographic knowledge), go to **www.tomnod.com**. On the Website, you'll see different ongoing activities that Tomnod is being used for. One example at the time of writing allowed users to view satellite imagery from Garamba to help stop elephant poaching. You can examine

Digitalglobe satellite imagery of many different unexplored areas and place a tag on the image if you identify helicopters, vehicles, or camps (all potential evidence of poaching). Take a look at some of the current and ongoing Tomnod campaigns.

Expansion Questions

- What kinds of current activities is Tomnod being used for to crowdsource information through the use of satellite imagery?

- What are some of the tips provided by Tomnod to help in using satellite imagery to identify various features of the campaign?

Digitalglobe makes high-resolution satellite imagery available online and gives end-users the ability to examine and make edits to the data to offer their own insights to emergencies. For instance, when the Malaysian Airlines flight MH370 went missing in 2014, Digitalglobe posted satellite imagery to Tomnod of the suspected areas where the plane may have gone down. Over two million people examined the hundreds of thousands of satellite images, marking things that they thought might have been evidence of the missing plane. Tomnod is active today with a variety of questions that rely on your time and expertise with satellite imagery to help answer (see *Hands-On Application 11.7: Crowdsourcing Satellite Imagery*).

Chapter Wrapup

Satellite remote sensing is a powerful tool for everything from covert surveillance to the monitoring of land use and land cover. The next chapter delves into a new set of geospatial satellite tools—the Earth Observing System, a series of orbital environmental satellites that continuously gather data about global processes. *Geospatial Lab Application 11.1* returns to the subject of using MultiSpec, and it has you doing more with Landsat 8 imagery and the multispectral capabilities of the OLI sensor. Before you get to the lab, check out the resources in *Satellite Imagery Apps* for some representative mobile apps as well as *Satellite Imagery in Social Media* for Facebook pages and Twitter accounts related to many of this chapter's concepts as well as a few YouTube videos that show off various satellite remote sensing applications.

Important note: The references for this chapter are part of the online companion for this book and can be found at **www.macmillanhighered.com/shellito/catalog**.

Satellite Imagery Apps

Here's a sampling of available representative satellite imagery apps for your phone or tablet. Note that some apps are for Android, some are for Apple iOS, and some may be available for both.

- **DigitalGlobe FirstLook:** An app designed for examining satellite imagery of events before and after emergency situations
- **NASA Earth as Art:** An app for viewing stunning imagery from Landsat satellites, EO-1, and other NASA remote sensing satellites

Satellite Imagery in Social Media

Here is a sampling of available social media outlets related to this chapter's concepts.

On Facebook, check out the following:

- **European Space Imaging:**
  www.facebook.com/europeanspaceimaging
- **DigitalGlobe:**
  www.facebook.com/DigitalGlobeInc
- **Landsat:**
  www.facebook.com/NASA.Landsat
- **PlanetLabs:**
  www.facebook.com/PlanetLabs
- **Skybox Imaging:**
  www.facebook.com/skyboximaging
- **Tomnod:**
  www.facebook.com/Tomnod
- **UN-SPIDER:**
  www.facebook.com/UNSPIDER

Become a Twitter follower of:

- **Airbus Defence and Space:** @AirbusDS_GEO
- **Blackbridge:** @BlackBridgeCorp
- **DigitalGlobe:** @digitalglobe
- **European Space Agency—Earth Observation:** @ESA_EO

- **International Charter:** @DisastersChart
- **NASA updates on Landsat:** @NASA_Landsat
- **PlanetLabs:** @planetlabs
- **Skybox:** @skyboximaging
- **Tomnod:** @tomnod
- **UN-SPIDER:** @UN_SPIDER
- **USGS updates on Landsat:** @USGSLandsat

You Tube On YouTube, watch the following videos:

- **A Planetary Perspective: With Landsat and Google Earth Engine** (a Google video featuring Landsat and its many applications): **www .youtube.com/watch?v=Ezn1ne2Fj6Y**
- **Amazon Deforestation: Timelapse** (a Google video showing how Landsat imagery is used to track deforestation in the Amazon over time): **www.youtube.com/watch?v=oBIA0lqfcN4**
- **Digitalglobe** (Digitalglobe's YouTube channel showing several different videos about their various satellites): **www.youtube.com/user /digitalglobeinc**
- **GeoEye-1 Imagery Collection Animation** (a video showing a simulation of how Geo-Eye-1 collects imagery): **www.youtube.com /watch?v=yLZR5pg8Amw**
- **NASA | Landsat's Orbit** (a NASA video showing the orbital path and 16-day revisit cycle of Landsat): **www.youtube.com/watch?v=P-lbujsVa2M**
- **NASA | Landsat 8 Celebrates First Year in Orbit** (a NASA video about Landsat 8 and its imagery): **www.youtube.com/watch?v=0XVnQEJ6w_Y**
- **NASA | TIRS: The Thermal InfraRed Sensor on LDCM** (a NASA video showing the applications of the new thermal instrument onboard LDCM): **www.youtube.com/watch?v=c6OmfYEgzA0**
- **NASA | What Doesn't Stay in Vegas? Sprawl** (a NASA video showing the growth of Las Vegas from 1972 to 2010 using Landsat imagery): **www.youtube.com/watch?v=xFzdyxwx50M**
- **Skybox Imaging** (a YouTube channel showing off videos and imagery captured by the Skysat satellites): **www.youtube.com/channel /UCRJ1IiCsukaep0BGInSbEgg**
- **The View from Space: Landsat's Role in Tracking Forty Years of Global Changes** (a USGS video featuring many Landsat applications): **www.youtube.com/watch?v=IJ2VWKlraSw**

Key Terms

across-track scanning (p. 377)
ALI (p. 391)
along-track scanning (p. 376)
Corona (p. 373)
Cubesat (p. 394)
EO-1 (p. 389)
EROS (p. 377)
ETM+ (p. 382)
Flock-1 (p. 394)
GeoEye-1 (p. 392)
geostationary orbit (p. 374)
GloVis (p. 386)
Hyperion (p. 391)
hyperspectral sensor (p. 391)
IKONOS (p. 392)
International Charter (p. 397)
Landsat (p. 381)
Landsat 7 (p. 382)
Landsat 8 (p. 382)
Landsat scene (p. 385)
LandsatLook (p. 386)
LDCM (p. 382)
MSI (p. 389)
MSS (p. 382)
multispectral sensor (p. 382)
nanosatellite (p. 394)
NASA (p. 373)
near-polar orbit (p. 374)
off-nadir viewing (p. 380)

OLI (p. 383)
pan-sharpening (p. 378)
panchromatic sensor (p. 378)
Pléiades (p. 395)
QuickBird (p. 392)
radiometric resolution (p. 379)
Rapideye (p. 396)
Sentinel (p. 388)
Sentinel-2 (p. 389)
Skysat (p. 394)
SLC (p. 386)
small satellite (p. 394)
spatial resolution (p. 378)
spectral resolution (p. 380)
SPOT (p. 395)
sun-synchronous orbit (p. 376)
swath width (p. 374)
tasking (p. 391)
temporal resolution (p. 380)
TIRS (p. 383)
TM (p. 382)
Tomnod (p. 397)
UN-SPIDER (p. 397)
WorldView-1 (p. 392)
WorldView-2 (p. 393)
WorldView-3 (p. 393)
WorldView-4 (p. 393)
Worldwide Reference System
  (p. 385)

Landsat 8 Imagery

This chapter's lab application builds on the remote sensing basics of the previous chapter and returns to using the MultiSpec program. In this exercise, you'll be starting with a Landsat 8 scene and creating a subset of it to work with. During the lab, you'll examine the uses for several Landsat 8 band combinations in remote sensing analysis.

Objectives

The goals for you to take away from this exercise are:

▶ To familiarize yourself further and work with satellite imagery in MultiSpec

▶ To create a subset image of a Landsat 8 scene

▶ To examine different Landsat 8 bands in composites and compare the results

▶ To examine various landscape features in multiple Landsat 8 bands and compare them

▶ To apply visual image interpretation techniques to Landsat 8 imagery

Using Geospatial Technologies

The concepts you'll be working with in this lab are used in a variety of real-world applications, including:

▶ Research for the World Wildlife Federation, which utilizes Landsat imagery to study deforestation patterns and their impact on climate and species

▶ Environmental sciences, which have utilized the multispectral characteristics of Landsat imagery to map the locations of coral reefs around the world

[Source: Max Milligan/AWL Images RM/Getty Images]

Obtaining Software

The current version of MultiSpec (3.4) is available for free download at **https://engineering.purdue.edu/~biehl/MultiSpec.**

Important note: Software and online resources can change fast. This lab was designed with the most recently available version of the software at the time of writing. However, if the software or Websites have significantly changed between then and now, an updated version of this lab (using the newest versions) will be available online at **www.macmillanhighered.com/shellito/catalog.**

Lab Data

Copy the folder Chapter11—it contains a Landsat 8 OLI/TIRS satellite image (called "landsat8sept272014.img") from 9/27/2014 of northern Ohio, which was constructed from data supplied via LandsatLook. The Landsat 8 image bands (bands 1–9 are sensed by OLI while bands 10 and 11 are sensed by TIRS) refer to the following portions of the electromagnetic (EM) spectrum in micrometers (mm):

- ▶ Band 1: Coastal (0.43 to 0.45 μm)
- ▶ Band 2: Blue (0.45 to 0.51 μm)
- ▶ Band 3: Green (0.53 to 0.59 μm)
- ▶ Band 4: Red (0.64 to 0.67 μm)
- ▶ Band 5: Near infrared (0.85 to 0.88 μm)
- ▶ Band 6: Shortwave infrared 1 (1.57 to 1.65 μm)
- ▶ Band 7: Shortwave infrared 2 (2.11 to 2.29 μm)
- ▶ Band 8: Panchromatic (0.50 to 0.68 μm)
- ▶ Band 9: Cirrus (1.36 to 1.38 μm)

▶ Band 10: Thermal infrared 1 (10.60 to 11.19 μm)

▶ Band 11: Thermal infrared 2 (11.50 to 12.11 μm)

Keep in mind that Landsat 8 imagery has a 30-meter spatial resolution (except for the panchromatic band, which is 15 meters).

Localizing This Lab

This lab focuses on a section of a Landsat scene from northeast Ohio. Landsat imagery is available for free download via LandsatLook at **http://landsatlook .usgs.gov.** This site will provide raw data that will have to be imported into MultiSpec and processed to use in the program. Information is available online for using this free Landsat data for use in MultiSpec at **https://engineering .purdue.edu/~biehl/MultiSpec/tutorials/MultiSpec_Tutorial_5.pdf.**

11.1 Opening Landsat Scenes in MultiSpec

1. Start MultiSpec.

2. MultiSpec will start with an empty text box, which will give you updates and reports of processes in the program. You can minimize the text box for now.

3. To get started with the Landsat image, select Open Image from the File pull-down menu. Alternatively, you can select the Open icon from the toolbar:

[Source: Purdue Research Foundation]

4. Navigate to the Chapter11 folder and select landsat8sept272014 as the file to open, then click Open.

5. A new dialog box will appear to let you set the display specification for the landsat8sept272014 image.

Set Display Specifications for:

landsat8sept272015.img

Area to Display

	Start	End	Interval
Lines	1	15921	2
Columns	1	15661	2

Display

Type: 3-Channel Color

Channels:

Red: 5 ☐ Invert

Green: 4 ☐ Invert

Blue: 3 ☐ Invert

Channel Descriptions...

Magnification: 0.083

Enhancement

Bits of color: 24

Stretch: Linear

Min-max: Clip 2% of Tails

Treat '0' as: Data

Number of display levels: 256

☐ Load New Histogram

Cancel   OK

[Source: Purdue Research Foundation]

6. Under **Channels,** you will see the three color guns available to you (red, green, and blue). Each color gun can hold one band (see the **Lab Data** section for information on which bands correspond to which parts of the electromagnetic spectrum). The number listed next to each color gun represents the band being displayed with that gun.

7. Display band 5 in the red color gun, band 4 in the green color gun, and band 3 in the blue color gun.

8. Accept the other defaults for now and click **OK.**

9. A new window will appear, and the landsat8sept272014 image will load in it.

11.2 Using Landsat 8 Imagery Bands

The Landsat OLI/TIRS image has several different bands (see Table 11.1 for which band represents which wavelengths), each with their own use. For instance, looking at the entire Landsat scene now (the 5-4-3) combination, you've got a broad overview of a large slice of northern Ohio using the near-infrared, red, and green bands.

1. Vegetated areas (such as grass or trees) are reflecting a lot of near-infrared light in the red color gun, causing those areas to appear in shades of red. However, there are a lot of other things in the image as well. Examine the Landsat scene, zooming in on some of the cyan areas on the lakeshore of Lake Erie, then answer Question 11.1.

> **Question 11.1** The features on the image in cyan are largely urbanized and developed areas. Why are they displayed in a cyan color on this image with the 5-4-3 band combination?

2. Open the landsat8sept272014 image again—but this time use band 10 in the red color gun, band 10 again in the green color gun, and also band 10 once more in the blue color gun. This will show band 10 in all three guns, so you will only see this band in greyscale. This version of the landsat8sept272014 image will load in a separate window.

3. Arrange the two windows (landsat8sept272014 in the 5-4-3 combination and landsat8sept272014 in the 10-10-10 combination) side by side to see both of them together. Keep in mind that band 10 in the Landsat 8 imagery is one of the thermal bands sensed by TIRS. Examine both of the Landsat scenes and answer Question 11.2.

> **Question 11.2** What do the brighter places on the 10-10-10 image correspond with? Why do these places mostly appear brighter than their surroundings in the 10-10-10 image?

4. Close the 10-10-10 version of landsat8sept272014.

5. Open the landsat8sept272014 image again, this time using a 9-9-9 combination (i.e., load the image with band 9 in the red gun, band 9 in the green gun, and band 9 in the blue gun). A new window will open with this image, which will be band 9 in greyscale. Place this 9-9-9 image side by side with your original 5-4-3 image.

6. Band 9 in the Landsat 8 imagery is designed for detecting cirrus clouds in imagery. Answer Question 11.3.

> **Question 11.3** Where are the cirrus clouds in this section of northern Ohio in the image? Why are they so hard to see in the regular 5-4-3 image?

7. Close the 9-9-9 image when you're done so you're only working with the 5-4-3 image.

11.3 Subsetting Images and Examining Landsat Imagery

Right now, you're working with an entire Landsat 8 scene, which is an area roughly 170 kilometers long by 183 kilometers wide (as shown in the graphic below—you are using the scene encompassed by path 19, row 31). For this lab, we want to focus only on the area surrounding downtown Cleveland. You will have to create a subset—in essence, "clipping" out the area that you're interested in and creating a new image from that.

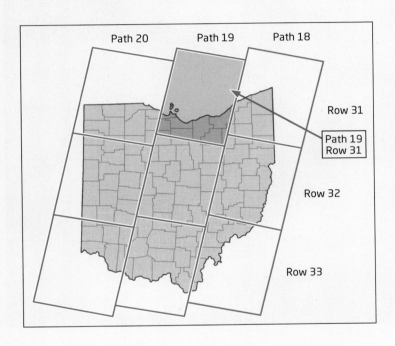

1. Zoom to the part of the landsat8sept272014 image that shows Cleveland (as in the following graphic):

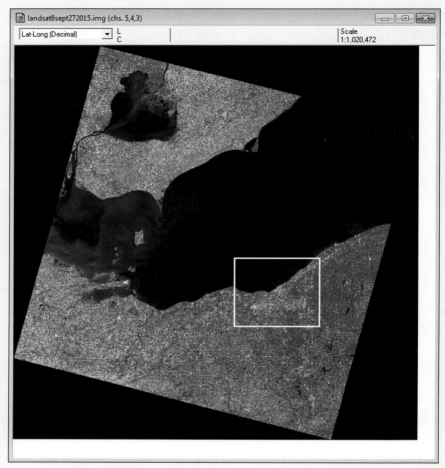

[Source: Purdue Research Foundation]

2. In the image, you should be able to see many features that make up downtown Cleveland—the waterfront area, a lot of urban development, major roads, and water features.

3. In order to create a new image that shows only Cleveland (a suggested region is shown in the graphic above), select Reformat from the Processor pull-down menu, then select Change Image File Format.

4. You can draw a box around the area you want to subset using the cursor, and the new image that's created will have the boundaries of the

box you've drawn on the screen. However, for the sake of consistency in this exercise, use the following values for the Area to Reformat:

a. Lines:
- Start 8977
- End 10729
- Interval 1

b. Columns:
- Start 9193
- End 11161
- Interval 1

Leave the other defaults alone and click OK.

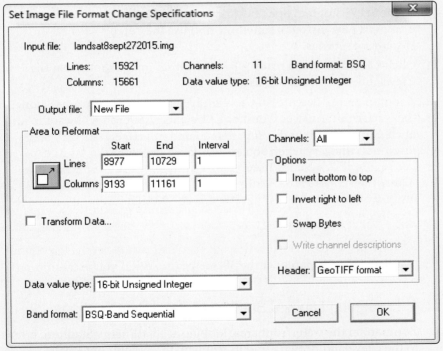

[Source: Purdue Research Foundation]

5. In the Save As dialog box that appears, save this new image in the Chapter11 folder (call it clevsub.img). Choose Multispectral for the Save as Type option (from the pull-down menu options next to Save as Type). When you're ready, click Save.

6. Back in MultiSpec, minimize the window containing the landsat8sept272014 image.

7. Open the clevsub image you just created in a new window (in the Open dialog box, you may have to change the Files of Type that it's asking about to "All Files" to be able to select the clevsub image option).

8. Open the clevsub image with a 5-4-3 combination (band 5 in the red gun, band 4 in the green gun, and band 3 in the blue gun).

9. Use the other defaults for the Enhancement options (stretch is Linear and range is Clip 2% of tails).

10. In the Set Histogram Specifications dialog box that opens, select the Compute new histogram method, and use the default Area to Histogram settings.

11. Click OK when you're ready. The new subset image shows that the Cleveland area is ready to use.

12. Zoom in on the downtown Cleveland area, especially the areas along the waterfront. Answer Questions 11.4 and 11.5. (You will want to also open Google Earth and compare the Landsat image to its very crisp resolution imagery when you answer these questions.)

> **Question 11.4** What kinds of features on the Cleveland waterfront cannot be distinguished at the 30-meter resolution you're examining?

> **Question 11.5** Conversely, what specific features on the Cleveland waterfront are apparent at the 30-meter resolution you're examining?

11.4 Examining Landsat Bands and Band Combinations

1. Zoom in on the waterfront area, looking at FirstEnergy Stadium, home to the Cleveland Browns, and its immediate surrounding area.

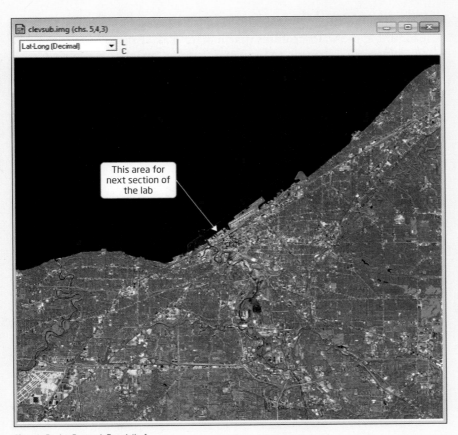

[Source: Purdue Research Foundation]

2. Open another version of the clevsub image using the 4-3-2 combination.

3. Arrange the two windows on the screen so that you can examine them together, expanding and zooming in as needed to be able to view the stadium in both three-band combinations.

Question 11.6 Which one of the two-band combinations best brought the stadium and its field to prominence?

Question 11.7 Why did this band combination best help in viewing the stadium and the field? (Hint: You may want to do some brief online research into the nature of the new stadium and its field.)

4. Return to the view of Cleveland's waterfront area. You'll now be examining the water features (particularly Lake Erie and the river). Paying careful attention to the water features, open four display windows, then expand and arrange them side by side to look at differences between them. Create the following image composites (one for each window):

   a. 4-3-2

   b. 7-6-5

   c. 5-3-2

   d. 2-2-2

   > **Question 11.8** Which band combination(s) is best for letting you separate water bodies from land? Why?

5. Return to the view of Cleveland's waterfront area. Focus on the urban features and vegetated features (zoom and pan where necessary to get a good look at urbanization). Open new windows with some new band combinations, and note how things change with each one as follows:

   a. 7-5-3

   b. 2-3-4

   c. 5-4-3

   > **Question 11.9** Which band combination(s) is best for separating urban areas from other forms of land cover (that is, vegetation, trees, etc.)? Why?

Closing Time

This exercise wraps up working with Landsat information as well as working with imagery in MultiSpec. Chapter 12 focuses on a whole different set of remote sensing satellites (part of the Earth Observing System), and its exercise will return to Google Earth to examine this imagery.

Exit MultiSpec by selecting Exit from the File pull-down menu. There's no need to save any work in this exercise.

12

Studying Earth's Climate and Environment from Space

NASA's Earth Observing System Program, Terra, Aqua, Aura, Suomi NPP, Other Earth Observing Missions, and NOAA Satellites

Climate change and global warming have become hot-button issues in today's world, affecting everything from a country's energy policies, to its economy, to its citizens' choices of homes and vehicles. The study and monitoring of the types of environmental processes that affect the land, oceans, and atmosphere occur on a global scale, and a continuous flow of information is needed. Monitoring changes in global sea surface temperature requires information captured from the world's oceans on an ongoing basis. In order to acquire information about the whole planet in a timely fashion, remote sensing technology can be utilized. After all, satellites are continuously orbiting the planet and collecting data—they provide the ideal technology for monitoring Earth's climate conditions and changes.

There are a lot of other remote sensing satellites up in orbit besides Landsat and the high-resolution satellites we discussed in Chapter 11. In fact, there's an entire set of science observatories orbiting hundreds of miles over your head that are dedicated to continuously monitoring things that may be critical to the planet's health, like sea surface temperature, the condition of glaciers, trace gases in the atmosphere, humidity levels, and Earth's heat radiation.

NASA's Earth Science Division operates a fleet of remote sensing missions dedicated to studying Earth's climate and environment. Among these is the Earth Observing System (**EOS**). Each satellite's capabilities are dedicated to an aspect of environmental monitoring or data collection related to Earth's surface, oceans, atmosphere, or biosphere. These EOS satellites often have coarse spatial resolution (for instance, some instruments have a spatial resolution of 1000 meters), but they consequently have a much larger swath width—and thus the ability to "see" an area up to 2300 kilometers across.

Their instruments are also set up to monitor very specific wavelengths tied to the purpose of their particular sensors. Their revisit times are usually

EOS NASA's Earth Observing System mission program

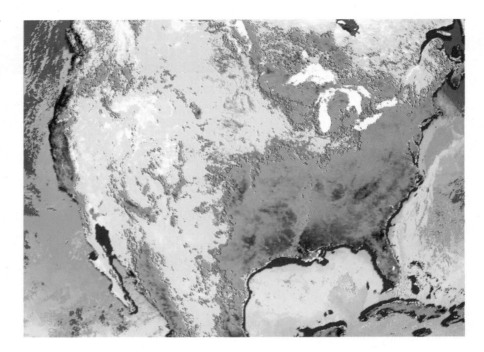

FIGURE 12.1 The health of land vegetation and the temperature of the sea surface measured through EOS remote sensing instruments. *[Source: MODIS Instrument Team, NASA Goddard Space Flight Center]*

relatively short, imaging the whole Earth in a couple of days to provide constant data streams of observation data about whatever environmental system they're examining.

There are a lot of different uses for EOS data, and many different products are derived from the information the satellites provide. This chapter can't hope to cover all of them; we will just demonstrate some sample applications. See **Figure 12.1** for an example of the kind of data and imagery that can be collected by EOS—the image was collected by a sensor onboard an EOS satellite over eight days in April 2000. It shows the health of vegetation (the "greening" of springtime) in North America and the sea surface temperature of its surrounding waters. The overall mission of EOS is to examine global environmental phenomena to advance knowledge and understanding of Earth's systems as a whole. The EOS missions were begun during the 1990s—there are several ongoing NASA Earth Science missions (with more planned for the future) with four key satellites being Terra, Aqua, Aura, and Suomi NPP.

What Is Terra and What Can It Do?

Terra the flagship satellite of the EOS program

Terra serves as the "flagship" satellite of the EOS program. It was launched in 1999 as a collaboration between the United States (NASA's Goddard Space Flight Center, NASA's Jet Propulsion Laboratory, and the NASA Langley

Research Center) and agencies in Canada and Japan, and it continues its remote sensing mission today. Terra's orbit is sun synchronous, and it flies in formation as one of four satellites called the **Morning Constellation**. This series of satellites gets this name because Terra crosses the Equator at 10:30 a.m. each day (it was originally named EOS-AM-1 because of this morning crossing). The other satellites in this morning constellation are Landsat 7 (which passes at 10:01 a.m.), EO-1 (which passes at 10:02 a.m.), and another satellite called SAC-C.

"Terra" means "Earth" in Latin. It's an appropriate name for the satellite, as its instruments provide numerous measures of (among many other functions) the processes involved with Earth's land and climate. There are five instruments onboard Terra, each with its own purpose and environmental monitoring capabilities. The data and imagery collected by Terra's instruments are processed into a number of specific products that are then distributed for use.

The first of Terra's instruments is **CERES** (Clouds and Earth's Radiant Energy System), a system designed to provide regular energy measurements (such as heat energy) from Earth's surface to the top of the atmosphere. CERES is used to study clouds and their effect on Earth's energy. CERES provides critical information about Earth's climate and temperature. As NASA notes, clouds play a defining role in the heat budget of Earth, but different types of clouds affect Earth in different ways—for instance, low clouds reflect more sunlight (and thus, help cool the planet), while high clouds reflect less but effectively keep more heat on Earth. By combining CERES data with data from other EOS instruments, scientists are able to examine cloud properties and gather information about Earth's heat and radiation (see **Figure 12.2** for an example of using CERES for examining Earth's radiated energy).

MISR (Multi-angle Imaging SpectroRadiometer) is the second of Terra's instruments. It has nine sensors that examine Earth, but each sensor is set up

Morning Constellation a set of satellites (including Terra, Landsat 7, SAC-C, and EO-1), each one passing the Equator in the morning during its own orbit

CERES the Clouds and Earth's Radiant Energy System instruments onboard Terra and Aqua

MISR the Multi-angle Imaging SpectroRadiometer instrument aboard Terra

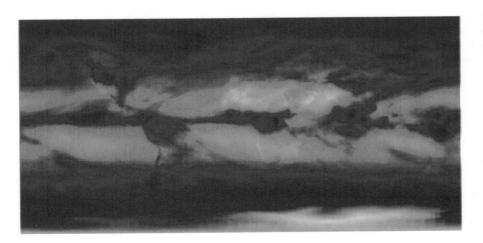

FIGURE 12.2 A global composite of CERES imagery from May 2015, showing the radiated energy from Earth (blue areas are colder, red areas are warmer).
[Source: NASA]

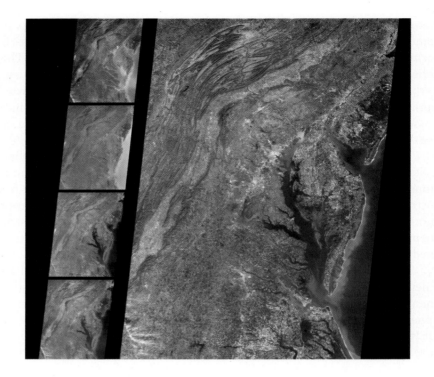

FIGURE 12.3 The operation of Terra's MISR instrument, showing the Appalachian Mountains, Chesapeake Bay, and Delaware Bay as captured from looking straight down (on the right) and from a variety of other angles (the four images on the left). *[Source: NASA/ GSFC/JPL, MISR Science Team]*

at a different angle. One of these sensors is at nadir, four of them are set at forward-looking angles, and the remaining four are set at the same angles but look backwards. MISR is a multispectral sensor, meaning that objects on the ground get viewed nine times (from different angles) in multiple bands, which means that Terra can effectively cover the entire planet in about nine days. By viewing things at multiple angles, stereo imagery (see Chapter 14) can be constructed, allowing scientists to measure things like cloud heights and the heights of plumes of smoke given off by wildfires or volcanoes. In addition, by showing the same area nine times in rapid succession, the images can provide a series of very quick temporal snapshots of the same phenomenon (**Figure 12.3**).

MISR data aids in environmental and climate studies by enabling scientists to better examine how sunlight is scattered in different directions, particularly by aerosols and atmospheric particles. For instance, a NASA article by Rosemary Sullivant details how MISR can study air pollution in the form of tiny particulates (caused by things like automobile or factory emissions, or by smoke and dust from wildfires, volcanic eruptions, and other natural phenomena) that can lead to increased health risks. Detailed information on concentrations of aerosols and atmospheric particles can help scientists understand the level of human exposure to these types of pollutants. MISR is extremely sensitive to measuring these types of aerosols, and its data can help scientists tell the difference between what's on the ground and what's in the atmosphere. By using MISR (in conjunction with other instruments) to study aerosols, scientists can help determine areas of high aerosol concentrations and provide data to aid in cleaning up air pollution.

A different Terra instrument used for measuring airborne pollutants is **MOPITT** (Measurements of Pollution in the Troposphere), which is designed for monitoring pollution levels in the lower atmosphere. MOPITT's sensors collect information to measure the amount of air pollution concentrations across the planet, creating an entire global image about every three days. Specifically, MOPITT monitors carbon monoxide concentrations in Earth's troposphere. Carbon monoxide, a global air pollutant, can be produced naturally as a result of volcanic eruptions and wildfires, or by human causes (such as vehicle emissions or industrial output).

Being able to track long-term carbon monoxide sources can aid in determining areas generating large amounts of pollutants and then monitoring how those pollutants are being circulated through the atmosphere. **Figure 12.4** shows an example of this kind of tracking in two MOPITT images of South America. The first, from March 2000, shows relatively low amounts of carbon monoxide. The second image, from September 2000, shows the dimensions of a large mass of carbon monoxide that was caused by biomass burning in the Amazon and the movement of carbon monoxide across the ocean to South America from fires in Africa.

Probably the most widely used instrument onboard Terra is **MODIS** (the Moderate Resolution Imaging Spectroradiometer), a sensor that

MOPITT the Measurements of Pollution in the Troposphere instrument onboard Terra

MODIS the Moderate Resolution Imaging Spectroradiometer instrument onboard Terra and Aqua

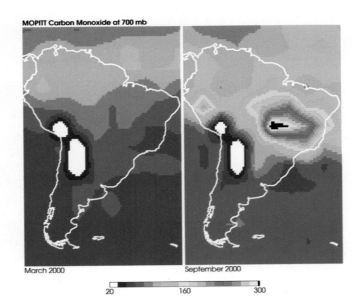

MOPITT Carbon Monoxide at 700 mb

March 2000

September 2000

20   160   300

FIGURE 12.4 MOPITT imagery, captured six months apart, showing growing carbon monoxide concentration levels in South America. *[Source: David Edwards and John Gille, MOPITT Science Team, NCAR/NASA]*

produces over 40 separate data products related to environmental monitoring. Being able to measure 36 spectral bands (including portions of the visible, infrared, and thermal) makes MODIS an extremely versatile remote sensing instrument, with multiple applications of its imagery—for instance, Figure 12.1 on page 414 shows vegetation health and water temperatures. MODIS products are used to examine numerous types of environmental features, including (among many others) snow cover, sea surface temperature, sea ice coverage, ocean color, and volcanoes. In addition, MODIS's thermal capabilities can track the heat emitted by wildfires, allowing for mapping of active fires across the globe (see **Figure 12.5** for an example of MODIS imagery of fires and their aftermath recorded in Wilsons Promontory National Park in Australia).

MODIS imagery has very coarse spatial resolution (most images have 1-kilometer resolution). However, the swath width of MODIS is 2330 kilometers, making it possible to examine broad-scale phenomena of large geographic regions. MODIS views most of the globe in one day, with complete global coverage being completed on a second day. With this

FIGURE 12.5 MODIS imagery examining the dimensions of a forest fire and resulting smoke plume at Wilsons Promontory National Park. The two smaller MODIS images show the park before and after the fire.
[Source: NASA/GSFC, MODIS Rapid Response]

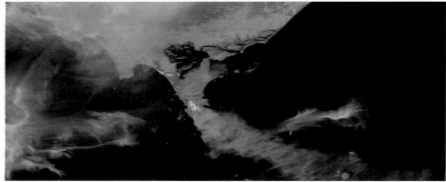

April 2, 2005 - Fire in Wilsons Promontory National Park, Victoria, Australia

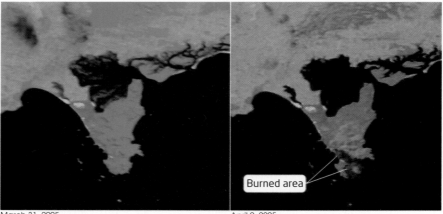

March 31, 2005   April 9, 2005

quick revisit time, MODIS allows scientists to look at changes in sea surface temperature or snow cover over time, make rapid and long-term
global leaf-area index studies, and track active large-scale fires, all on a
regular basis. For instance, the red and near-infrared bands sensed
by MODIS can be used to regularly compute global measures of NDVI
(see Chapter 10), thus allowing for the constant monitoring of the health
of the planet's trees and vegetation. Another application of MODIS lets
researchers study the health of phytoplankton in the oceans. As NASA
notes, when phytoplankton are unhealthy or under heavy stress, they take
the sunlight they have absorbed and re-emit it as fluorescence, which can
then be observed by MODIS. In this way, scientists can track potential
harmful algal (phytoplankton) blooms around the world via MODIS. See
Figure 12.6 for an example of MODIS imagery of Egypt and the River
Nile—new lakes in the desert are visible adjoining the Nile. Also, check
out *Hands-On Application 12.1: MODIS Rapid-Fire Online* for examples
of current MODIS imagery online.

The last Terra instrument is **ASTER** (Advanced Spaceborne Thermal
Emission and Reflection Radiometer). ASTER has relatively high spatial
resolution and is often referred to as a sort of "zoom lens" for Terra's other
instruments. With its thermal and infrared measurements, ASTER is also

ASTER the Advanced
Spaceborne Thermal
Emission and Reflection
Radiometer instrument
onboard Terra

FIGURE 12.6 MODIS
imagery showing new
lakes along the River Nile.
*[Source: Image by Robert
Simmon, Reto Stöckli, and Brian
Montgomery, NASA GSFC]*

HANDS-ON APPLICATION 12.1

MODIS Rapid-Fire Online

Recent and archived MODIS data is available online at NASA's MODIS Rapid Response System. Several MODIS images (not the raw data, but processed images) can be viewed at **http://rapidfire.sci.gsfc.nasa.gov/gallery**. Check out some of the images from recent dates and see what kind of phenomena NASA is tracking. Click "Display most recent images" on the Web page for the most recent MODIS acquisition data available. Use the imagery from the most recent available date. In this

chapter's Geospatial Lab Application, you'll examine more MODIS imagery using some other sources available from NASA.

Expansion Questions
- What kinds of phenomena are currently being monitored or examined by MODIS?
- How does the large swath width of MODIS imagery help in monitoring these kinds of phenomena?

used to obtain information about land surface temperature. In addition, through ASTER's thermal capabilities, natural hazards such as wildfires can be monitored, and phenomena such as volcanoes can be observed and data obtained on their health and lava flows (see **Figure 12.7** for an example of ASTER imagery of the lava flow from the Bezymianny Volcano on Russia's Kamchatka Peninsula). See *Hands-On Application 12.2: ASTER Applications* for further examples of ASTER imagery.

FIGURE 12.7 A composite image of the Bezymianny Volcano, located on Russia's Kamchatka Peninsula, and its lava flow as seen by ASTER's sensors. *[Source: NASA/GSFC/MITI/ERSDAC/JAROS, and U.S./Japan ASTER Science Team, University of Pittsburg]*

HANDS-ON APPLICATION 12.2

ASTER Applications

For a look at a variety of other applications of ASTER, go to **http://asterweb .jpl.nasa.gov/gallerymap.asp**—this is the ASTER Web Image Gallery, a collection of imagery detailing how ASTER is used in a wide variety of real-world situations. You can select a geographic location from the map—for instance, select some of the options nearest to where you are now, or choose from the categories on the left-hand menu. Check out how ASTER is used for archeology, geology, and hydrology, and also for various environmental studies.

Expansion Questions

* What are some of the specific uses of ASTER's capabilities for monitoring volcanoes and natural hazards?

* What are some of the specific uses of ASTER's capabilities for studying glaciers?

What Is Aqua and What Does It Do?

Launched in 2002 as a joint mission between NASA and agencies in Brazil and Japan, **Aqua** is another key EOS satellite. In Latin, "aqua" means "water," indicating the main purpose of the satellite: to examine multiple facets of Earth's water cycle. Analysis of water in all its forms—solid, liquid, and gaseous—is the key element of all six instruments onboard Aqua. Terra and Aqua are designed to work in concert with one another—their sun-synchronous orbits are set up similarly so that while Terra is on a descending path, Aqua is ascending (and vice versa). Because of this setup, Terra crosses the Equator in the morning while Aqua crosses in the afternoon (Aqua was originally called EOS-PM to complement Terra's original EOS-AM-1 name). This connection is further strengthened as both satellites carry a MODIS and a CERES instrument, in essence doubling the data collection performed by these two tools.

Beyond duplicate MODIS and CERES instruments, Aqua carries four others that are unique to its mission of examining Earth's water cycle: AMSU-A, HSB, AIRS, and AMSR-E. The **AMSU-A** (Advanced Microwave Sounding Unit) instrument is used to create profiles of the temperature in the atmosphere. AMSU-A uses 15 microwave bands and is referred to as a "sounder" because its instruments are examining a three-dimensional atmosphere, similar to the way "soundings" were used by ships to determine water depths. AMSU-A's data provides estimates not only of temperature data but also of precipitation and atmospheric water vapor. Similarly, the four microwave bands of the **HSB** (Humidity Sounder for Brazil) measured atmospheric water vapor levels (in other words, humidity) in the atmosphere until the instrument stopped operating in 2003.

Aqua's fifth instrument is **AIRS** (Advanced Infrared Sounder), whose uses include measuring temperatures and humidity levels in the atmosphere as well as information about clouds. AIRS data works in close conjunction with the

Aqua an EOS satellite whose mission is to monitor Earth's water cycle

AMSU-A the Advanced Microwave Sounding Unit instrument onboard Aqua

HSB the Humidity Sounder for Brazil instrument onboard Aqua (which ceased operation in 2003)

AIRS the Advanced Infrared Sounder instrument onboard Aqua

FIGURE 12.8 A global temperature map measured by AIRS showing the "polar vortex" of 2014. *[Source: NASA/JPL]*

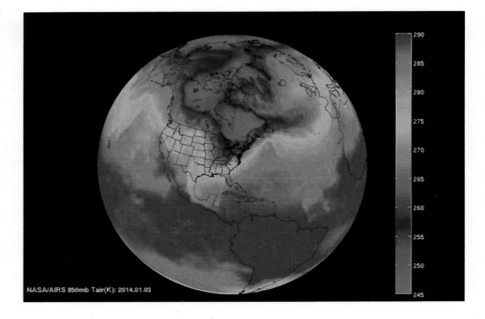

data from AMSU-A (and formerly with HSB) to create a combined sounding system for Aqua (sometimes referred to as the "AIRS Suite" for studying climate and helping to improve weather forecasting). However, even alone AIRS provides a wealth of climate-related data, from daily water vapor and carbon monoxide maps of the globe to being able to measure the daytime air temperature of the entire planet on a daily basis (see **Figure 12.8** and *Hands-On Application 12.3: Tracking Earth's Climate and Temperature with AIRS*).

AMSR-E (the Advanced Microwave Scanning Radiometer for EOS) is the last of the Aqua instruments. It uses 12 microwave bands to cover most of the planet in one day and completes a global dataset in the second day. Unfortunately, due to mechanical issues, AMSR-E stopped producing

AMSR-E the Advanced Microwave Scanning Radiometer for EOS instrument onboard Aqua

HANDS-ON APPLICATION 12.3

Tracking Earth's Climate and Temperature with AIRS

AIRS is capable of monitoring a wide range of climate-related conditions for Earth. To see some recent examples, go to **http://airs.jpl.nasa.gov/index.html**. From there, choose the option for Resources and then for Today's Earth Maps. A new set of maps will appear, showcasing the kind of data products available through AIRS, such as air temperature, water vapor, carbon monoxide, and ozone levels. The maps will be displayed of the most recently available data (likely of today's date and a recent time period). Examine each of them to see how AIRS is used for climate and environmental monitoring.

Expansion Questions

- Currently, what geographic areas have the warmest and coldest daytime air temperature?

- Currently, what geographic areas are seeing the highest levels of carbon monoxide in the atmosphere?

HANDS-ON APPLICATION 12.4

The Earth Observatory and 10 Years of Aqua

Aqua's instruments have been collecting imagery of Earth since 2002. NASA has set up an online retrospective of Aqua's accomplishments as part of its Earth Observatory Website (see page 430). This is sort of a "greatest hits" of Aqua over a decade of observing. To check them out, go to **http://earthobservatory.nasa.gov/Features /Gallery/aqua.php.** You'll see a slideshow of numerous different applications of Aqua imagery—clicking on a thumbnail will bring up a larger image as well as a full description of what type of climate or environmental condition is being studied.

Expansion Questions

- What types of environmental features are Aqua's instruments monitoring?

- How could a fully functioning AMSR-E be used to measure sea surface temperature in relation to hurricane locations? See the study of Hurricane Irene on the Website for more information.

- How can AIRS be used to monitor air quality conditions related to sulfur dioxide?

imagery in October 2011. AMSR-E monitored a variety of environmental factors that affect global climate conditions, including sea ice levels, global sea surface temperature, water vapor, wind speed, and amounts of global rainfall. For example, by assessing the amount of rain across the planet on a near-daily basis, AMSR-E could provide measures of how much precipitation storms can produce as they move across land or oceans. See *Hands-On Application 12.4: The Earth Observatory and 10 Years of Aqua* for more about AMSR-E and the other Aqua instruments.

What Is Aura and What Does It Do?

The **Aura** EOS satellite was designed as a collaboration between NASA and agencies in Finland, the Netherlands, and the United Kingdom. In Latin, "aura" means "breeze," which helps describe the satellite's mission—examination of elements in the air, especially the chemistry of Earth's atmosphere. Aura orbits in formation with Aqua and other EOS satellites to form what is referred to as the **A-Train** of satellites.

This organization of satellites is called the A-Train for two reasons: because it serves as the **Afternoon Constellation** of satellites (to complement the Morning Constellation that Terra flies in) and also because two of the key satellites (Aqua and Aura) begin with the letter "A." The newest member of the A-Train, OCO-2, a satellite designed to measure global carbon dioxide, became the new lead after its launch in 2014. A satellite called GCOM-W1 (the Global Change Observation Mission-Water, also referred to as "SHIZUKU") flies second in the constellation. Aqua comes next, CALIPSO follows closely after, then CloudSat follows about 103 seconds later, and

Aura an EOS satellite dedicated to monitoring Earth's atmospheric chemistry

A-Train another term for the Afternoon Constellation

Afternoon Constellation a set of satellites (including Aqua and Aura) that pass the Equator in the afternoon during their orbits

FIGURE 12.9 The configuration of EOS satellites that make up the A-Train. *[Source: NASA]*

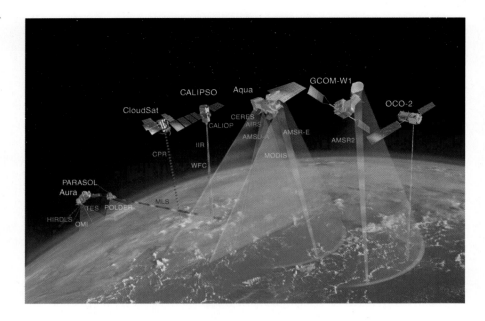

finally Aura brings up the rear (**Figure 12.9**). A satellite called PARASOL used to orbit before Aura, but it was removed from the formation on December 2, 2009. Other satellites are currently planned to become part of the A-Train after their launches. The combined data from the A-Train of satellites give scientists a rich data set for analysis of climate change questions.

Aura carries four instruments onboard, each utilized in some type of atmospheric observation (such as ozone concentrations or air quality) as it relates to global climate change. The first, **HIRDLS** (High Resolution Dynamics Limb Sounder), measures temperature, levels of water vapor, ozone, and other trace gases to examine qualities such as the transportation of air from one section of the atmosphere to another. As NASA notes, HIRDLS data is also used to examine pollution to see what is naturally occurring (from ozone) and what is generated by human activity. Similarly, Aura's **MLS** (Microwave Limb Sounder) instrument senses microwave emissions in five bands to examine carbon monoxide and ozone in the atmosphere. MLS data can be used as an aid in measuring ozone destruction in the atmosphere. Aura's third instrument, **TES** (Tropospheric Emission Spectrometer), measures things related to pollution, including ozone and carbon monoxide. Since TES is capable of sensing from the land surface up into the atmosphere, its data can be used to assess air-quality levels in urban areas by measuring the levels of pollutants and ozone in cities.

Aura's final instrument is **OMI** (Ozone Monitoring Instrument), which (as the name implies) is dedicated to keeping an eye on changes in ozone levels in the troposphere. OMI is a hyperspectral sensor that views sections of the visible-light spectrum as well as the ultraviolet portion of the

HIRDLS the High Resolution Dynamics Limb Sounder instrument onboard Aura

MLS the Microwave Limb Sounder instrument onboard Aura

TES the Tropospheric Emission Spectrometer instrument onboard Aura

OMI the Ozone Monitoring Instrument onboard Aura

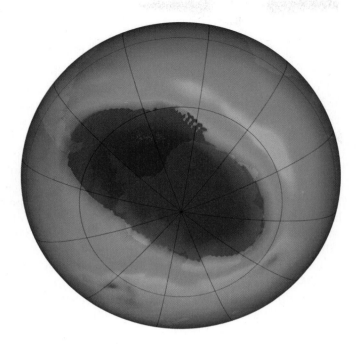

FIGURE 12.10 OMI imagery showing a hole in the ozone layer above Antarctica. *[Source: NASA/ Goddard Space Flight Center Scientific Visualization Studio]*

electromagnetic spectrum. In Chapter 10 we discussed how the atmosphere absorbs a lot of electromagnetic radiation and also how ozone is the primary greenhouse gas that absorbs harmful ultraviolet light. OMI data aids scientists in measuring the amount of ultraviolet radiation penetrating to Earth by examining clouds and ozone levels. **Figure 12.10** shows an example of imagery from OMI, showing the thinning (or hole) in the ozone layer above Antarctica. See *Hands-On Application 12.5: The Earth Observatory and 10 Years of Aura* for more examples of Aura's instruments' applications.

HANDS-ON APPLICATION 12.5

The Earth Observatory and 10 Years of Aura

Aura has been monitoring the Earth since 2004. Like the collection of Aqua imagery used in *Hands-On Application 12.4*, NASA has set up a "greatest hits" collection of impressive Aura applications via the Earth Observatory. Open your Web browser and go to **http:// earthobservatory.nasa.gov/Features/Gallery/ aura.php.** You'll see a slideshow of numerous different applications of Aura imagery. Clicking on a thumbnail will bring up a larger image as well as a full description of what type of climate or atmospheric condition is being studied.

Expansion Questions

- What specific effects are being monitored or examined using Aura's instrumentation?

- What kind of findings have there been on monitoring ozone levels in recent years with OMI?

- How has OMI been used for monitoring dust storms?

What Is Suomi NPP and What Does It Do?

Suomi NPP a joint satellite mission of NASA, NOAA, and the U.S. Department of Defense

NOAA the National Oceanic and Atmospheric Administration, a U.S. federal agency focused on weather, oceans, and the atmosphere

ATMS the Advanced Technology Microwave Sounder instrument onboard Suomi NPP

CrIS the Cross-track Infrared Sounder instrument onboard Suomi NPP

OMPS the Ozone Mapping Profiler Suite instrument onboard Suomi NPP

VIIRS the Visible Infrared Imaging Radiometer Suite instrument onboard Suomi NPP

Suomi NPP is one of the 'next generation' of Earth observing satellites, launched in 2011, and designed in a joint partnership between NASA, the U.S. Department of Defense, and **NOAA** (National Oceanic and Atmospheric Administration). NOAA operates and maintains a series of weather satellites (see page 431) and the "NPP" part of the name stands for the National Polar-orbiting Partnership, a reference to Suomi NPP's origins as a bridge between older and newer model NOAA weather satellites. Suomi NPP was named after the late Dr. Verner E. Suomi, a scientist at the University of Wisconsin–Madison who was considered "the father of satellite meteorology."

Suomi NPP carries five different instruments for monitoring Earth's climate and atmosphere, some of which have similar characteristics to those from earlier EOS missions, but now with more upgraded capabilities. The first of these is another CERES instrument, like those onboard Aqua and Terra (see page 415). The second instrument is **ATMS** (Advanced Technology Microwave Sounder), which is designed to provide profiles of atmospheric temperature and moisture. The third instrument is **CrIS** (Cross-track Infrared Sounder), which measures water vapor and temperature in the atmosphere. This data from ATMS and CrIS is of great aid in weather forecasting models. The fourth instrument is **OMPS** (Ozone Mapping Profiler Suite), which is designed to measure the levels of ozone across the globe. OMPS's ozone monitoring capabilities help scientists in determining whether the ozone layer is recovering.

The fifth (and probably most versatile) instrument onboard Suomi NPP is **VIIRS** (Visible Infrared Imaging Radiometer Suite). With a swath width of 3040km, VIIRS can image the entire Earth twice each day at 750km resolution. Like MODIS (see pages 417–419), VIIRS has a wide variety of applications and is able to sense 22 different bands at once. VIIRS imagery is used for studying (among other things) cloud properties, ocean color, sea surface temperature, land surface temperature, and fires across the planet. VIIRS is also capable of examining global chlorophyll content of the oceans (see **Figure 12.11**).

VIIRS also senses in what's referred to as the day-night band, which allows VIIRS to accurately capture imagery of Earth at night. Satellite imagery data is not often taken at night, but VIIRS' day-night band is specifically designed to do just that. The day-night band captures the low light of reflected moonlight from clouds, terrain, and water, as well as the light cast from urban features, producing impressive imagery of Earth at night (see **Figure 12.12**). For more about working with VIIRS imagery, see *Hands-On Application 12.6: The VIIRS View Spinning Marble.*

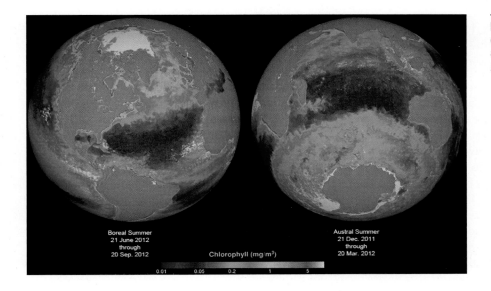

FIGURE 12.11 Global ocean chlorophyll content as measured by VIIRS. *[Source: NASA]*

FIGURE 12.12 The U.S. West Coast at night, as seen by VIIRS. *[Source: NASA]*

HANDS-ON APPLICATION 12.6

The VIIRS View Spinning Marble

The Raytheon Corporation has set up a free online tool that lets you examine a virtual globe of different applications of VIIRS imagery called the VIIRS View Spinning Marble. To use this utility, go to **www.raytheon.com/media /spinningmarble/web.html** (you will also have to install and run the Unity player for the Spinning Marble to work properly). When the Spinning Marble opens, you'll have your choice of three different VIIRS global images, accessible by icons at the bottom of the screen: regular visible light imagery, ocean chlorophyll content, and nighttime imagery. These three layers can be made more or less prominent by spinning their globe icons and thus adjusting the lines showing their display intensities at the bottom of the screen. You can pause and zoom in on the globe; after a brief time it will begin to spin again.

Expansion Questions

- With adjusting the nighttime imagery up to its maximum and the daytime visible imagery to its minimum levels, examine some areas around the globe (such as North America and Europe) in comparison to others (such as Africa and South America). How does the nighttime imagery help in determining urban population centers?

- By examining the ocean chlorophyll layer at its maximum level, where are the highest concentrations of chlorophyll found around the globe?

THINKING CRITICALLY WITH GEOSPATIAL TECHNOLOGY 12.1

How Can EOS Data Be Used in Studying and Monitoring Climate Change?

Climate change studies are going to be heavily reliant on accurate data related to Earth's climatic conditions and temperature changes as well as to the kinds of environmental conditions that influence alterations in Earth's processes. Changes in Earth's surface temperature are often used as indicators of climate change. What other indicators are used to study global climate change? Which instruments onboard EOS satellites can be used as data sources for each of these types of indicators?

What Other Earth Observing Satellites Are Out There?

There are a lot of other ongoing NASA Earth Science satellite missions beyond the four mentioned in this chapter, each dedicated to different aspects of Earth's environment and climate. For instance, data from the Cloudsat mission is used in studying questions related to clouds and climate change, while the SORCE mission studies energy from the sun. Recent Earth Observing missions include SMAP, a mission to study soil moisture from space; OCO-2, a mission to study carbon dioxide levels; and Aquarius, an instrument aboard the SAC-D satellite, which studies the

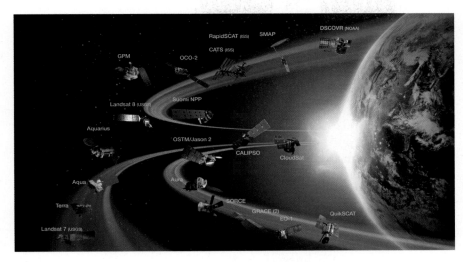

FIGURE 12.13 The various satellite missions of the NASA Earth Science Division *[Source: NASA]*

salinity of oceans. See **Figure 12.13** for all of these current missions, and also *Hands-On Application 12.7: NASA Eyes on the Earth* for a very cool NASA Website filled with interactive information and data about all of the EOS satellites.

HANDS-ON APPLICATION 12.7

NASA Eyes on the Earth

NASA has set up a Web application that allows you to view the real-time positions of Terra, Aqua, Aura, and the other satellite missions, and to examine 3D data maps of the products and imagery they capture. Open your Web browser and go to **http://eyes.nasa.gov/earth/index.html** (you may have to download and install a special plugin for all of the functions to work correctly). When Eyes on the Earth opens, a separate window will open that allows you to view the orbital paths of all the EOS satellite missions (which includes Terra, Aqua, and Aura, along with other missions such as CALIPSO or Cloudsat). Pressing the "Real Time" button will show you the current positions and paths of the satellites. Clicking on a particular satellite in orbit allows you to access information about that mission, view its ground path, and compare the size of the satellite with other objects.

By selecting one of the icons at the top of the screen, you can view EOS imagery and investigate environmental applications based on topics such as global temperature, sea level, and ozone concentrations.

After selecting a category, you can choose a range of dates of imagery to view as animation or as a series of images, or you can just view the most recently available data map.

Expansion Questions

- View the latest data map of sea level variations (available via the Sea Level icon). What areas of the globe are seeing the greatest rise in sea level (based on the OSTM imagery that loads)?

- View the latest data map of carbon monoxide levels (available via the Carbon Monoxide icon). What areas of the globe are experiencing the highest levels of carbon monoxide (based on the AIRS imagery that loads)?

- View the latest data map of daytime air temperatures (available via the Global Temperature icon). What areas of the globe are seeing the warmest temperatures during the day (based on the AIRS imagery that loads)?

With all of this Earth observation data and imagery collected on a near-daily basis, you might expect there would be a mechanism to get this information out into the hands of the public. Much of the available EOS data is processed and compiled into a series of data products rather than as raw imagery or values. NASA has a number of Web-based platforms in place to distribute these products (as well as EOS imagery) so that they can be viewed or analyzed:

Visible Earth a Website operated by NASA to distribute EOS images and animations of EOS satellites or datasets

▶ **Visible Earth** is a NASA-operated Website from which you can download pictures and animations related to EOS missions (online at **http://visibleearth.nasa.gov**). Visible Earth also features popular detailed composite images such as NASA's **Blue Marble** (showing the entire Earth from space).

Blue Marble a composite MODIS image of the entire Earth from space

NASA NEO the NASA Earth Observations Website, which allows users to view or download processed EOS imagery in a variety of formats, including a version compatible for viewing in Google Earth

▶ **NASA NEO** (NASA Earth Observations) is a Website from which you can download global EOS imagery from multiple dates in a format compatible with Google Earth, effectively "wrapping" the Google Earth virtual globe in EOS imagery (online at **http://neo.sci.gsfc.nasa.gov**). You'll be using NASA NEO in *Geospatial Lab Application 12.1*.

Earth Observatory a Website operated by NASA that details how EOS is utilized in conjunction with numerous global environmental issues and concerns

▶ **Earth Observatory** is a NASA Website devoted to global environmental and climate issues and how EOS contributes to our understanding of those issues (online at **http://earthobservatory.nasa.gov**). Think of it as a sort of online "magazine" with articles and EOS imagery from 1998 to the present (see **Figure 12.14** for an example of one of the "Images of the Day," as well as some interactive components to work with EOS imagery (see *Hands-On Application 12.8: Using the Earth Observatory to Work Interactively with EOS Imagery*).

FIGURE 12.14 Ice flowing to the Leidy Glacier in Greenland—one of the striking Images of the Day available via Earth Observatory. *[Source: NASA Earth Observatory image by Jesse Allen, using data from NASA/GSFC/METI/ERSDAC/JAROS, and U.S./Japan ASTER Science Team]*

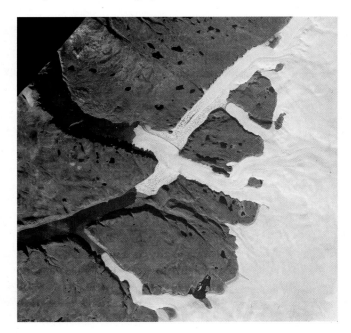

HANDS-ON APPLICATION 12.8

Using the Earth Observatory to Work Interactively with EOS Imagery

The Earth Observatory features the ICE (Image Composite Explorer) tool, which allows users to analyze applications of remotely sensed data (we used this tool back in *Hands-On Application 10.3: Examining NDVI with NASA ICE*). To get started using EOS data in one of the ICE scenarios, go to **http://earthobservatory.nasa .gov/Experiments/ICE/Channel_Islands**. This will allow you to examine MODIS products as they relate to conditions in California's Channel Islands. The Website supplies background material on the Channel Islands and shows how MODIS is used for monitoring phytoplankton conditions. Within the

page is a link to launch ICE, which will allow you to observe imagery related to sea surface temperature, chlorophyll content, and fluorescence.

Expansion Questions

*   How does MODIS measure chlorophyll content and fluorescence, and how can EOS imagery and data be used to monitor phytoplankton content in the oceans?

*   Examine the MODIS imagery for chlorophyll, fluorescence, and sea surface temperature. What connections can be drawn between these three subjects?

Also, keep in mind that NASA doesn't have a monopoly on broad-scale environmental remote sensing satellites. For example, NOAA operates and maintains a series of satellites dedicated to numerous environmental monitoring functions, including weather observations. The **GOES** (Geostationary Operational Environmental Satellite) program is a series of geostationary satellites tasked with monitoring weather conditions—each satellite constantly monitors the weather at the same location. The **POES** (Polar Orbiting Environmental Satellites) are polar orbiting and part of a long-running series of satellites. These NOAA satellites carry the **AVHRR** (Advanced Very High Resolution Radiometer) sensor, which views in six bands (early versions viewed only in five). See *Hands-On Application 12.9: Examining NOAA Satellite Imagery Applications* for some examples of NOAA satellite imagery as well.

> **GOES** Geostationary Operational Environmental Satellite; a series of geostationary satellites tasked with monitoring weather conditions
>
> **POES** Polar Orbiting Environmental Satellites; a series of polar-orbiting NOAA satellites
>
> **AVHRR** Advanced Very High Resolution Radiometer; the six-band sensor onboard NOAA satellites

HANDS-ON APPLICATION 12.9

Examining NOAA Satellite Imagery Applications

NOAA's remote sensing program is a critical part of global data collection as it relates to climate. An example of the use of NOAA imagery for environmental monitoring is available online at **http://coastwatch.glerl.noaa.gov**, which takes you to the NOAA CoastWatch Great Lakes Node. At this Website, you can view very recent (and near-real-time) AVHRR and GOES imagery.

Expansion Questions

*   For the most recent day and time available under the AVHRR imagery section, what sections of Lake Michigan and Lake Huron are seeing their warmest daytime sea surface temperatures?

*   What sections of the same lakes are seeing the warmest nighttime sea surface temperatures?

> **SOS** Science on a Sphere, a NOAA initiative used in projecting images of Earth onto a large sphere

Another example of enhanced geospatial visualization is NOAA's Science on a Sphere (**SOS**) device. SOS features a six-foot sphere that has animated images projected onto its surface, turning the sphere into a global model of Earth. With NOAA's SOS tool, viewers can view a simulated version of Earth's surface that allows them to see geospatial data such as ocean currents, sea surface temperature, land-use information, and global fires, as well as data products derived from satellites in the Earth Observing System (see **Figure 12.15**). Think of global EOS datasets that are projected onto an actual spherical surface several feet tall. Museums and science centers around the world are using SOS technology for geospatial education related to Earth and its climate and environment.

Chapter Wrapup

The satellites of the Earth Observing System provide a constant source of remotely sensed data that can be utilized in numerous studies related to the climate and environment of our planet. Whether they're used in monitoring wildfires, ozone concentrations, potential algae blooms, or tropical cyclones, the applications described in this chapter just scratch the surface of the uses of the EOS, and they give only a modest outline of how these geospatial tools affect our lives. This chapter's lab will have you start working with EOS imagery and Google Earth to get a better feel for some of the applicability of the data. Also, check out *Earth Observing Mission Apps* for some apps for your mobile device that use data and imagery from various satellites and *Earth Observing Missions in Social Media* for some Facebook, Twitter, and Instagram accounts, as well as some blogs and YouTube videos showcasing the many satellites of the EOS and their applications.

In the next chapter, we'll dig into a new aspect of geospatial technology: modeling and analyzing landscapes and terrain surfaces.

Important note: The references for this chapter are part of the online companion for this book and can be found at **www.macmillanhighered .com/shellito/catalog.**

Earth Observing Mission Apps

Here's a sampling of available representative EOS apps for your phone or tablet. Note that some apps are for Android, some are for Apple iOS, and some may be available for both.

- **Earth-Now:** An app from NASA's Jet Propulsion Lab featuring global climate data and information from EOS missions
- **SatCam:** An app that allows you to capture the conditions of the ground and sky when Terra, Aqua, or Suomi NPP passes overhead
- **VIIRS View Spinning Marble:** An app from Raytheon that allows you to view global Suomi NPP's VIIRS imagery on your mobile device
- **WxSat:** An app that displays weather imagery obtained from a geostationary satellite

The Earth Observing Missions in Social Media

Here's a sampling of some representative Facebook, Twitter, and Instagram accounts, along with some YouTube videos and blogs related to this chapter's concepts.

On Facebook, check out the following:

- **NASA Earth Observatory:**
  www.facebook.com/NASAEarthObservatory
- **NOAA Satellite and Information Service:**
  www.facebook.com/NOAANESDIS
- **NPP—Suomi National Polar-orbiting Partnership:**
  www.facebook.com/NASA.NPP

Become a Twitter follower of:

- **NASA Climate:** @EarthVitalSigns
- **NASA Earth Data:** @NASAEarthData
- **NASA Earth Observatory:** @NASA_EO
- **NASA OCO-2 Mission:** @IamOCO2
- **NASA SMAP Mission:** @NASASMAP
- **NOAA Satellites:** @NOAASatellites
- **Suomi NPP Satellite:** @NASANPP

Become an Instagram follower of:

- **NASA Goddard** (includes remotely sensed imagery from EOS satellites): **@nasagoddard**

On YouTube, watch the following videos:

- **Airsnews** (a YouTube channel featuring climate science and the AIRS instrument aboard Aqua):
  www.youtube.com/user/airsnews
- **Earth Observing Fleet with NPP and Aquarius** (a video showing the orbital paths of the various EOS satellites):
  www.youtube.com/watch?v=WT-WhWHPyk4
- **NASA | Aqua MODIS: Science and Beauty** (a NASA video that shows several applications of the MODIS instrument on Terra and Aqua):
  www.youtube.com/watch?v=1jqFxZI_2XY
- **NASA Aura** (a NASA YouTube channel with several videos showcasing the applications of Aura's instruments):
  www.youtube.com/user/NasaAura
- **NASA Earth Observatory** (a YouTube channel with videos from the Earth Observatory):
  www.youtube.com/user/NASAEarthObservatory
- **NASA | NPP and the Earth System** (a video about Suomi NPP and what it can do):
  www.youtube.com/watch?v=F-ejWjVRoIM
- **OCO-2 to Shed Light on Global Carbon Cycle** (a video about OCO-2 and its capabilities):
  www.youtube.com/watch?v=X5VWayNrtWk
- **Science on a Sphere** (a NOAA YouTube channel containing many videos showcasing remotely sensed imagery via SOS):
  www.youtube.com/user/scienceonasphere

For further up-to-date info, read up on these blogs:

- **Earth Matters** (a NASA Earth Observatory blog showcasing applications of the Earth Observing System satellites): **http://earthobservatory.nasa.gov/blogs/earthmatters**

- **EarthNow** (a blog concerning Science on a Sphere applications and visualizations): **http://sphere.ssec.wisc.edu**

Key Terms

A-Train (p. 423)
Afternoon Constellation (p. 423)
AIRS (p. 421)
AMSR-E (p. 422)
AMSU-A (p. 421)
Aqua (p. 421)
ASTER (p. 419)
ATMS (p. 426)
Aura (p. 423)
AVHRR (p. 431)
Blue Marble (p. 430)
CERES (p. 415)
CrIS (p. 426)
Earth Observatory (p. 430)
EOS (p. 413)
GOES (p. 431)
HIRDLS (p. 424)

HSB (p. 421)
MISR (p. 415)
MLS (p. 424)
MODIS (p. 417)
MOPITT (p. 417)
Morning Constellation (p. 415)
NASA NEO (p. 430)
NOAA (p. 426)
OMI (p. 424)
OMPS (p. 426)
POES (p. 431)
SOS (p. 432)
Suomi NPP (p. 426)
Terra (p. 414)
TES (p. 424)
VIIRS (p. 426)
Visible Earth (p. 430)

Earth Observing Missions Imagery

This chapter's lab will introduce you to some of the basics of examining imagery from three different Earth Observing Missions: Terra, Aqua, and Suomi NPP. You will be examining data from MOPITT as well as many types of imagery from MODIS and VIIRS. You will also be using online resources from NASA and others in conjunction with Google Earth.

Objectives

The goals for you to take away from this lab are:

▶ To utilize Google Earth as a tool for examining remotely sensed EOS imagery

▶ To examine the usage and function of MODIS imagery

▶ To examine the usage and functions of the day-night band imagery from VIIRS

▶ To use Terra and Aqua imagery for environmental analysis

Using Geospatial Technologies

The concepts you'll be working with in this lab are used in a variety of real-world applications, including:

▶ Forestry, which uses MODIS imagery on a near-daily basis to aid in active wildfire and forest fire monitoring

▶ Climatology, which uses MODIS imagery to track and monitor hurricanes and tropical storms and aid in emergency preparedness programs

[Source: © Karen Wattenmaker/www.forestphoto.com]

Obtaining Software

The current version of Google Earth Pro (7.1) is available for free download at **www.google.com/earth/explore/products/desktop.html**

Important note: Software and online resources can change fast. This lab was designed with the most recently available version of the software at the time of writing. However, if the software or Websites have significantly changed between then and now, an updated version of this lab (using the newest versions) is available online at **www.macmillanhighered.com /shellito/catalog**.

Lab Data

Copy the folder **Chapter12**—it contains the following KML datasets (for Google Earth) from NASA's Earth Observatory:

▶   russia_tmo_2012170.kml, a MODIS image showing fires in Siberia

▶   ge_44478.kml, a MODIS image showing a phytoplankton bloom off the coast of Iceland

▶   irene_amo_2011238.kml, a MODIS image showing Hurricane Irene

▶   isaac_tmo_2012241.kml, a MODIS image showing Hurricane Isaac

▶   black_marble.kml, a VIIRS image showing Earth at night

▶   Bakken_2012317.kml, a VIIRS image showing a section of northwestern North Dakota at night

Chapter12 also contains the following KML datasets from NASA/ Goddard Space Flight Center and NASA's Scientific Visualization Studio:

▶ a002900.kml, an animation of MOPITT imagery from Terra

▶ a003255.kml, an animation of MODIS imagery from Aqua

Localizing This Lab

This lab uses EOS data from a variety of locations across the globe, but there are several ways to examine some EOS imagery at a more local level.

In Section 12.1 take a look through the NASA Earth Observatory Website and archived information, and look for some related phenomena (such as other fires or storms) that are either relatively close to your location or whose MODIS imagery overlaps into your area.

In Section 12.2, examine one year of MOPITT data to determine conditions in your local area, rather than on a global scale, and keep track of the carbon monoxide (CO) levels for your region.

In Section 12.3, view the VIIRS nighttime imagery of your own local region, rather than examining Africa and South America.

In Sections 12.4 and 12.5, examine some of the Terra and Aqua imagery products to determine the climatic interactions in your region, rather than on a global scale. Open and examine images from multiple dates in Google Earth for your area, rather than one month's worth of data.

12.1 Viewing MODIS Imagery with Google Earth

1. Start Google Earth Pro.

2. We'll begin by examining a MODIS image of wildfires burning in Siberia. More information about this is available through NASA's Earth Observatory at **http://earthobservatory.nasa.gov/IOTD/view.php?id=78305**. The MODIS image comes in a format called KML, which can be read by Google Earth and overlaid on top of the normal Google Earth imagery. From the File pull-down menu, select Open. Navigate to the Chapter12 folder, select the russia_tmo_2012170.kml file, and click Open. A new item called "Siberia Burns" will be added to Google Earth's Temporary Places, and the view will rotate to the location of the image. Note that you can turn the image display on and off by clicking a checkmark in the "Siberia Burns" box.

3. Owing to the large extent of the MODIS image, you'll have to zoom out to see the whole thing and zoom in to different parts of it to examine the fires themselves.

> **Question 12.1** How are the fires being shown in this MODIS imagery?

> **Question 12.2** How is the extent of the fires being tracked via MODIS? (Hint: What else is visible in the MODIS scene aside from the fires themselves?)

4. Next, we'll examine a MODIS image of a phytoplankton bloom off the coast of Iceland. More information about this is available through NASA's Earth Observatory at **http://earthobservatory.nasa.gov/IOTD /view.php?id=44478.** Open a the KML called ge_44478.kml. Google Earth will rotate around to Iceland and a new MODIS scene will appear (and a new item called "Phytoplankton Bloom off Iceland" will be added to your Temporary Places).

5. Again, you may have to first zoom out to see the extent of the MODIS image and then zoom in to examine some of the details.

> **Question 12.3** According to the Earth Observatory, this is a MODIS image of phytoplankton in the North Atlantic. Why is this a phenomenon that is important enough to be tracked by MODIS, and what does the imagery show an observer about this phenomenon?

6. Lastly, we'll be examining MODIS imagery of large destructive storms, particularly Hurricane Irene from 2011 and Hurricane Isaac from 2012. More information about monitoring these two storms is available at **http://earthobservatory.nasa.gov/NaturalHazards/view.php?id =51931** (for Irene) and **http://earthobservatory.nasa.gov/IOTD /view.php?id=79008** (for Isaac). Open a new KML file called irene _amo_2011238.kml to view the MODIS imagery of Hurricane Irene. Google Earth will rotate to the East Coast of the United States. As with the other MODIS images, you'll have to zoom out to see the extent of the imagery of Irene.

7. Once you've examined the MODIS image of Irene, open a new KML file called isaac_tmo_2012241.kml, which is of Hurricane Isaac. Zoom out from the image. You'll see the two hurricane images overlap, so you'll have to turn one image off and the other on to examine both of them.

> **Question 12.4** How are the scope and capabilities of the MODIS instrument used for monitoring weather or storm formations?

8. At this point you can turn off all four of your KML files.

12.2 Viewing VIIRS Imagery with Google Earth

1. Next, we'll take a look at some global imagery from VIIRS's day-night band. More information about this is available through NASA's Earth Observatory at **http://earthobservatory.nasa.gov/NaturalHazards /view.php?id=79793**. Open the file called black_marble.kml. A new item called "Black Marble" will be added to your Temporary Places. This is the nighttime equivalent of the "Blue Marble" captured by MODIS imagery. In this nighttime view of Earth, you'll see the city lights of urban areas shining brightly in the darkness.

2. Rotate the globe and zoom out so that you can view the continent of Africa and examine the imagery. You should also put a checkmark in the Borders and Labels option in the Layers box before answering Question 12.5 and 12.6.

> **Question 12.5** From looking at the continent as a whole through the VIIRS imagery, in which countries are the major urban population centers in Africa? (*Hint*: With the Borders and Labels layer turned on, you may have to zoom in a bit to see the labels appear with the country names.)

3. Next, change your view so that you can view all of Canada and examine the imagery.

> **Question 12.6** From looking at the continent as a whole through the VIIRS imagery, in which provinces are the major urban population centers in Canada? (*Hint*: With the Borders and Labels layer turned on, you may have to zoom in a bit to see the labels appear with the provinces' names.)

4. Turn off the black_marble.kml layer. Next, add the bakken_2012317. kml file to GE. A new layer called Gas Drilling, North Dakota, will be added to your Temporary Places, and the globe will zoom in to a section of northwestern North Dakota. Although North Dakota is sparsely populated, there are a lot of lights that can be seen by VIIRS at night. These represent gas and oil drilling sites of the Bakken shale play in the region. More information about this is available through NASA's Earth Observatory at **http://earthobservatory.nasa.gov/IOTD/view .php?id=79810**

> **Question 12.7** How does the VIIRS imagery help determine where shale and oil drilling is occurring in this section of North Dakota?

5. Turn off the bakken_2012317.kml layer when you're done.

12.3 Using NASA Scientific Visualization Studio Imagery in Google Earth

Data from NASA's Scientific Visualization Studio (SVS) provides imagery or animations pertaining to several dates, and these images are placed together as a time series. Some of these animations have been formatted to work with Google Earth as KML files.

1. Open a new KML file called a002900.kml. A new item will be added to Google Earth's Temporary Places called "Global Atmospheric Carbon Monoxide in 2000," which is a series of MOPITT images from March through December 2000. More information about this can be found at the SVS Website at **http://svs.gsfc.nasa.gov/vis/a000000 /a002900/a002900**. You'll see that a legend has been added to Google Earth as well, showing which colors on the image correspond to what levels of carbon monoxide. If necessary, use the Google Earth tools to adjust the view so that all of Earth is shown in the Viewer.

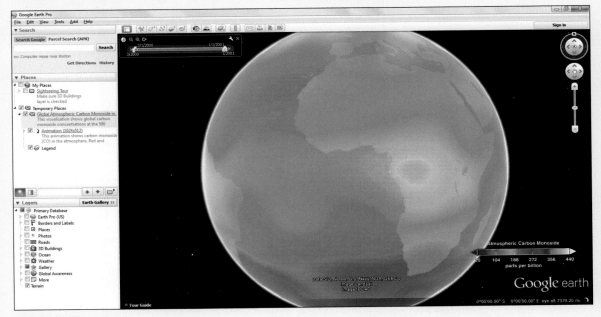

[Source: Google]

2. A slider bar has also been added to the top of the Google Earth display. You can use this slider bar to step through the MOPITT imagery day by day.

3. To animate the data, click on the Wrench icon on the slider bar.

[Source: Google]

4. In the Date and Time Options box that opens, select 3/1/00 as the Start date/time and 1/1/01 as the End date/time.

5. Reduce the Animation speed slider from fastest back to the slowest option on the left.

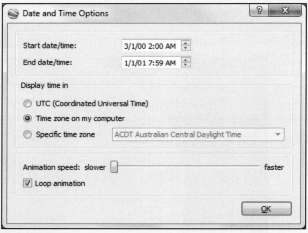

[Source: Google]

6. Place a checkmark in the Loop animation box. Click OK to close the dialog.

[Source: Google]

7. Click on the Play animation button on the slider bar to begin the animation.

8. Some frames might come up blank or the animation may not play at first—Google Earth seems to need to run completely through the animation once for it to load all the MOPITT imagery. On the second run-through of the animation, there should be no blank spots in the data.

9. You can press the Play animation button (it will now have a pause symbol) at any time to stop the animation. Answer Questions 12.5 and 12.6. (You will have to rotate the globe while the animation is playing to answer the questions. Also keep in mind the times and dates that are running in the upper-right-hand side of the screen; you may want to pause or rewind to examine certain dates in the past.)

> **Question 12.8** At what dates are the highest concentrations of carbon monoxide being monitored by MOPITT in Africa and South America?

> **Question 12.9** Beyond the carbon monoxide levels in Africa and South America (which were likely generated by wildfires), from what other geographic areas can you see high concentrations of carbon monoxide developing and spreading?

10. Turn off the Global Atmospheric Carbon Monoxide in the 2000 KML file when you're done.

11. Next, open another Scientific Visualization Studio KML called a003255.kml. A new item will be added to Google Earth's Temporary Places called "Aqua MODIS imagery of Hurricane Katrina," and Google Earth will rotate to the Gulf Coast of the United States. More information about this MODIS imagery can be found at **http://svs .gsfc.nasa.gov/vis/a000000/a003200/a003255**.

12. Set up the animation as you did with the MOPITT carbon monoxide animation, and play this new animation. It will show several days of imagery from the MODIS instrument on Aqua, during which time Hurricane Katrina was building before making landfall on the Mississippi Gulf Coast. You may have to pause or replay the animation to view all of the imagery.

> **Question 12.10** Given what you've seen of MODIS imagery and the information on the SVS Website, should MODIS be used as the sole instrument to monitor a massive storm formation like Hurricane Katrina? Why or why not?

13. Turn off the Aqua MODIS imagery of Hurricane Katrina when you're done.

12.4 Using the NASA Earth Observations (NEO) Web Resources

1. Open your Web browser and go to **http://neo.sci.gsfc.nasa.gov.** This is the Website for NEO (NASA Earth Observations), an online source of downloadable Earth observation satellite imagery. In this portion of the lab, you'll be using NEO's EOS imagery in conjunction with Google Earth in a KMZ format (which is a compressed version of KML).

[Source: NASA NEO]

2. Click on the Atmosphere option at the top of the page.

3. Select the option for Carbon Monoxide.

4. This is an image of global carbon monoxide concentration for one month, collected using the MOPITT instrument aboard Terra (see the section on About this dataset for more detailed information).

5. Rather than providing you with a view of a flat satellite image, NEO gives you the option of examining the imagery draped across a virtual globe.

6. Select 2014 for the year and then choose the option for October2014 (this will display imagery from October 2014, which is what we want to examine). Select Google Earth from the Downloads File Type pull-down menu, then click on the option for 1440 × 720 to begin the download of the KMZ file. Locate the downloaded KMZ file on your computer and double-click it to open the file in Google Earth.

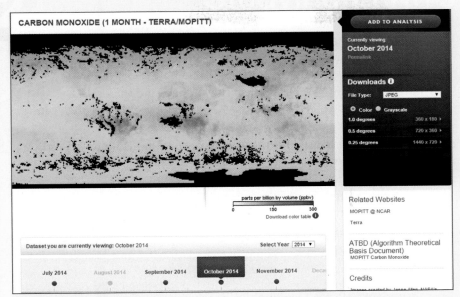

CARBON MONOXIDE (1 MONTH - TERRA/MOPITT)

ADD TO ANALYSIS

Currently viewing:
October 2014
Permalink

Downloads ⓘ
File Type: JPEG ▼

○ Color ● Grayscale
1.0 degrees   360 x 180 ›
0.5 degrees   720 x 360 ›
0.25 degrees   1440 x 720 ›

parts per billion by volume (ppbv)
0   150   300
Download color table ⓘ

Related Websites
MOPITT @ NCAR

Terra

ATBD (Algorithm Theoretical
Basis Document)
MOPITT Carbon Monoxide

Credits

Dataset you are currently viewing: October 2014   Select Year 2014 ▼

July 2014   August 2014   September 2014   October 2014   November 2014   Dec

[Source: NASA NEO]

7. Rotate Google Earth to examine the MOPITT imagery draped over the globe.

> **Question 12.11** Geographically, where were the highest concentrations of carbon monoxide for this month?

8. Back on the NEO Website, select the **Energy** tab, then select **Land/ Surface Temperature [Day]**.

9. Choose the option for 2014, then select **September**.

10. Use **Google Earth** for the **Downloads File Type** option and click on the 1440 × 720 option.

11. In Google Earth, turn off the MOPITT image (it will be in your Temporary Places listings).

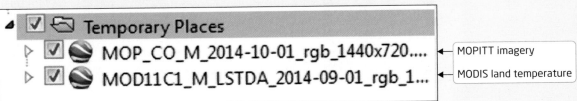

Temporary Places
☑ 🌐 MOP_CO_M_2014-10-01_rgb_1440x720.... ← MOPITT imagery
☑ 🌐 MOD11C1_M_LSTDA_2014-09-01_rgb_1... ← MODIS land temperature

[Source: Google]

12. Rotate Google Earth to examine the new MODIS image.

> **Question 12.12** Geographically, where were the lowest daytime land temperatures for this month?

13. Back on the NEO Website, select the Life tab, and then select Vegetation Index [NDVI] 2000+ MODIS. Choose 2014 for the year, then select July.

14. NDVI is a metric used for measuring vegetation health for a pixel. As we discussed in Chapter 10, the higher the value of NDVI, the healthier the vegetation at that location.

15. Use Google Earth for the Downloads File Type option and click on the 1440 × 720 option.

16. In Google Earth, turn off your other two Terra images.

17. Rotate Google Earth and zoom in to examine the new MODIS composite image.

> **Question 12.13** Geographically, where were the areas with the least amount of healthy vegetation on the planet for this month?

18. In Google Earth, turn off this last MODIS image.

12.5 EOS Imagery from the National Snow and Ice Data Center

1. NASA NEO is not the only source of EOS data for viewing in a virtual globe format. One example is the National Snow and Ice Data Center (NSIDC). Go to their Website at **http://nsidc.org/data/google_earth.**

2. Go to the View More Snow, Ice, Glaciers, Permafrost, and Sea Ice on the Globe section of the Website. Look for the section called Global Snow Cover from MODIS.

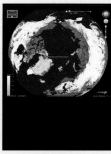

Global Snow Cover from MODIS

View global, monthly average snow cover by month for 10 years (2000-2009)

Note: This time series runs from October 2000 to May 2009. The following months are missing: March 2002 and December 2003

2009 snow cover (KML, 670 KB)   2004 snow cover (KML, 670 KB)
2008 snow cover (KML, 670 KB)   2003 snow cover (KML, 670 KB)
2007 snow cover (KML, 670 KB)   2002 snow cover (KML, 670 KB)
2006 snow cover (KML, 670 KB)   2001 snow cover (KML, 670 KB)
2005 snow cover (KML, 670 KB)   2000 snow cover (KML, 670 KB)

Data Source: MODIS/Terra Snow Cover Monthly L3 Global 0.05Deg CMG, Version 5

[Source: Google/NASA]

3. Read about what the imagery is showing, and then open the KML file for 2009 Snow Cover. The KML file will open in Google Earth, like the ones from NEO.

4. In Google Earth, turn off your other three NEO images.

5. In Google Earth's Temporary Places, there will be a new expandable item for NSIDC. By expanding this item and looking through the options, you'll see another expandable item called Snow Cover. This shows that snow data for each month of 2009 is available as an option that can be clicked on and off. To examine just one month of data, turn off all of the other months. You can also use the slider bar that appears at the top of the Google Earth view to adjust the snow data in Google Earth from month to month.

6. Rotate Google Earth and zoom in to examine the new MODIS image.

Question 12.14 Geographically, where were the greatest concentrations of snow cover in the southern hemisphere in June 2009?

Question 12.15 Geographically, where were the greatest concentrations of snow cover in the northern hemisphere in December 2009?

7. At this point, you can exit Google Earth Pro by selecting Exit from the File pull-down menu.

Closing Time

This exercise showed off a number of different ways of viewing and visually analyzing data available from EOS satellites. The next chapter's going to change gear, and we'll start looking at the ground that those satellites are viewing.

13 | Digital Landscaping

Topographic Maps, US Topos, Contours, Digital Terrain Modeling, Digital Elevation Models (DEMs), Lidar, 3DEP, and Applications of Terrain Data

There's one big element missing in geospatial technology that we haven't yet dealt with: the terrain and surface of Earth. All of the maps, imagery, and coordinates discussed have dealt with two-dimensional (2D) images of either the surface or the developed or natural features on the land. We haven't yet described how the actual landscape can be modeled with geospatial technology. This chapter delves into how terrain features can be described, modeled, and analyzed using several of the geospatial tools (like GIS and remote sensing) that we're already familiar with. Whether you are modeling topographic features for construction or recreational opportunities, or you're just interested in seeing what the view from above looks like, the landforms on Earth's surface are a principal subject for geospatial analysis (see **Figure 13.1** on page 450 for an example of visualizing Mount Everest with Google Earth).

First, when we're examining landforms on Earth's surface, we will have to assign an elevation to each location. In terms of coordinates, each x/y pair will now have a **z-value** that will indicate that location's elevation. These elevations have to be measured relative to something—when a point has an elevation of 900 feet, this indicates that it is 900 feet above something. The "something" represents a baseline, or **vertical datum,** that is the zero point for elevation measurements. Some maps of landforms indicate that the vertical datum is taken as mean sea level (represented by the National Vertical Datum of 1929). When this vertical datum is used and an elevation value on a map indicates 1000 feet, that number can be read as 1000 feet above mean sea level. Much geospatial

z-value the elevation assigned to an x/y coordinate

vertical datum a baseline used as a starting point in measuring elevation values (which are either above or below this value)

FIGURE 13.1 Mount Everest modeled as a digital terrain landscape and shown in Google Earth. *[Source: Google Earth]*

NAVD88 the North American Vertical Datum of 1988, the vertical datum used for much geospatial data in North America

data in North America utilize the North American Vertical Datum of 1988 (**NAVD88**) as this origin point for elevations. The U.S. National Geodetic Survey has announced they are currently developing a new vertical datum to take the place of NAVD88, but it will not be ready until 2022. Coastal terrain models may also use mean high water as their vertical datum.

How Can Terrain Be Represented on Topographic Maps?

topographic map a map created by the USGS to show landscape and terrain as well as the location of features on the land

A common (and widely available) method of representing terrain features is the **topographic map**. As its name implies, a topographic map is a printed paper map designed to show the topography of the land and the features on it. The U.S. topographic mapping program ran from 1945–1992 and the topographic maps it produced were published by the USGS (United States Geological Survey) at a variety of map scales. A common topographic map was a 1:24000 quadrangle, also referred to as a 7.5 minute topographic map, insofar as it displays an area covering 7.5 minutes of latitude by 7.5 minutes of longitude. Topographic maps were also available in smaller map scales, such as 1:100000 and 1:2000000.

DRG Digital Raster Graphic—a scanned version of a USGS topographic map

Topographic maps are available in a digital format as a **DRG** (Digital Raster Graphic). DRGs are scanned versions of topographic maps that have been georeferenced (see Chapter 3) so that they'll match up with other geospatial data sources when used in GIS. Usually, DRGs are available in **GeoTIFF** file format (see Chapter 7 for more information about TIFFs), which allows for high-resolution images while also providing spatial reference.

However, DRGs are just scanned images. To use the specific features on a topographic map (such as the streets or rivers) as separate GIS data sets, those features would have to be digitized (see Chapter 5) into their own geospatial layers. This is what a Digital Line Graph (**DLG**) is—the digitized features from a topographic map. DLGs are vector datasets representing transportation features (such as streets, highways, and railroads), hydrography features (such as rivers or streams), or boundaries (such as state, county, city, or national forest borders (see **Figure 13.2** for an example of several DLGs and their corresponding DRG).

Topographic maps (and thus, DRGs) model the landscape through the use of **contour lines**—imaginary lines drawn on the map that join points of equal elevation (see **Figure 13.3** on page 452). These contour lines are available in DLG format as a hypsography layer. A particular elevation on the surface is represented by a contour line drawn on the map. However, because elevation is really a continuously variable

> **GeoTIFF** a graphical file format that can also carry spatial referencing
>
> **DLG** Digital Line Graph—the features (such as roads, rivers, or boundaries) digitized from USGS maps
>
> **contour line** an imaginary line drawn on a map to connect points of common elevation

FIGURE 13.2 1:24,000 DLG data showing roads, railroads, boundaries, and hydrologic features of Youngstown, Ohio, along with the corresponding 1:24000 DRG of Youngstown. [*Source: (left) Libre Map Project (right) USGS*]

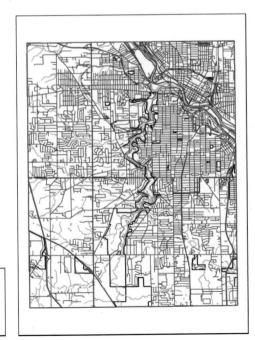
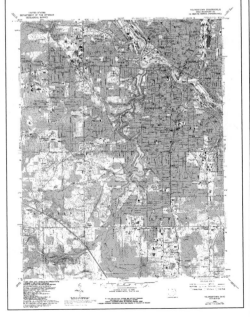

Youngstown DLGs
- Roads DLG
- Rails DLG
- Hydro DLG
- Boundaries DLG

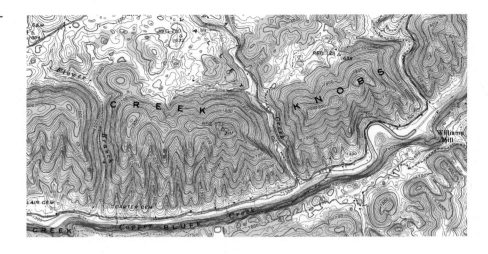

FIGURE 13.3 Contour lines as shown on a DRG (at a 20-foot contour interval). *[Source: USGS]*

phenomenon, it's often difficult to represent every change in elevation (such as from eight feet of elevation to ten feet of elevation) without overloading the map to the point of uselessness with contour lines. Thus, contour lines are drawn a certain elevation distance apart (say, a new contour line for every 50 feet of elevation).

This elevation or vertical difference between contour lines is called the **contour interval** and is set up according to the constraints of the map and the area being measured. In general, a wider contour interval is selected when mapping more mountainous terrain (since there are numerous higher elevations), and a narrower contour interval is used when mapping flatter terrain (since there are fewer changes in elevation and more details can be mapped). Small-scale maps (see Chapter 7) tend to use a wider contour interval because they cover a larger geographic area and present more basic information about the terrain, whereas large-scale maps generally utilize a narrow contour interval for the opposite reason (they show a smaller geographic area and thus can present more detailed terrain information).

A big thing to keep in mind when dealing with topographic maps is that they are no longer being produced by the USGS. The original USGS topographic mapping project lasted until 1992, with sporadic revisions made to a small percentage of the maps for a few years after that. The topographic map program was replaced by The National Map (see Chapter 1) to deliver digital geospatial data (including contour line data) for the United States. Since these topographic maps are no longer being produced, neither are their corresponding DRGs or DLGs.

Even though printed or scanned topographic maps haven't been produced for several years, the next generation of digital topographic mapping is the **US Topo** series. Like the old topographic maps, the US Topos are delivered in 7.5 minute quad sections, but in GeoPDF file

contour interval the vertical difference between two adjacent contour lines drawn on a map

US Topo a digital topographic map series created by the USGS to allow multiple layers of data to be used on a map in GeoPDF file format

format (see Chapter 7). A US Topo has multiple layers of data stored on the same map, which can be turned on and off according to what features you want to display on the map. US Topo maps include contours, recent transportation features (roads, railroads, airports), transportation names, contours, boundaries, and orthophoto images, all as separately accessible layers. A US Topo also contains a lot of data (including the scale, the dates of map creation, and similar information) in the white border that surrounds the map. This information-filled border is referred to as the map's **collar**, and like the other layers in the US Topo it can be displayed or hidden. See **Figure 13.4** for an example of a US Topo.

collar the white, information-filled border around a US Topo

FIGURE 13.4 A 2013 US Topo of Youngstown, Ohio, showing the available layers for the map. *[Source: USGS]*

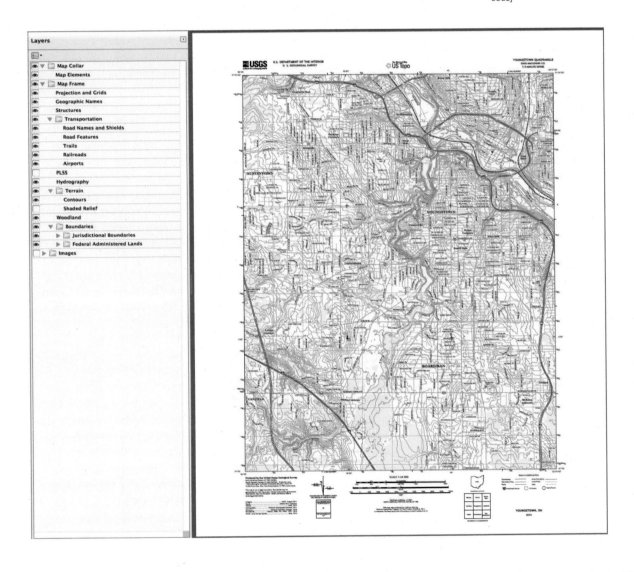

The data contained in the US Topos (such as the contours or transportation layers) is derived from national GIS data. The USGS produces over 100 US Topos per day and each US Topo is on a three-year revision cycle. Also, historic topographic maps are being converted over to the GeoPDF format as part of the Historic Topographic Map Collection (HTMC) and freely distributed online.

As a GeoPDF, each US Topo is a stand-alone product that can be used on a computer or on a mobile device. The GeoPDF can also store coordinate information for each location on the map, and a set of tools available from a special free plug-in allows you to measure distances, calculate the size of areas, and also connect the US Topo to a GPS device for obtaining your current location information (see *Hands-On Application 13.1: US Topos as GeoPDFs* for how to obtain and work with US Topos).

HANDS-ON APPLICATION 13.1

US Topos as GeoPDFs

Before working with US Topos, you'll need to install the TerraGo Toolbar in order to take advantage of all of the GeoPDF features available with the US Topo. The TerraGo Toolbar can be downloaded directly from **http://info.terragotech.com/download/terrago-toolbar**. With the GeoPDF downloaded and the toolbar installed, now it's time to get some US Topos to work with.

US Topos can be easily obtained through The National Map (see Chapter 1) in the same way you would get other data layers. An alternative way to browse current and historic US Topos is through the USGS Store at **http://store.usgs.gov**. Click on the link for Map Locator & Downloader and you'll be taken to a separate Website to search for available maps. At this point, you can first type the name of a place to search for available US Topos. For instance, search for East Lansing, Michigan, and the display map will shift to show that. Next, click on the red marker in the center of the East Lansing view and a new pop-up box will appear to show you all of the available maps for that location. For each of the maps labeled as 7.5 × 7.5 GRID, click on the circular icon with the blue and white cross on it to add that map to the download cart.

As you add the 7.5 maps (these will be US Topos or Historic Topographic Map Collection maps) to the download cart, you'll see your cart begin to fill. Once you have the 7.5 × 7.5 GRID maps for all available dates, click on Download All Cart Items. A zip file will download to your computer containing all of the GeoPDFs that you have added to the cart. When the zip file finishes downloading, unzip the US Topos and open the one with the most recent date. For further information on accessing and using US Topos, see the Quickstart guide from the USGS, online here: **http://nationalmap.gov/ustopo/quickstart.pdf**.

Expansion Questions

- What dates of US Topos and historic topographic maps are available for East Lansing, Michigan?

- What is the approximate elevation of Spartan Stadium on the Michigan State University campus? (Hint: You can use the information on the Orthoimage and Geographic Names layers to locate the campus and stadium and the Contours layer for elevation.)

- What is the contour interval of the US Topo? What is the highest elevation shown on the map?

THINKING CRITICALLY
WITH GEOSPATIAL TECHNOLOGY 13.1

If Everything's Digital, Do We Still Need Printed Topographic Maps?

A similar question was raised back in Chapter 2: If all of the historic USGS topographic map products are available digitally and in easily accessible formats that can be used on a desktop, laptop, or tablet, is there still a need to have printed copies (at a variety of map scales) on hand? Keeping in mind that the topographic mapping program ran through 1992 and only a small percentage of the maps were updated after that (and the estimated median date for the currentness of a map was 1979), are the available printed maps up to date enough to be of use? Do surveyors, geologists, botanists, archeologists—or any other professionals who require topographic information for fieldwork—need to carry printed topo quads with them? Is there a need for geographers to have printed versions of several quads when seamless digital copies (which can be examined side by side) are so easily available? Are there situations in which a printed topo map is necessary?

How Can Geospatial Technology Represent Terrain?

In geospatial technology, terrain and landscape features can be represented by more than just contour lines on a map. A digital terrain model (**DTM**) is the name given to a model of the landscape that is used in conjunction with GIS or remotely sensed imagery. The function of a DTM is both to accurately represent the features of the landscape and to be a useful tool for analysis of the terrain itself. The key to a DTM is to properly represent a z-value for x and y locations. With a z-value, the model can be shown in a perspective view to demonstrate the appearance of the terrain, but this doesn't necessarily make it a three-dimensional model. In fact, a DTM is usually best described as a two-and-a-half dimensional model. In a 2D model, all coordinates are measured with x and y values, without a number for z at these coordinates. In a **two-and-a-half-dimensional (2.5D) model**, a single z value can be assigned to each x/y coordinate as a measure of elevation at that location. In a full **three-dimensional (3D) model**, multiple z-values can be assigned to each x/y coordinate. Most DTMs have one elevation value measured for the terrain height at each location, making them 2.5D models (**Figure 13.5**).

> **DTM** a representation of a terrain surface calculated by measuring elevation values at a series of locations
>
> **two-and-a-half-dimensional (2.5D) model** a model of the terrain that allows for a single z-value to be assigned to each x/y coordinate location
>
> **three-dimensional (3D) model** a model of the terrain that allows for multiple z-values to be assigned to each x/y coordinate location

FIGURE 13.5 A comparison of (a) 2D, (b) 2.5D, and (c) 3D models of terrain. [Source: Esri]

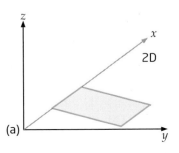
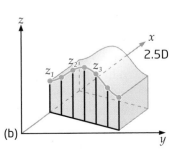
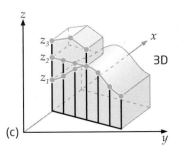

FIGURE 13.6 A TIN representation of the terrain around Gate City, Virginia. *[Source: Esri]*

An example of a type of DTM that's used in terrain modeling is a **TIN** (Triangulated Irregular Network), in which selected elevation points of the terrain, those that the system deems the "most important," are used in constructing the model. Points are joined together to form nonoverlapping triangles, representing the terrain surfaces (see **Figure 13.6**). While TINs are often used in terrain modeling in GIS, another very widely used type of digital terrain model is the DEM.

TIN Triangulated Irregular Network—a terrain model that allows for non-equally spaced elevation points to be used in the creation of the surface

 ## What Is a DEM?

A **DEM** is a digital elevation model, a specific type of model of the terrain and landscape that's produced by the USGS and others. A DEM is based on regularly spaced point data of elevations, but it can be converted to a raster grid representation (see Chapter 5) for use with other geospatial data, such as satellite imagery or GIS layers. In grid format, the elevation values are represented by the grid cell value, while the grid resolution represents the size of the cell being measured. Thus, DEM resolution is measured in much the same way that we measure the resolution of a remotely sensed image (see Chapter 10). If a DEM has 30-meter resolution, then each of the DEM's raster grid cells is set up at 30 meters in size.

DEM digital elevation model—a representation of the terrain surface, created by measuring a set of equally spaced elevation values

DEMs have been created through a variety of methods, including using DLG contour information, the use of photogrammetry, as well as remotely sensed data and stereo imagery (see Chapter 14). The ASTER sensor onboard the Terra satellite (see Chapter 12) was used to create a global DEM at 30-meter resolution. Another source of terrain data is the Shuttle Radar Topography Mission (**SRTM**), which originated as a part of a mission of the Space Shuttle Endeavor in February 2000. For 11 days, Endeavor used a special radar system to map Earth's terrain and topographic features from orbit, and the result was a highly accurate digital elevation model. At the mission's close, roughly 80% of Earth was examined and modeled as 90-meter DEMs (30-meter DEM data is also available for the United States). In 2014, the U.S. government announced that it would be releasing SRTM-2 (level 2 data, or approximately 30-meter resolution elevation data) for the entire globe.

SRTM the Shuttle Radar Topography Mission, flown in February 2000, which mapped Earth's surface from orbit for the purpose of constructing digital elevation models of the planet

An additional remote sensing method of terrain mapping is called **lidar** (a mashup of the words "light" and "radar" but also used as an acronym for Light Detection and Ranging). Rather than firing a microwave pulse at a ground target as a radar system would, lidar uses a laser beam to measure the terrain. In lidar, a plane is equipped with a system that fires a series of laser beams (between 30,000 and 250,000 pulses per second) at the ground. The laser beams are reflected from the ground back to the plane, and based on the distance from the plane to targets on the ground, the elevation of the landscape (as well as objects on the surface of the terrain) can be determined. GPS (see Chapter 4) is used in part to determine where the beams are striking the ground. See **Figure 13.7** for an example of how lidar data is collected.

The data collected by lidar is often referred to as **point cloud** data, owing to the high volume of point locations measured (you could potentially measure millions of locations in a single flight). After the data is collected and processed, the end result of a lidar mission is a highly accurate set of x/y locations with a z-value (see **Figure 13.8** on page 458 for an example of a lidar-derived elevation model). These "bare earth" elevation models of open terrain have a high degree of accuracy for their vertical measurements (between 15–30 cm), and today lidar equipment is also attached to UAS drones and can achieve 3–5 cm vertical accuracy.

However, Earth is covered by more than just open terrain. Lidar points will also measure the heights of features such as tree canopies or roofs of buildings and can be used to create realistic looking 3D models of these objects on Earth's surface. We'll get to 3D visualization in Chapter 14, but these digital surface models (**DSMs**) can be developed using lidar to measure the heights of all types of things on Earth's surface, not just the ground. The heights of objects like trees or buildings are

lidar a process in which a series of laser beams fired at the ground from an aircraft is used both to create highly accurate elevation models and also to measure the height of objects from the ground

point cloud the name given to the massive number of elevation data measurements collected by lidar

DSM digital surface model; a measurement of the heights of ground elevations as well as the objects on top of the ground as captured by lidar

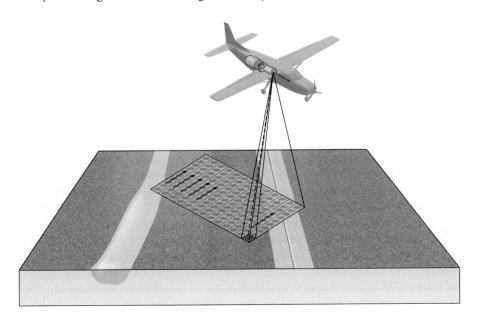

FIGURE 13.7 Gathering point cloud data: the process behind measuring terrain heights using lidar.

FIGURE 13.8 Mount
St. Helens as viewed from
lidar data. *[Source: USGS]*

removed to create an elevation model of the actual Earth's surface. As such, lidar has become a very versatile remote sensing tool that is used extensively in both government and private sector fields. A **LAS** file is the industry standard for storing lidar data, and many software packages (such as ArcGIS) are able to generate elevation models and calculate object heights using this data.

LAS the industry standard data format used for lidar data

Digital elevation data is available from the USGS through the **3DEP** (3D Elevation Program) initiative. 3DEP data is available for free download from The National Map (see *Hands-On Application 1.3: The National Map Viewer* on page 11 for instructions on how to obtain various geospatial datasets). The purpose of 3DEP is twofold: to have a digital elevation product that covers the entire United States at the best possible resolution, and to eliminate any gaps in the data. 3DEP is designed to be "seamless," insofar as its data uses the same datum, projection, and elevation unit. With 3DEP, users can select which sections of the national elevation dataset they need and then download those areas. 3DEP data is available at 1 arc second (about 30-meter), 1/3 arc second (about 10-meter), and 1/9 arc second (about 3-meter) resolutions, depending on the region. At the time of writing, some areas around the United States also had 1-meter resolution 3DEP data available. To see what's available for 3DEP data for your own area, see *Hands-On Application 13.2: U.S. Elevation Data and the National Map.*

3DEP the 3D Elevation Program, a U.S. government program that provides digital elevation data for the entire United States

3DEP draws from a variety of different sources for its elevation data. Lidar is a key method used to derive the DEMs that comprise 3DEP data, as well as radar remote sensing in areas such as Alaska. Point cloud data for

HANDS-ON APPLICATION 13.2

U.S. Elevation Data and The National Map

The USGS distributes 3DEP data through The National Map. To see what types of elevation data are available, go to the TNM download viewer platform at **http://viewer.nationalmap.gov /basic**. Under the options for Data on the left side, put a checkmark in the box next to Elevation Products 3DEP. This will allow to you see all of the available elevation resolutions for 3DEP. Under this, click on the option for Product Availability and a series of radio buttons will appear, allowing you to see what 3DEP elevation data resolutions are available at areas across the United States. For instance, clicking the radio button for DEM 1-meter will show you which areas of the country have available 1-meter 3DEP elevation data.

Next, you can go to a particular location to see what elevation data is available. For instance, in the Search location box, type in Topeka, Kansas. The Viewer will center on Topeka, but you may also need to zoom out a little bit from the initial search area to see the entire area. Select each of the radio buttons to see the kinds of 3DEP data available for download for Topeka.

Expansion Questions

- What resolutions of 3DEP elevation data are available for Topeka, Kansas?
- For your own local area, what types of elevation data and which resolutions of data are available?

areas as well as derived DSMs are also part of the 3DEP data sets. For 15 years, the main resource for digital elevation data within the United States was the **NED** (National Elevation Dataset) that provided bare-earth elevation data for the entire country. However, NED was retired in 2015, and its data has been rolled into the 3DEP elevation layers.

> **NED** the National Elevation Dataset, which provided digital elevation coverage of the entire United States, and is now part of 3DEP

How Can Digital Terrain Models Be Utilized?

With terrain data, many different types of terrain analysis can be performed. DTMs are used in geospatial technology for creating **viewsheds**—maps that show what can be seen (or not seen) from a particular vantage point. A viewshed is used for determining how far a person's visibility is (that is, what they can see) from a location before his or her view is blocked by the terrain. A viewshed can be created by selecting a vantage point and then computing how far a person's line of sight extends in all directions around them until something blocks the view. DTMs are also used for a variety of hydrologic applications, like calculating the accumulation of water in an area or delineation of stream channels or watersheds.

> **viewshed** a data layer that indicates what an observer can and cannot see from a particular location due to terrain

DTMs can be used to derive a new dataset of **slope** information—rather than elevation values, slope represents the change of elevations (and the rate of change) at a location by calculating the rise (vertical distance) over the run (horizontal distance). When a slope surface is created from a DTM, information can be derived not just about heights, but also about the steepness

> **slope** a measurement of the rate of elevation change at a location, found by dividing the vertical height (the rise) by the horizontal length (the run)

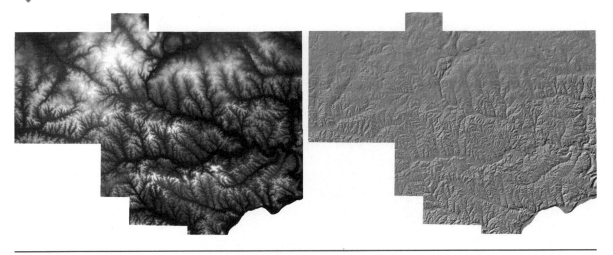

FIGURE 13.11 A DEM of Columbiana County, Ohio, and a hillshade of the DEM made using a Sun elevation of 45° and a Sun azimuth of 315°. *[Source: USGS/Esri]*

a Sun altitude of 45 degrees and a Sun azimuth of 315 degrees is shown in **Figure 13.11**.

Hillshading provides a good shaded map of what the terrain will look like in various lighting conditions, but there are plenty of features on the landscape (like roads and land cover) that aren't shown with a hillshade. In order to see these types of features, we can use a process called **draping**, which essentially shows the terrain model with a remotely sensed image (or another dataset) on top of it. **Figure 13.12** shows an example of a Landsat TM image draped over a DEM. Draping is achieved by first aligning the image with its corresponding places on the terrain model, then assigning the z-values from those locations on the terrain (in Esri terminology, these are referred to as **base heights**) to those locations on the image. In essence, locations on the image are assigned a z-value that correspond with the terrain model.

> **draping** a process in which an image is given z-values to match the heights in a digital terrain model
>
> **base heights** the z-values of a digital terrain model that can then be applied to an image in the process of draping

FIGURE 13.12 A DEM of the Laurelville/Hocking region of Ohio and a Landsat TM image draped over the DEM. *[Source: USGS/Esri]*

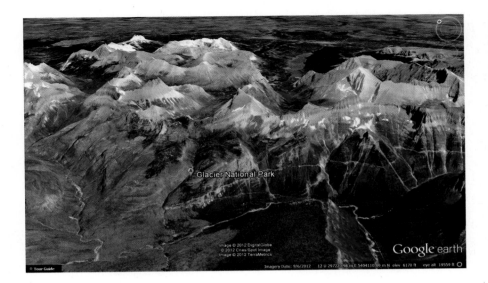

FIGURE 13.13
Examining imagery on a
digital terrain model of
Glacier National Park in
Google Earth. *[Source: Google
Earth]*

Draping is a common technique to show remotely sensed imagery on terrain features (the terrain model of Mount Everest in Figure 13.1 on page 450 appears in this way). Programs such as Google Earth can show pseudo-3D landscapes by draping imagery over the underlying terrain models (see **Figure 13.13**). By creating a new draped image in perspective view, we can get new visual information about the appearance of the landscape that's not directly obtainable through an examination of contours or non-perspective DTMs (see *Hands-On Application 13.3: Terrain and Imagery Examples in Google Earth* for more information). For example, draping a dataset of contours over a DTM and looking at it in perspective can visually demonstrate how contour lines match up with the elevations that they represent.

HANDS-ON APPLICATION 13.3

Terrain and Imagery Examples in Google Earth

Start up Google Earth and search for Glacier National Park in Montana. Use the zoom slider and the other Google Earth navigation tools to change the view so that you're examining the mountains and terrain of Glacier in a perspective view. Remember that what you're examining here is imagery that has been draped over a digital terrain model representing the landscape of this section of the country. Fly around the Glacier area (see Figure 13.13 for an example) and get a feel for navigation over draped imagery. You'll be doing more of this (among many other things) in *Geospatial Lab Application 13.1.*

Expansion Questions

- How does the addition of draped remotely sensed imagery on the terrain add to your sense of visualizing the features of the terrain?

- Note that as you fly through Glacier National Park, many of the land cover features (such as trees) are not rendered in 3D along with the terrain. Why is this?

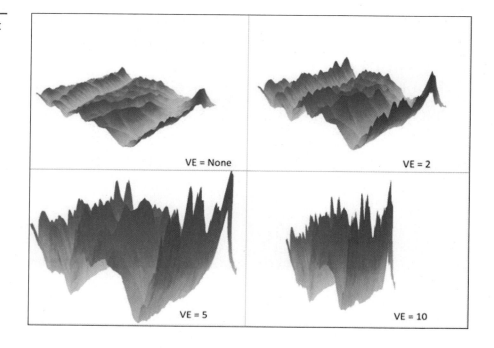

Even with hillshading or draping to improve the terrain's appearance, there's no getting around the fact that some sections of terrain don't have much variety in terms of changing elevations and features. In these cases, differences in landscape elevations or slopes may be difficult to see when viewing or interacting with a digital terrain model. The solution is to artificially enhance the differences between elevations so that the landscape can be better visualized. **Vertical exaggeration** is the process of artificially altering the terrain model for visualization purposes so that the vertical scale of the digital terrain model is larger than the horizontal scale. If the model's vertical exaggeration is "5x," then the vertical scale (the z-values) will be five times that of the horizontal scale. These types of artificial changes can really enhance certain vertical features (for example, valleys appear deeper and peaks appear higher) for visual purposes. The downside to vertical exaggeration is that it alters the scale of the data and should be used only for visualization. For a comparison of different vertical exaggerations applied to a DTM, see **Figure 13.14**.

vertical exaggeration a process whereby the z-values are artificially enhanced for purposes of terrain visualization

Chapter Wrapup

This chapter has explored both methods of modeling and ways of visualizing the landforms on Earth's surface. In the next chapter, we'll do a lot more with the 3D aspect of visualization. After all, there's more than just the terrain that can be viewed in 3D—there are plenty of structures, buildings, and natural growth that can be added to a pseudo-3D terrain model to improve the realism of the scene. Chapter 14, which picks up where this one leaves off, continues pursuing new avenues of 3D visualization.

Before then, check out *Terrain and Topography Apps* for some downloadable apps that will allow you to use topographic maps on your mobile device. Also take a look at *Digital Terrain in Social Media* for information from Twitter and Instagram accounts as well as some YouTube videos related to this chapter's concepts. This chapter's lab will allow you to implement the chapter topics using Google Earth Pro.

Important note: The references for this chapter are part of the online companion for this book and can be found at **www.macmillanhighered .com/shellito/catalog**.

Terrain and Topography Apps

Here's a sampling of available representative terrain apps for your phone or tablet. Note that some apps are for Android, some are for Apple iOS, and some may be available for both.

- **ElevMaps:** An app that provides 3D versions of landscapes and topographic maps for outdoors use
- **MyTopoMaps—Trimble Outdoors:** An app that gives you access to seamless USGS topographic maps

Digital Terrain in Social Media

Here's a sampling of some representative Twitter and Instagram accounts, along with some YouTube videos related to this chapter's concepts.

Become a Twitter follower of:

- **Open Topography:** @OpenTopography
- **USGS (United States Geological Survey):** @USGS
- **USGS News about Mapping:** @USGSNewsMapping

Become an Instagram follower of:

- **USGS:** @usgs

> **You Tube On YouTube, watch the following videos:**
>
> - **Elevation** (a USGS video about the National Elevation Dataset):
>   **www.youtube.com/watch?v=HMQQgPj1mX4**
> - **Exploring US Topo GeoPDFs** (a USGS video about using US Topos):
>   **www.youtube.com/watch?v=xigXqORgsTE**
> - **How does LiDAR remote sensing work?** (a basic introduction to lidar):
>   **www.youtube.com/watch?v=EYbhNSUnIdU**
> - **US Topo** (a USGS video concerning the development of US Topos):
>   **www.youtube.com/watch?v=hv0jxsW3qgY**

Key Terms

3DEP (p. 458)
base heights (p. 462)
collar (p. 453)
contour interval (p. 452)
contour line (p. 451)
DEM (p. 456)
DLG (p. 451)
draping (p. 462)
DRG (p. 450)
DSM (p. 457)
DTM (p. 455)
GeoTIFF (p. 451)
hillshade (p. 461)
LAS (p. 458)
lidar (p. 457)
NAVD88 (p. 450)
NED (p. 459)
perspective view (p. 460)

point cloud (p. 457)
pseudo-3D (p. 460)
slope (p. 459)
slope aspect (p. 460)
SRTM (p. 456)
Sun altitude (p. 461)
Sun azimuth (p. 461)
three-dimensional (3D) model (p. 455)
TIN (p. 456)
topographic map (p. 450)
two-and-a-half-dimensional (2.5D) model (p. 455)
US Topo (p. 452)
vertical datum (p. 449)
vertical exaggeration (p.464)
viewshed (p. 459)
z-value (p. 449)

Digital Terrain Analysis

This chapter's lab will introduce you to some of the basics of digital terrain modeling—working with DTMs, slope, viewsheds, and imagery draped over the terrain model. You'll be using the free Google Earth Pro for this lab.

Objectives

The goals for you to take away from this lab are:

▶ To examine pseudo-3D terrain and navigate across it in Google Earth

▶ To examine the effects of different levels of vertical exaggeration on the terrain

▶ To create a viewshed and analyze the result

▶ To create an animation that simulates flight over 3D terrain in Google Earth

▶ To create an elevation profile for use in examining a DTM and slope information

Using Geospatial Technologies

The concepts you'll be working with in this lab are used in a variety of real-world applications, including:

▶ Civil engineering, in which slope calculations from digital elevation models are utilized to aid in determining the direction of overland water flow, a process that contributes to calculating the boundaries of a watershed

▶ Military intelligence, which makes use of terrain maps and digital elevation models in order to have the best possible layout of real-world areas (as opposed to two-dimensional maps) when planning operations involving troop deployments, artillery placements, or drone strikes

[Source: PATRICK HERTZOG/AFP/Getty Images]

Obtaining Software

The current version of Google Earth Pro (7.1) is available for free download at **www.google.com/earth/explore/products/desktop.html.**

Important note: Software and online resources can change fast. This lab was designed with the most recently available version of the software at the time of writing. However, if the software or Websites have significantly changed between then and now, an updated version of this lab (using the newest versions) will be available online at **www.macmillanhighered.com /shellito/catalog**.

Lab Data

There is no data to copy in this lab. All data comes as part of the Google Earth data that is installed with the software or is streamed across the Internet when using Google Earth.

Localizing This Lab

The lab can be performed using areas close to your location, as Google Earth's terrain and imagery covers the globe.

13.1 Examining Landscapes and Terrain with Google Earth Pro

1. Start **Google Earth Pro (GE)**. By default, GE's Terrain option will be turned on for you. Check in the Layers box to be sure that there is a checkmark next to Terrain (and don't uncheck that for the remainder of the lab). Having this layer turned on activates the DTM that underlies the imagery within GE.

2. Next, as this lab is focused on digital terrain modeling, we'll want to use the most detailed DTM as possible. From the Tools pull-down menu, select Options. In the Google Earth Options dialog box that appears, choose the 3D View tab and place a checkmark in the box next to Use high quality terrain (disable for quicker resolution and faster rendering). Next, click Apply and then click OK to close the dialog box and return to GE.

3. For this lab, we'll be using GE to examine a terrain model of Zion National Park in Utah—in the Search box, type Zion National Park, UT. GE will zoom down to the area of Zion National Park. In the Layers box, expand the More option and place a checkmark next to Parks/Recreation Areas. This will show you the outline of Zion in green. Zoom out so that you can see the entire layout of Zion in the view. For more information about Zion National Park, check out **www.nps.gov /zion/index.htm**.

4. You will see a question mark symbol labeled as "Visitor Center" next to a label for "Park Headquarters." Center the view on this area of Zion.

5. From the Tools pull-down menu, choose Options. In the dialog box that appears, click on the Navigation tab and make sure that the box that says Automatically tilt when zooming has its radio button filled in. This will ensure that GE will zoom in to areas while tilting to a perspective view. Next, click Apply and then click OK to close the dialog box and return to GE.

6. Use the Zoom Slider to tilt the view down, so that you have a good perspective view on Zion (see *Geospatial Lab Application 1.1* for more info on using this tool). Basically, push the "plus" button on the Zoom Slider down, and your view will zoom in and tilt down to a perspective where you can see the sides of the mountains and canyons in Zion in pseudo-3D. You can also hold down the Ctrl key on the keyboard while moving the mouse forward or backwards to tilt the view.

7. Use your mouse wheel to move the view in and out (do this instead of using the Zoom Slider, because otherwise you'll keep tilting backwards as well) as well as holding the Ctrl key until you can position the view as if you were at the park headquarters/visitor center and facing south (see below for an example).

[Source: Google/DigitalGlobe/USDA Farm Service Agency, Image NMRGIS]

8. There are two references to heights or elevation values in the bottom portion of the view:

▶ The heading marked "elev" shows the height of the terrain model (that is, the height above the vertical datum) where the cursor is placed.

▶ The heading marked "Eye alt" shows Google Earth's measurement for how high above the terrain (or the vertical datum) your vantage point is.

Maneuver your view so your "Eye alt" is a good level above the terrain (perhaps about 4100–4200 feet), and you'll be able to see the mountain area that surrounds the park headquarters in Zion.

> **Question 13.1** How does the pseudo-3D view from this position and altitude aid in bringing out the terrain features of Zion (compared to what you originally saw in the overhead view earlier)?

13.2 Vertical Exaggeration and Measuring Elevation Height Values in Google Earth Pro

Google Earth Pro also allows you to alter the vertical exaggeration of the terrain layer. As we discussed on page 464, vertical exaggeration changes the vertical scale but keeps the horizontal scale the same, so it should be used for visualization purposes only.

1. To look at different levels of vertical exaggeration, select Options from the Tools pull-down menu. Select the 3D View tab.

2. In the box marked Elevation Exaggeration, you can type a value (between 0.01 and 3) for vertical exaggeration of GE's terrain. Type a value of 2, click Apply and OK, and then re-examine the area surrounding the park headquarters in Zion.

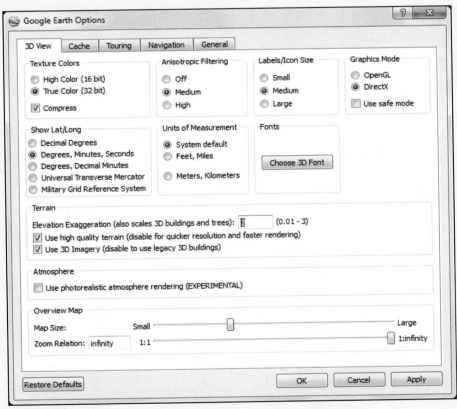

[Source: Google]

Question 13.2 How did the vertical exaggeration value of 2 affect the view of Zion?

472

> **Question 13.3** Try the following values for vertical exaggeration: 0.5, 1, and 3. How did each value affect the visualization of the terrain? In addition to the value of 2 you examined in Question 13.2, which value of vertical exaggeration was the most useful for a visual representation of Zion and why?

3. Reset the Elevation Exaggeration to a value of 1 when you're done.

4. From here, we'll examine the elevation values of the terrain surface itself. Although Google Earth will always show the imagery over the terrain model, elevation information of each location is available. Wherever you move the cursor on the screen, a new value for elevation is computed in the "elev" option at the bottom of the view. By zooming and placing the cursor on its symbol on a spot on the screen, you can determine the elevation value for that location.

> **Question 13.4** At what elevation is the height of the park headquarters/visitors center?

13.3 Working With Viewsheds in Google Earth Pro

Now that you've got a pretty good idea of how the landscape of Zion looks in the areas nearby the park headquarters, we'll now create a viewshed that will allow you to see what is visible and what is blocked from sight at a certain location.

1. To begin, we'll see what areas of Zion can be seen and cannot be seen from the park headquarters. Zoom in closely on the question mark symbol that GE uses to mark the park headquarters. This symbol is what we'll use as the location for creating a viewshed.

2. Select the Add Placemark icon from the toolbar and put the placemark right on the park headquarters question mark symbol (see Geospatial Lab Application 1.1 for how to do this).

3. In the New Placemark dialog box, name this new placemark Park HQ. Click OK in the New Placemark dialog box to accept the new name and close the dialog.

Google Earth - New Placemark	☒
Name: Park HQ	⚑
Latitude: 37°12'33.32"N	
Longitude: 112°58'49.10"W	

[Source: Google]

4. You'll see the new Park HQ placemark in the Places box. Right-click on it and select Show Viewshed. When prompted about the placemark being too low, click on Adjust automatically.

5. GE will compute the viewshed. Zoom out a bit to see the extent of the viewshed—all of the areas covered in green are the places that can be seen from the location of the Park HQ placemark and the areas not in green cannot be seen from there. After zooming out, look at the areas immediately south and southeast of the Park HQ placemark.

> **Question 13.5** Can the ranger station, the campground, or the two picnic areas just south and southeast of the park headquarters be seen from the Park HQ vantage point?

6. Click on Exit viewshed to remove the viewshed layer.

7. About two miles to the northeast of the park headquarters is a scenic overlook. Its symbol in GE is a small green arrow pointing to a green star. Move over to that location so that you can see it in the view and zoom in closely on the overlook symbol.

8. Put a new placemark directly on the green star and name this placemark Zion Overlook.

9. Create a viewshed for the Zion Overlook point and answer Questions 13.6 and 13.7. Exit the viewshed once you're done.

> **Question 13.6** Can the area labelled as parking to the immediate north of your Zion Overlook point be seen from the overlook?

> **Question 13.7** What is blocking your view of Sand Beach trail from the overlook? Be specific in your answer. *Hint*: You may want to zoom in closely to your Zion Overlook placemark and position the view as if you were standing at that spot and looking in the direction of Sand Beach trail.

13.4 Flying and Recording Animations in Google Earth

Google Earth Pro allows you to record high-definition video of the areas you view in GE. In this section, you'll be capturing a video of the high-resolution imagery draped over the terrain to record a virtual tour of a section of Zion. Before recording, use the mouse and the Move tools to practice flying around the Zion area. You can fly over the terrain, dip in and out of valleys, and skim over the mountaintops. Don't forget you can also hold down the Ctrl key and move the mouse to tilt your view. It's important to have a good feel of the controls, as any movements within the view will be recorded to the video. When you feel confident in your skills for flying over 3D terrain in Google Earth, move on to the next step.

1. The tour you will be recording will start at the park headquarters (your Park HQ placemark), move to the scenic overlook (your Zion Overlook placemark), and then finish at the lodging/parking area about a mile to the north of the overlook. We'll do a short "dry run" of this before you record. To begin, double-click on the Park HQ placemark in the Places box and the view will shift there.

2. Next, double-click on the Zion Overlook placemark in the Places box and the view will jump to there.

3. Lastly, use the mouse and Move tools to manually fly over the terrain a mile north of the overlook to the parking and lodging area and end the tour there.

4. If you need to, try this dry run of maneuvering between the three points so you feel comfortable prior to recording. When you're ready to make your tour video, double-click on the Park HQ placemark in the Places box to return to the starting point of the tour. Also, take away the checkmarks next to the Park HQ and Zion Overlook placemarks in the Places box so that the two placemarks will not appear in the view (and thus not appear in the video) so that all you will see is the GE imagery and terrain.

5. On Google Earth's toolbar, select the Record a Tour button.

[Source: Google]

6. A new set of controls will appear at the bottom of the view:

[Source: Google]

7. To start recording the video, press the red record button.

Important note: If you have a microphone hooked up to your computer, you can get really creative and narrate your tour—your narration or sounds will be saved along with your video.

8. After showing the park headquarters in the view for a couple seconds, double-click on the Zion Overlook placemark in the Places box to jump to there. Show the overlook for a few seconds, then use the mouse and move commands to fly to the lodging/parking area to the north.

9. When you're done, press the red record button again to end your recording of the tour.

10. A new set of controls will appear at the bottom of the screen, and Google Earth will begin to play your video.

[Source: Google]

11. Use the rewind and fast-forward buttons to skip around in the video, and also the play/pause button to start or stop. The button with the two arrows will repeat the tour or put it on a loop to keep playing.

12. If the tour is how you want it, save the tour by pressing the disk icon (Save Tour) button. Call it Ziontour in the dialog box that opens. The saved tour will be added to your Places box (just like all other GE layers). If the tour is not looking how you want it, return to Step 4 and remake the tour.

13. Right now, the tour can only be played in GE. You'll want to export your tour to a high-definition video that can be watched by others or shared over the Web. To start with this process, first close the DVR control box in the view by clicking on the x in the upper-right-hand corner of the controls.

14. Next, from the Tools pull-down menu, select Movie Maker.

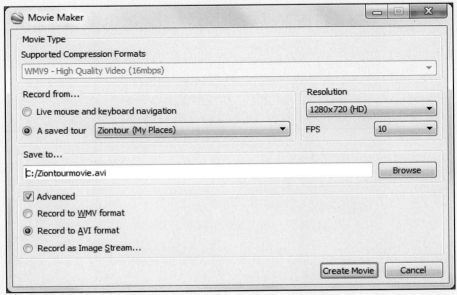

[Source: Google]

476

15. In the Movie Maker dialog box, choose 1280 × 720 (HD) for Resolution. This will create a high-definition (HD) video of your tour.

16. Next, under Record from... choose the radio button for A saved tour and choose Ziontour (My Places). This will be the source of your video.

17. Put a checkmark in the Advanced box, and then select the radio button for Record to AVI format.

18. Under Save to… use the browse button to navigate to the drive where you want to save your video to and call it Ziontourmovie.

19. Leave the other defaults alone and click Create Movie. A dialog box will appear showing the process of the conversion process to create your video.

20. When prompted if you want to watch your movie, click Yes.

> **Question 13.8** Show (or submit) your final .avi file of your video tour of Zion to your instructor, who will check it over for its quality and completeness for you to get credit for this question.

13.5 Measuring Profiles and Slopes in Google Earth Pro

1. In the last part of this lab, you will be examining profiles of the digital terrain model and examining the slope information. To begin, return the view to the Park HQ placemark.

2. Zoom out so that you can see both the Park HQ and the Zion Overlook placemarks clearly in the view. What we'll do next is construct a profile (a two-dimensional view) of the digital terrain between the two placemarks and then examine the elevations and slope of the terrain.

3. To examine the terrain profile, you must first draw a path between the two points. To begin, choose the ruler tool from the GE toolbar.

[Source: Google]

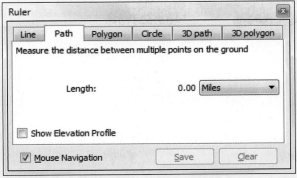

[Source: Google]

4. In the Ruler dialog box that opens, select the **Path** tab.

5. In the view, you'll see that your cursor has changed to a crosshairs symbol. Left-click once on the Park HQ placemark location and then click the left mouse button once on the Zion Overlook placemark position. You'll see a yellow line drawn between them and the Length of this line will be shown in the Ruler dialog.

6. In the Ruler dialog, click **Save**.

7. A new dialog will open and allow you to give this path you've drawn a name. Call it **ParkHQOverlookPath** and click **OK** in the naming dialog.

8. You'll see a new item added to the Places box called PathHQOverlookPath. Right-click on it and select **Show Elevation Profile**.

9. A new window will open below the view, showing the elevation profile between the two points. This profile will show you, in two dimensions, the outline and elevations of the terrain in between the two points.

14 See the World in 3D

3D Geovisualization, 3D Modeling and Design, Prism Maps, SketchUp, and Google Earth in 3D

Up until the previous chapter, all the topics in this book dealt with geospatial technology in a two-dimensional format, whether as geospatial data, measurements, maps, or imagery. Chapter 13 began to go beyond two dimensions and start on a third, presenting terrain in a perspective and pseudo-3D view. This chapter looks entirely at presenting geospatial information in three dimensions, as well as designing and visualizing 3D data and concepts.

There's no doubt that 3D visualization of data is impressive—as technology has improved and computers have become faster and more powerful, 3D rendering has become more familiar to us. Video games and simulators are extremely impressive to watch, and computer-animated movies have made huge strides in recent years. While geospatial technologies haven't (yet) reached the technical level of some of the state-of-the-art CGI we see in some high-budget movies, many 3D modeling and visualization techniques are capable of creating perspective views, 3D maps, and realistic-looking 3D objects. For instance, Bing Maps and Google Earth both support viewing and interacting with 3D geospatial data. See **Figure 14.1** for an example of realistic-looking 3D geospatial visualization in Google Earth.

Before proceeding, keep in mind the discussion from Chapter 13 concerning two-and-a-half-dimensional (2.5D) data versus three-dimensional (3D) data. If only one height value can be assigned to an x/y location, then the data is considered 2.5D; if multiple height values can be assigned to an x/y location, then the data is fully 3D (3D data also has volume). For ease of reading and usage, this chapter uses the term "3D" throughout to refer to all geospatial phenomena that incorporate a third dimension into their design or visualization, although technically, some examples will be 2.5D or pseudo-3D.

FIGURE 14.1 Lower Manhattan (New York City), rendered in 3D.
[Source: Google/Sanborn]

 ## What Is 3D Modeling?

3D modeling (in geospatial terms) refers to the design of spatial data in a three-dimensional (or pseudo-3D) form. Designing a 3D version of the buildings in a city's downtown district or a 3D version of the Eiffel Tower would be examples of 3D modeling of geospatial data. Like all of our spatial data, the 3D models should be georeferenced to take advantage of real-world coordinate systems and measurements. If you're designing a 3D model of your house, you'll want to start with a georeferenced base so that you can make accurate measurements of your house's footprint. An ortho-photo, high-resolution satellite image, or architectural diagram with spatial reference are all good starting points for a georeferenced base to begin modeling from. Digitizing the house's footprint will give you a polygon (with two dimensions, x and y), but a 3D-style object has to have a third dimension—a z-dimension. This **z-value** will be the height of the object above whatever surface it's standing on—in the case of your house, the ground it's built on.

In order to create a 3D version of the polygon, it will have to be extruded to reach the height specified by the z-value. **Extrusion** is the process of giving height to an object. If your house is 15 feet high, the footprint polygon can be extruded to a height of 15 feet. Extrusion will change the objects—extruding a polygon will transform it into a **block.** In this case, the building footprint will change into a block 15 feet high (see **Figure 14.2** on page 482 for the comparison between the two-dimensional polygon footprint and the extruded block).

3D modeling designing and visualizing data that contains a third dimension (a z-value) in addition to x- and y-values

z-value the numerical value representing the height of an object

extrusion the extending of a flat object to have a z-value

block a flat polygon that has been extruded to transform it into an object with a z-value

FIGURE 14.2 A flat polygon versus an extruded block. *[Source: Sketchup/Trimble]*

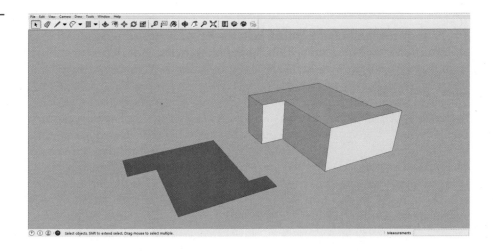

Any object can be extruded to a particular height by assigning it a z-value. Keep in mind that these items extrude from the ground level up and some objects can't be designed this way. For example, suppose you want to design a 3D model of a 10-foot-high pedestrian walkway that's 15 feet over a busy road. If you digitize the bridge's footprint and extrude it to 10 feet, it will look like a wall blocking the road rather than a bridge over it. To get around this, you'll have to apply an **offset**, or a z-value, where the initial object is placed, before you do the extrusion. In this case, the digitized polygon representing the bridge will have to be offset to a height of 15 feet (its height from the ground as it crosses over the road), then extruded to a height of 10 feet (to capture the dimensions of the walkway). **Figure 14.3** illustrates the difference between a regular extruded block and an extruded block that has been offset from the surface.

offset a value applied to an object that is not attached to the ground, but is entirely above ground level

FIGURE 14.3 An offset polygon that has been extruded versus a single extruded block. *[Source: Sketchup/Trimble]*

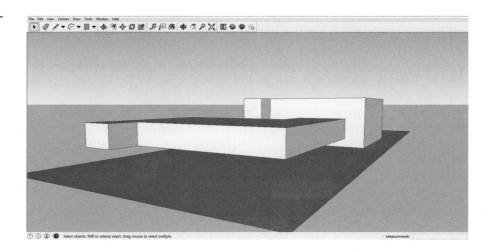

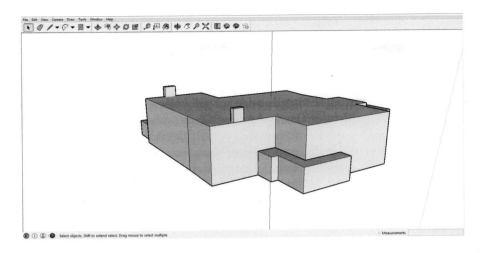

FIGURE 14.4 Multiple extruded blocks making up a single building (a very early version of Weller House, on Youngstown State University's campus).
[Source: Sketchup/Trimble]

A building being modeled will likely consist of multiple blocks (or other shapes), each representing a different portion of the building. Multiple levels of a house or different-sized parts of a building can each be represented by blocks. Just as a GIS dataset can consist of multiple polygons, a 3D representation can consist of multiple extruded shapes and blocks (see **Figure 14.4**). Each of these blocks can then be treated as its own object for further modeling.

In 3D modeling it's not enough just to place an object in its correct georeferenced location—it must also conform correctly to the real-world terrain and to real-world elevations. If your house is located 900 feet above sea level, you don't want to base your house at a ground level of zero feet. If you do, your house will be shown not on Earth's surface, but 900 feet underground when you merge your 3D model with other geospatial data. Terrain modeling was discussed back in Chapter 13, but the key element to remember with 3D modeling is the concept of **base heights** (to use Esri terminology). These values represent the elevation of the terrain above a vertical datum.

When you're designing a 3D object (like a model of your house) that you intend to place on an area of the terrain, you should make sure that the terrain's base heights are applied to the object you're designing. In this way, the software will understand that the base of your house begins at an elevation of 900 feet, and extruding your house's footprint to a height of 15 feet means its rooftop will be 915 feet above sea level, rather than 15 feet above ground level (zero feet). Many geospatial software packages enable the user to combine 3D modeling techniques with terrain modeling to help with the design of more realistic 3D visualizations (some specific programs that do this are described later in this chapter).

Once the footprints and elevations are correctly set in place, it's time to start making those gray blocks look like the thing you're trying to model. Every block has several **faces**, each face representing one side of the object. For example, an extruded rectangular block will have six faces: top, bottom,

base heights the elevation values assigned to the terrain upon which the objects will be placed

face one side of a block

FIGURE 14.5 A version of Youngstown State University's Weller House with features added and colors, textures, and additional details applied to the block faces (compare with Figure 14.4). [Source: Sketchup/Trimble]

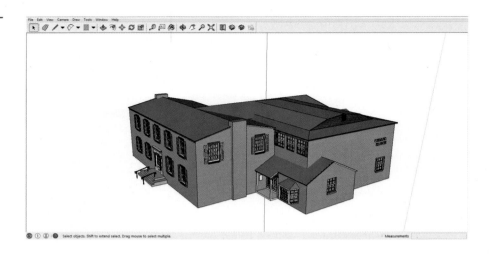

and four sides. Each face can be "painted" to give it a more realistic appearance. This painting can take a variety of forms—doing something as simple as changing the color of one face to, say, yellow (thus repainting one side of your house to make it yellow) can make a considerable difference to its appearance.

More realistic appearances can be achieved by applying a **texture** to a face. A texture is a graphic designed to simulate the appearance of materials (like brick, wood, stone, glass, etc.). With textures, you don't have to draw individual bricks on the side of a building—you just apply a red-brick texture and the face's appearance is changed to red bricks. Similarly, windows can be designed by applying a translucent texture to a window shape drawn onto a face, and various roofing materials can be applied to the faces that make up the roof of your house (see **Figure 14.5** for an example of a 3D building with various textures applied to the block faces).

Since a texture is just a graphic, it's possible to apply other graphics to a face. For instance, you could take a digital camera picture of your home's

> **textures** graphics applied to 3D objects to give them a more realistic appearance

THINKING CRITICALLY
WITH GEOSPATIAL TECHNOLOGY 14.1

What's the Advantage of Using 3D Design?

The use of 3D visualization has become increasingly widespread as a means of viewing and using geospatial data. As processors have gotten faster and graphics and animation have improved, it's become easier than ever to render large amounts of geospatial data in realistic 3D form. Why is visualization of geospatial data so critical to conveying a message to the user or viewer of the data? Beyond the "cool" factor, what makes 3D visualization a better conveyor of data than a simple 2D map? What advantages does 3D visualization have over regular 2D for the presentation of data and communication of geospatial information?

particular brick and apply it to a face of your model. If it's accuracy you want, you might even take a picture of an entire side of your house, resize it, and "paste" that image onto the corresponding entire face of your model. However, it's not just buildings and objects that can be visualized using 3D modeling techniques. 3D maps can also quickly communicate spatial information to the viewer.

How Are 3D Maps Made?

A 3D-style map can be made by applying the kind of 3D modeling concepts that we've just discussed. However, this time you won't be working with extruding polygons that represent building footprints on the ground, but with the shapes and objects on a map and what they can represent. A regular choropleth or thematic map (see Chapter 7) can be transformed into a 3D-style map by extruding the polygons that are displayed on the map. The extruded polygons are called raised **prisms,** and a map that features these extrusions is referred to as a **prism map.** In a prism map, the shapes on the map are extruded to a height that reflects the data value assigned to them. For instance, if a thematic map of the United States is intended to show population by state, then each state's shape will be extruded to a "height" of the number of units equal to that state's population. Thus, the polygon shape of California will be a raised prism extruded to a value of 36.7 million, while neighboring Nevada's shape will only be extruded to a value of 2.6 million.

 Figure 14.6 shows an example of a prism map dealing with world population. Each country's polygon has a value of the population statistics for that country. Those countries with the highest raised prisms have the largest

> **prisms** the extruded shapes on a prism map
>
> **prism map** a 3D thematic map from whose surface shapes (like the polygons that define states or nations) are extruded to different heights that reflect different values for the same type of data

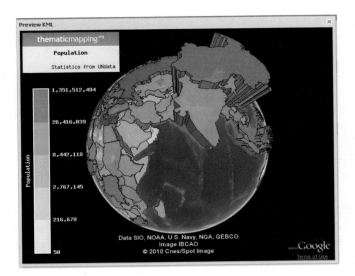

FIGURE 14.6 A prism map showing the 2010 populations of every country in the world.
[*Source: thematicmapping.org, Google, Data SIO, NOAA, U.S. Navy, NGA, GEBCO, Image IBCAO, 2010 Cnes/Spot Image*]

HANDS-ON APPLICATION 14.1

Creating Prism Maps Online

The Thematic Mapping Engine (also used back in Chapter 7) enables you to create prism maps and to display them using Google Earth. Open your Web browser and go to **http://thematicmapping.org/engine**. Select an indicator that you want to map (such as population, life expectancy, or mobile phone subscribers) and a year, then select Prism for the technique. Press Download to open the prism map in Google Earth. Try several of the classification combinations to see how they alter the raised prisms in the data.

Expansion Questions

- Create a prism map of population of the year 2010—these will be regular count

values, not normalized ones (like on a choropleth map). How does a prism map help in displaying and interpreting these kinds of count data?

- Using the indicator CO2 emissions per capita for the year 2004 (using the Quantiles method of classification, with five classes), create both a prism map and a regular 2D choropleth map and download both of them into Google Earth. Switch back and forth between the two layers—what differences in visual communication does a prism map offer when mapping this dataset?

number of persons (in this example, the map shows China and India's huge populations relative to neighboring countries). The polygon shape of each country is extruded relative to the number of units being measured.

See Hands-On Application 14.1: Creating Prism Maps Online for an online tool for creating prism maps.

How Can 3D Modeling and Visualization Be Used with Geospatial Technology?

There are numerous 3D design and modeling programs on the market today, and many commercial geospatial programs offer some sort of capability of visualizing or working with 3D data (there's no way this chapter can get into all of them, but we will discuss a few). For instance, Esri's ArcGIS for Desktop program contains the 3D Analyst extension (which enables numerous 3D features related to terrain and modeling), as well as two other interfaces for working with 3D data—**ArcGlobe** and **ArcScene.** As its name implies, ArcGlobe allows for 3D visualization on a "global" level. If you're examining 3D surfaces and objects in Los Angeles, you can fly north across the globe to view other 3D spatial data in San Francisco, then continue north to look at the landscape in Seattle. ArcGlobe functions like a virtual globe program and works with all types of Esri file formats. ArcScene contains a full set of 3D visualization tools for ArcGIS and allows for the same type of

ArcGlobe a 3D visualization tool used to display data in a global environment, which is part of Esri's ArcGIS for Desktop program

ArcScene the 3D visualization and design component of Esri's ArcGIS for Desktop program

3D modeling, except on a "local" scale. If you were modeling one section of a 3D Los Angeles, that's the extent of the data you'd be working with.

Google Earth also supports 3D visualization. Back in *Geospatial Lab Application 1.1: Introduction to Geospatial Concepts and Google Earth,* we examined a 3D representation of places in South Dakota. Google Earth contains many more 3D objects and features representations of 3D buildings and structures across the world (buildings in New York City, London, and other cities around the world). It's easy to create 3D objects, structures, buildings, landforms, and more through the use of the **SketchUp** software program and then view them in Google Earth. The regular version of SketchUp (called "SketchUp Make") is made freely available for download by Trimble (SketchUp was formerly owned by Google), while a Pro version with additional features is available for purchase (you'll be downloading and using the free version of SketchUp in this chapter's Geospatial Lab Application). SketchUp is a very versatile and intuitive-to-use program that allows for a full range of 3D design capabilities. Structures created with SketchUp can be as realistic and detailed—or as basic—as you want them to be. See **Figure 14.7** for an example of a football stadium designed in SketchUp.

Shapes can be molded, extruded, bent, and moved to build any type of structure you can envision. Realistic textures can be applied to faces for a better appearance, and digital photos can also be resized and applied to a face for a more photorealistic appearance. SketchUp also features a process called Match Photo, which enables you to design the polygon dimensions of a building straight from a digital photo. From there, the photo can be directly applied to the faces for a realistic-looking 3D design.

SketchUp also enables the use of **components** (or pre-made 3D objects) for added realism and detail in designing 3D models. Components give you a quick way to insert trees, cars, furniture, landscaping, and similar features into 3D models (see **Figure 14.8** on page 488 for examples of components used in SketchUp). Like other 3D objects, components are constructed

SketchUp a 3D modeling and design software program distributed online by Trimble

components pre-designed (or user-designed) objects in SketchUp

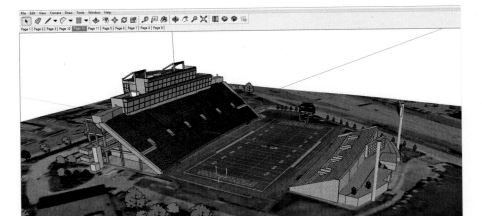

FIGURE 14.7 A version of Youngstown State University's football stadium complex, displayed in SketchUp Make. *[Source: Sketchup/ Trimble]*

FIGURE 14.8 Sample components, displayed in SketchUp. *[Source: Sketchup/ Trimble]*

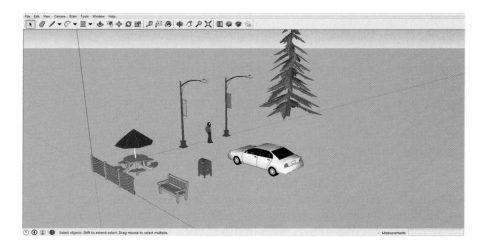

using shapes and textures, but can be stored for re-use throughout the construction of a model. For instance, if you're designing a 3D version of a car dealership, you can quickly add a couple dozen different car components and customize them, rather than design each 3D car model individually. Components are also useful for elements you'll use multiple times in a model. For example, if you're designing a 3D version of a school building, you could design a window once, store it as a component, and then easily re-use the same model of a window multiple times around the school.

Geospatial Lab Application 14.1 will introduce you to several aspects of using SketchUp for 3D design. You'll begin with a building's footprint, use tools to create a shape, design elements of the building (such as the entranceway), and adapt components to fit in with your own ideas. A file created with SketchUp (like the one you'll make in the lab) uses the "*.skp" file extension. The **SKP file** consists of information about the polygons that make up the structure, the textures applied, and the geometry of how these objects fit together. However, the SKP format is native to SketchUp. You (or whoever you share your data with) have to be running SketchUp to open and examine an SKP file. In addition, you can open only one SKP file at a time, which causes some difficulties if you've designed multiple buildings (each in its own SKP file) and want to see how they look next to one another.

One way around these issues is to convert the SKP files into another format called **KMZ**, which can be used in Google Earth. Each building becomes its own KMZ file, and many KMZ files can be opened at once in Google Earth. KMZ is a compressed version of another file format called **KML**, which is short for Keyhole Markup Language. KML and KMZ are file formats that can be interpreted using Google Earth. By converting an SKP file into a KMZ file, your SketchUp model can be viewed interactively in Google Earth. **Figure 14.9** shows an example of this—Youngstown State

SKP file the file format and extension used by an object created in SketchUp

KMZ the compressed version of a KML file

KML Keyhole Markup Language—the file format used for Google Earth data

FIGURE 14.9 Youngstown State University's Stambaugh Stadium complex, converted from an SKP file in SketchUp and displayed as a KMZ file in Google Earth. *[Source: Google Earth/Sketchup/Trimble]*

University's stadium complex (containing both sides of the stadium, stadium lights, scoreboard, and goal posts) was converted from an SKP file into a KMZ file. The KMZ file can be opened in Google Earth, which overlays the 3D model of the stadium over the Google Earth imagery and allows the user to rotate around the stadium, fly across the field, zoom into a bleacher seat, or utilize any of the other usual uses of Google Earth. KMZ has become a standard for creating objects or overlays to view in Google Earth—other file formats can be converted to KMZ (such as satellite imagery), and then be placed or draped across Google Earth (recall how you examined KMZ files of EOS imagery in Google Earth in *Geospatial Lab Application 12.1*).

The big question that should be asked at this point is: How does Google Earth know where properly to place that KMZ file? If Google Earth knows just where to place it, the Youngstown State University stadium must have had some sort of spatial reference attached to it, so that it ends up placed over the imagery of the stadium's location and not at some random point on the globe. The best way of going about this is to know the object's spatial location before you start to do any modeling—otherwise, you'll find yourself trying to georeference the 3D model's location after the fact. SketchUp works hand in hand with Google Earth to obtain this data. SketchUp has the capability to take a "snapshot" of imagery (like the area you can see with Google Earth), and import that snapshot into SketchUp (complete with a model of the terrain being shown and the relevant spatial reference information). With SketchUp, you'll have a color overhead image with spatial reference information to use as a starting point for your modeling efforts (see **Figure 14.10** on page 490 for examples of this geolocation). When you've completed your work in

FIGURE 14.10 A section of Theodore Roosevelt National Park and Medora in North Dakota in Google imagery (above) and in an image version (including terrain) in SketchUp (below). *[Source: Google Earth; Sketchup/Trimble]*

SketchUp, you can convert your SKP file (*.skp) into a KMZ file (*.kmz), which will then contain the necessary information to place the model in its proper spatial location in Google Earth. This is how you'll begin 3D modeling in *Geospatial Lab Application 14.1*—first by obtaining a "snapshot" of the overhead view of the building, and then by creating the building's footprint from this starting point.

With the popularity of SketchUp and Google Earth, it's become easy for people to create spatially referenced data in SketchUp and then place it in Google Earth. To help facilitate the use of the programs (and to help foster a geospatial 3D modeling community), Trimble maintains an online **3D Warehouse** as a central repository of 3D models from users around the

3D Warehouse Trimble's online repository of 3D objects created using SketchUp

HANDS-ON APPLICATION 14.2

Digging into Trimble's 3D Warehouse

The Trimble 3D Warehouse is full of all sorts of 3D models from around the world. To start digging through the 3D Warehouse, go to **https://3dwarehouse.sketchup.com**. In the 3D Warehouse, you'll see a lot of different 3D models, not just 3D Geospatial ones. For our purposes, look at some of the options under the Featured Geo Models category for numerous 3D models that users have uploaded to the Warehouse. Take a look at some of them, select their particular Warehouse page, and then download the model (you'll see the option for it on its page) into Google Earth, so that you can interact with it.

Expansion Questions

- Look for structures and buildings of your local area in the 3D Warehouse and download them into Google Earth. What types of structures have been created and uploaded to the 3D Warehouse for your own local area?

- Search for the model of "Christ the Redeemer," the famous statue in Rio de Janeiro, Brazil. Download the model and view it in Google Earth. Zoom in to view the statue and its surroundings overlooking Rio de Janeiro. What other features did the designer of the model create besides just the statue itself? Why is this?

world. In the 3D Warehouse, you can find new components, objects, and a wide variety of 3D buildings, ranging from state capital buildings to national monuments, sports arenas, and many other architectural constructions from around the world. SketchUp makes it easy to interact with the 3D Warehouse—the software allows you to upload your model to the warehouse and to download items from the warehouse either into SketchUp or directly into Google Earth. The 3D Warehouse is a great resource for models created by users of all levels of ability, and the site contains guidelines on how to model for inclusion into Google Earth (see *Hands-On Application 14.2: Digging into Trimble's 3D Warehouse* for more information about the 3D Warehouse).

With these types of 3D modeling and design techniques, all manner of 3D features can be created and utilized with other geospatial data. It's one thing to look at a flat map of a university campus, but it's another thing entirely to have it laid out in a realistic 3D environment that you can interact with. Prospective students are likely to get a better sense of place by viewing a 3D representation of their dorms and understanding where they are in relation to the urban areas surrounding the campus.

The same idea holds true for designing any area in interactive 3D. Future planning efforts can be better implemented when the planners can actually see the visual impact that various types of building plans would have on the neighborhood (with respect to height, texture, and landscaping). With 3D geospatial data, the locations of items and structures can be seen in relation to each other and to the surrounding terrain. For example,

FIGURE 14.11 The Eiffel Tower (and surrounding areas in Paris) as shown in 3D using Google Earth. *[Source: Google Earth]*

FIGURE 14.12 A section of the Las Vegas Strip as shown in 3D using Google Earth. *[Source: Google Earth]*

CityEngine a powerful rule-based 3D visualization program designed for quickly modeling 3D cities

in Google Earth, turning on the "3D Buildings" layer activates a wide range of 3D content tied to spatial locations that you can interact with (see **Figures 14.11** and **14.12** for examples of rendered 3D features in a geospatial technology environment, and *Hands-On Application 14.3: 3D Buildings in Google Earth* for how to interact with this data in Google Earth). 3D modeling techniques and other methods of shaping and designing geospatial data are advancing rapidly. Esri's **CityEngine** program uses a rule-based structure to enable users to work rapidly on the construction of entire rendered 3D cities without having to extrude and model individual

HANDS-ON APPLICATION 14.3

3D Buildings in Google Earth

Figures 14.11 and 14.12 are both images of 3D structures in Google Earth, but they're far from being the only places that have been built and placed in Google Earth. For this application, start up Google Earth, zoom to Boston, Massachusetts, and look around Boston Common and the downtown area. From the available options in the Layers Box (if necessary, see Chapter 1 for more information about Google Earth's layout), select the option for 3D Buildings and place a checkmark in it. Tilt your view down so you're looking at Boston from a perspective view, and you'll see various buildings and structures start to appear in 3D. Fly around the city and see how Boston has been put together in 3D.

When you're done, try looking at the design of 3D buildings in other cities, such as New York City,

Washington, D.C., and Las Vegas, Nevada (as well as non-U.S. destinations like London, United Kingdom, or Moscow, Russia). Check some more cities to see if their buildings have been designed in 3D for Google Earth.

Expansion Questions

*   Check out your own local city in Google Earth. Has the city been created in 3D for Google Earth? If not, what features of it have been added to Google Earth under the "3D Buildings" option?

*   Use Google Earth to examine the Las Vegas Strip using the "3D Buildings" option. What areas and features of the Strip had to have been designed as 3D models, to accompany the extruded buildings with textures added to them?

buildings. See *Hands-On Application 14.4: 3D CityEngine Web Scenes* for some interactive online examples of this. Other companies, such as PLW Modelworks, utilize remotely sensed imagery and aerial photography to create remarkably realistic large-scale 3D models (see **Figure 14.13** for an example).

FIGURE 14.13 A 3D model of the city of Naples, Italy, created by PLW Modelworks. *[Source: PLW ModelWorks]*

HANDS-ON APPLICATION 14.4

3D CityEngine Web Scenes

For some interactive examples of using CityEngine, go to **www.arcgis.com /home/group.html?owner=CityEngine&title =CityEngine%20Web%20Scenes.** All of the options on this ArcGIS Online Website are various city area Web Scenes generated using CityEngine. By selecting one of them, the CityEngine Web Viewer will open and load the 3D city data. Within the CityEngine Web Viewer, controls for navigating the scene are in the left-hand corner, various layers that can be turned off and on will be on the right-hand side of the screen, and a series of bookmarks will be along the bottom. By clicking on a bookmark, the view will shift to show the camera angle at that point. Check out several of the examples, particularly Desert City, Esri Campus, and Philadelphia Redevelopment.

Expansion Questions

*   How does the 3D visualization of Philadelphia in its current state and with the proposed redevelopment aid in showing which areas, spatially, will be impacted by the redevelopment plan?

*   How are various exterior and interior city features shown and visualized in both the Desert City and the Esri Campus Web Scenes?

 ## How Can Geospatial Data Be Visualized in 3D?

Even with all of these great 3D modeling and design programs, when their products are viewed they're still only seen in a two-dimensional environment on a computer screen. It's one thing to view a 2D choropleth map (like in Chapter 7), another thing to create a pseudo-3D representation of the mapped data (like a prism map), and another thing altogether to be able to interact with the data in an immersive environment. Examining a realistic geospatial landscape and flying across it (like in Chapter 13) is very neat, but what if you could simulate the full 3D effect of diving through canyons or banking over ridgelines? As computers get faster and develop the ability to handle geospatial data more quickly and easily, new methods of visualizing data are being developed. Immersive "virtual reality" technology allows you to actually become part of the data, move with the landscape, have yourself placed inside a structure, or move and see the surface of the map. This sense of immersion, or interactivity, with geospatial data adds immensely to our understanding of the nature of spatial data, whether we are researchers or students.

Another way of making geospatial data appear more realistic (and adding a sense of depth and immersion) is to view the data in stereo. Stereo imagery shows two images (for instance, two images of the same landscape), rendered apart from each other. When you look at something, your eyes are seeing the same thing from two slightly different points of view—which is why you see depth and distance (and view the real world in three dimensions). Stereo images use a similar principle: they are two representations of the same thing, taken from two slightly different positions of the camera or sensor. Remote sensing has long been making use of stereo imagery for measuring the heights and depths of areas. Overlapping aerial photography and images acquired from sensors (such as those onboard SPOT 5 or the ASTER or MISR sensors on Terra) can be used for creating stereo data.

A simple way of viewing imagery or data in stereo (which several geospatial software programs have the capability for) is to create a red and blue **anaglyph** of the data—in which the two images are slightly separated, but one is highlighted in red and the other in blue. By wearing a pair of 3D glasses with colored gels for lenses (one red and one blue), the anaglyph images appear to be raised or elevated off the screen in a "3D effect" (see **Figure 14.14**). This type of stereo effect (where one eye views the red and the other eye views the blue) creates a type of **stereoscopic 3D** that gives the illusion of depth in parts of the image. Some items may appear in the foreground and others in the background. Red and blue anaglyphs are certainly nothing new, but today, when stereoscopic 3D movies have become all the rage (including the top-grossing film of all time), there are more high-tech ways of using these stereo concepts with geospatial technology.

If you go to a 3D movie today, you're not going to be wearing red and blue lenses, but rather a pair of polarized glasses. The same type of 3D glasses can be used with the **GeoWall**, a computer tool that's used for examining imagery in an immersive three-dimensional environment with stereoscopic 3D viewing techniques. First developed by the members of the GeoWall Consortium in 2001, it has spread to more than 400 systems, which have been developed for use in schools and colleges (along with places like the EROS Data Center and NASA's Jet Propulsion Laboratory). The GeoWall can be used for viewing geospatial data in a stereoscopic 3D format.

anaglyph a new image created by slightly offsetting the features in an existing image from each other and highlighting them in two different colors

stereoscopic 3D an effect that creates the illusion of depth or immersion in an image by utilizing stereo imaging techniques

GeoWall a powerful computer tool for displaying data and imagery in stereoscopic 3D

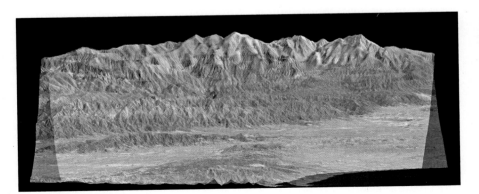

FIGURE 14.14 An anaglyph of a terrain model of the Los Angeles area of southern California. *[Source: NASA]*

The GeoWall consists of several components that can be purchased separately, to keep the cost of the whole setup affordable (around $10,000). First, two high-power Digital Light Processing (DLP) projectors project the two images that are necessary for stereo viewing. These projectors are hooked into a powerful computer with a dual-output video card (so that the computer can send its output into both of the projectors). The images are projected through special filters in front of the projectors' lenses, which polarize the light. The light is projected onto a special screen made of material that preserves the integrity of the polarized light (rather than scattering it). Lastly, the user wears a pair of polarized 3D glasses to view the imagery. See **Figure 14.15** for the components of the GeoWall projectors and their filters.

Wearing the 3D glasses, the viewer perceives the images in a 3D effect as if the items and images are appearing off the screen—similar to the effect created by a stereoscopic 3D movie. Viewing and interacting with imagery like this gives the viewer a sense of depth or immersion, as some items will be in the foreground of the image and some will be in the background. **Figure 14.16** shows an image of a GeoWall in operation, from the point of view of an outsider who is not wearing the polarized glasses (without the glasses, the images appear blurry and slightly askew from one another). GeoWall viewers position themselves in front of the screen, and are able to see the images projected in stereoscopic 3D.

Using the GeoWall for visualizing geospatial data carries a lot of benefits. The GeoWall enables the viewer to see maps draped over a surface model or terrain data displayed in stereoscopic 3D. Certainly, being able to dive inside a meteor crater or view volcanic features, coastlines, or river valleys in a stereoscopic 3D environment allows the viewer to get a realistic sense of the terrain features. 3D cities in Google Earth, landscapes, geologic processes, draped satellite imagery, draped topographic maps, stereoscopic photos, and tours of virtual environments are all ways of viewing types of data with a GeoWall setup.

FIGURE 14.16 Imagery projected onto the screen by the GeoWall system (as viewed without the glasses). *[Source: Courtesy Bradley A. Shellito]*

Chapter Wrapup

The goal of this chapter was to provide an introduction to some of the basic concepts of 3D visualization and how they can be incorporated into geospatial technology. *Geospatial Lab Application 14.1* will get your hands dirty working with 3D design and visualization using SketchUp and Google Earth. However, 3D modeling goes well beyond SketchUp—designing, programming, and working with 3D modeling and visualization comprise a whole industry on its own: video games, movies, television and Web animation, and rapidly developing simulators of many kinds are constantly serving up extremely impressive and ever-more realistic 3D designs. There's no way to cover all the possible applications, software, companies, and changes going on in this rapidly developing field in just one short chapter.

Check out *3D Visualization Apps* for some apps related to this chapter's contents, as well as *3D Visualization in Social Media* for some Facebook pages, Twitter accounts, Instagram accounts, YouTube videos, and blogs dealing with geospatial technology and 3D visualization.

3D visualization is a very appealing and quickly evolving area of geospatial technology, but there are many new developments yet to come as technology advances. Chapter 15 explores some of these new frontiers in geospatial technology.

Important note: The references for this chapter are part of the online companion for this book and can be found at **www.macmillanhighered .com/shellito/catalog**.

3D Visualization Apps

Here's a sampling of available representative 3D visualization apps for your phone or tablet. Note that some apps are for Android, some are for Apple iOS, and some may be available for both.

- **Google Earth:** An app for viewing Google Earth in 3D
- **Maps 3D and Navigation:** A mobile app for navigation and routing that also contains pseudo-3D versions of the maps

3D Visualization in Social Media

Here's a sampling of some representative Facebook, Twitter, and Instagram accounts, along with some YouTube videos and blogs related to this chapter's concepts.

On Facebook, check out the following:

- **CityEngine:**
  www.facebook.com/CityEngine
- **SketchUp:**
  www.facebook.com/sketchup

Become a Twitter follower of:

- **CityEngine:** @CityEngine
- **Mapzen** (makers of the open source Tangram, used for 2D and 3D modeling): @mapzen
- **PLW Modelworks:** @plwmodelworks3D
- **SketchUp:** @SketchUp

Become an Instagram follower of:

- **SketchUp:** @sketchup_official

On YouTube, watch the following videos:

- **CityEngine TV** (a YouTube channel featuring several different applications of CityEngine): **www.youtube.com/user/CityEngineTV**
- **Esri CityEngine Trailer** (a short demo of CityEngine's capabilities): **www.youtube.com/watch?v=aFRqSJFp-I0**

- **New 3D imagery for Google Earth for mobile** (a Google video showing new 3D city imagery available in Google Earth): **www.youtube.com/watch?v=N6Douyfa7l8**

- **PLW Modelworks** (the PLW Modelworks YouTube channel, which features numerous videos showcasing their realistic 3D city models): **www.youtube.com/user/plwmodelworks**

- **The SketchUp Channel** (a YouTube channel featuring videos about a variety of SketchUp applications and tutorials): **www.youtube.com/user/SketchUpVideo**

For further up-to-date info, read up on these blogs:

- **ArcGIS Resources—3D GIS Category** (blog posts concerning topics related to working with 3D GIS and 3D visualization): **http://blogs.esri.com/esri/arcgis/category/subject-3d-gis**

- **SketchUpdate Blog** (the official blog for SketchUp): **http://blog.sketchup.com**

Key Terms

3D modeling (p. 481)
3D Warehouse (p. 490)
anaglyph (p. 495)
ArcScene (p. 486)
ArcGlobe (p. 486)
base heights (p. 483)
block (p. 481)
CityEngine (p. 492)
components (p. 487)
Extrusion (p. 481)
face (p. 483)

GeoWall (p. 495)
KML (p. 488)
KMZ (p. 488)
offset (p. 482)
prism map (p. 485)
prisms (p. 485)
SketchUp (p. 487)
SKP file (p. 488)
stereoscopic 3D (p. 495)
textures (p. 484)
z-value (p. 481)

3D Modeling and Visualization

This chapter's lab will introduce you to concepts of modeling and design using three-dimensional objects. Working from an aerial image of a building with spatial reference, you will design a building in 3D using SketchUp, incorporating accurate height measurements, textures, and other effects, and place it in its proper location in Google Earth Pro. While there are many methods that can be used for designing a building (including Google's Match Photo tools), this lab will show you how to use a variety of SketchUp tools through the active process of creating a building.

Important note: Even though this lab uses a fixed example, it can be personalized by selecting another structure you're more familiar with—such as your home, workplace, or school—and using the same techniques described in this lab to design that structure instead.

As in *Geospatial Lab Application 7.1,* there are no questions to answer—the product you'll create in this lab will be a 3D representation of a building in SketchUp, and you'll place a version of it in Google Earth Pro. The answer sheet for this chapter will be a checklist of items to help you keep track of your progress in the lab.

Objectives

The goals for you to take away from this lab are:

▶ To obtain imagery to use in a design environment

▶ To familiarize yourself with the uses of the SketchUp software package, including its operating environment and several of its tools

▶ To apply textures to objects

▶ To utilize components in 3D design

▶ To transform your 3D model into the KMZ file format and place it in Google Earth Pro

Using Geospatial Technologies

The concepts you'll be working with in this lab are used in a variety of real-world applications, including:

▶ Urban planning, where 3D models make it possible to visualize new development projects, assess the visual impact of new buildings on the surrounding area, and develop projects that will be in character with the cityscape

▶ Archaeology, where 3D modeling can be used to recreate ancient cities and the layouts of archeological sites

[Source: photobank.kiev.ua/Shutterstock]

Obtaining Software

The current version of SketchUp (SketchUp Make) is available for free download at **www.sketchup.com**. The current version of Google Earth Pro (7.1) is available for free download at **www.google.com/earth/explore /products/desktop.html**.

Important note: Software and online resources can change fast. This lab was designed with the most recently available version of the software at the time of writing. However, if the software or Websites have significantly changed between then and now, an updated version of this lab (using the newest versions) will be available online at **www.macmillanhighered.com /shellito/catalog**.

Lab Data

There is no data to be copied for this lab. You will, however, use the "Bird's Eye" imagery from Bing Maps (see *Hands-On Application 9.6: Oblique Imagery on Bing Maps* on page 323).

Localizing This Lab

In this lab you will design a 3D model of a building on Youngstown State University's (YSU's) campus, and then view it in Google Earth Pro. In

Section 14.2, you will take a "snapshot" of the building to start with. Rather than examining a campus building at YSU, you can start at this point with a base image of any local building—such as the school building you're in, a nearby landmark or government building, or your house. You can use the same steps in this lab and the SketchUp tools to design any building or structure you so choose in place of the one used in this lab—the same lab techniques of using SketchUp will still apply; you'll just have to adapt them to your own 3D design needs.

14.1 Starting SketchUp

1. Start SketchUp. Select the option for Choose Template, and then from the available options select Simple Template—Feet and Inches. There are several different templates available, depending on what type of 3D design you want to do.

2. Next, select the Start using SketchUp button.

3. SketchUp will open into a blank working environment, with three intersecting lines at the center. These represent the axes you will use for 3D design—red and green are the horizontal (x and y) axes, while blue is the vertical (z) axis.

4. You'll see the SketchUp toolbar at the top of the screen. You'll use many of these tools during the lab—the tools that you'll use most often in this lab (from left to right on the toolbar) function as follows:

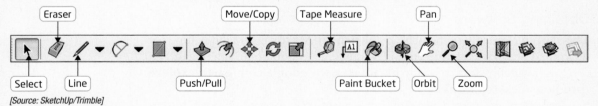

[Source: SketchUp/Trimble]

▶ The black cursor is the Select tool that lets you choose or select objects.

▶ The pink eraser is the Eraser tool that allows you to remove lines and objects.

▶ The pencil is the Line tool that lets you draw, or construct, lines.

▶ The box with the red arrow coming up from it is the Push/Pull tool that allows you to alter the shape of polygons.

▶ The four red arrows represent the Move/Copy tool that can be used to drag objects around the screen and also to create copies of objects.

▶ The extended yellow tape measure is the Tape Measure tool that allows you to measure the length or width of objects.

▶ The bucket spilling paint is the Paint Bucket tool that allows you to change the color or texture of objects.

▶ The swirling red and green lines around the pole represent the Orbit tool that allows you to change your view and maneuver around the scene.

▶ The hand is the Pan tool that allows you to move around the view.

▶ The magnifying glass is the Zoom tool that allows you to zoom in and out of the scene.

▶ There is another specialized tool (Add Location) that you'll make use of, but the lab will point this out to you when you need it.

14.2 Obtaining Google Imagery

SketchUp can use Google imagery as a starting point for your design work in SketchUp, and then you can transfer your 3D structures into Google Earth Pro (GE).

1. To begin, start Google Earth Pro.

   Important note: This lab will have you design a simplified representation of Lincoln Building (formerly known as Williamson Hall, the original College of Business building) on YSU's campus. All of the measurements of things like heights, window lengths, window measurements, and so on are either just approximations or values generated for this lab and for use only in this simplified model. To use real values for Lincoln (or your own buildings) you would have to make measurements of the structure or consult blueprints for actual heights and sizes.

2. Lincoln Building is located on YSU's campus at the following latitude/longitude decimal degrees coordinates:

   • Latitude: 41.103978

   • Longitude: −80.647777

3. Zoom GE into this area and you'll see the top of the building (see the following graphic). Fill the GE view with the entirety of the building and return to SketchUp. This should give you a good look at the overhead view of the Lincoln Building for reference in the next step. You'll use GE later, so minimize it for now.

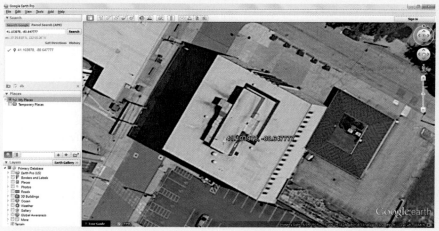

[Source: Google Earth]

4. Back in SketchUp, select the Add Location icon from the toolbar:

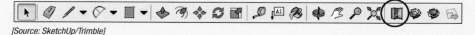

[Source: SketchUp/Trimble]

5. Add Location allows you to take a "snapshot" of Google imagery and import it into SketchUp (with information on the coordinates of the view as well as terrain information) for use as a digitizing source.

6. In the Add Location box that appears, type the coordinates of Lincoln Building and click Search. Use the zoom and pan tools (which are the same as the Google Earth tools) to expand the view, so that you can see all of Lincoln Building in the window (as you just did in Google Earth).

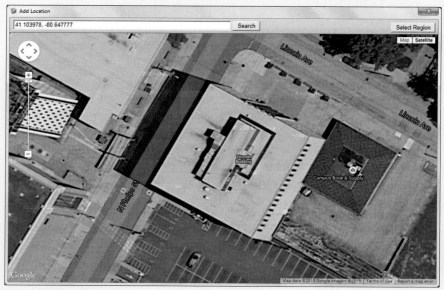

[Source: Google/SketchUp/Trimble]

7. When you've got the view set up how you want it, click the Select Region button.

8. A new image will appear—a bounding box with blue markers at the corners. Drag the corners of the box so that it stretches all the way across Lincoln Building (like in the following graphic). This area will be the "snapshot" that SketchUp will take of the imagery, which you'll use in the modeling process—so be sure to capture the whole area.

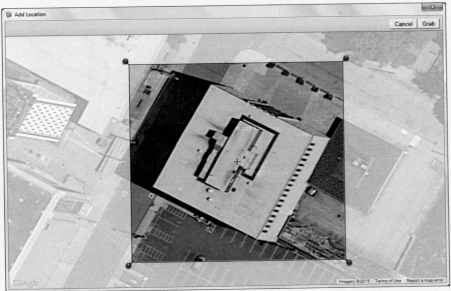

[Source: Google/SketchUp/Trimble]

9. When everything is ready, click Grab.

10. After the importing process is done, the image (in color) should appear in SketchUp, centered on the axes.

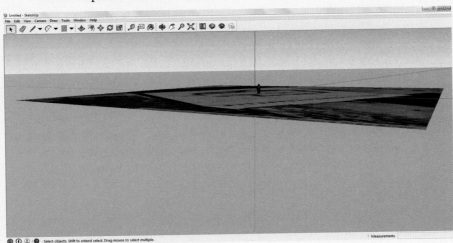

[Source: Google/SketchUp/Trimble]

11. In SketchUp, you can turn the terrain on and off within the snapshot. From the File pull-down menu, select Geo-location, and then select Show Terrain. You'll see a bit of warp and change added to the image, indicating that the terrain (that the image is draped on) is turned on. You can use this menu option to toggle the terrain on and off by placing a checkmark in that menu option. For now, toggle the terrain off (so the image is perfectly flat).

14.3 Digitizing the Building Footprint

1. You can digitize the dimensions of Lincoln Building directly from the imported aerial view. The easiest way to do this is to position the view so that it's looking directly down on the image. From the Camera pull-down menu, select Standard Views, then select Top.

2. The view will switch to show the image as seen from directly above. To zoom in or out so that you can see the entire building, select the Zoom icon from the toolbar (or if your mouse has a scroll wheel, you can use that to zoom as well). To zoom using the icon, select it, then roll the mouse forward or back to zoom in and out.

3. When you can see the outline of the entire building, it's time to start digitizing. Click on the Pencil icon (the Line tool) on the toolbar to start drawing.

4. Your cursor will now switch to a pencil. Start at the lower left-hand corner of the building and click the pencil onto the building's corner. A purple point will appear. Now, drag the pencil over to the lower right-hand corner and hover it there for a second. The words "on face in group" will appear. This is SketchUp's way of telling you just what surface you're drawing on (in this case, the face is the surface of the image).

5. Click the pencil on the lower right-hand corner, then at the upper right-hand corner, then at the upper left. The guide may change to "Perpendicular to the edge," indicating that you've drawn a perpendicular line. Finally, drag the pencil back to where you started. When you reach the beginning point, the cursor should turn green and the word "Endpoint" will appear. This indicates that you've closed the polygon of the building's footprint in on itself. Click the pencil at the endpoint and the polygon will appear, covering the entire building.

   Important note: If you have only lines and not a complete polygon, return to the lower corner and remake the footprint polygon.

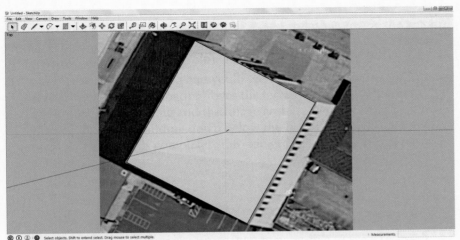

[Source: Google/SketchUp/Trimble]

14.4 Extruding the Building Footprint

1. With the building's footprint successfully digitized, the next step is to give it some height.

2. From the Camera pull-down menu, select Standard Views and select Iso. The view will switch to a perspective view.

3. From the toolbar, select the Push/Pull tool.

4. Your cursor will change again. Hover it over the footprint polygon and you'll see it slightly change texture. Click on the footprint and hold down the mouse button, then use the mouse to extrude the polygon by pushing forward.

5. In the lower-right-hand corner of SketchUp, you'll see the real-world height you're extruding the building to. Lincoln Building is approximately 72 feet high. During the extrusion operation, you can type 72' and hit the Enter key on the keyboard while holding down the mouse button, and the polygon being pushed/pulled will automatically extrude to that height.

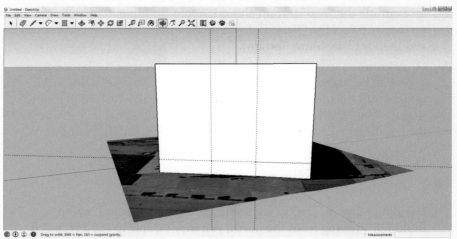

[Source: Google/SketchUp/Trimble]

7. If you find you're creating too many construction lines, you can always select them with the cursor icon and delete them by using the Eraser icon on the toolbar (in fact, anything you create can be removed with the eraser).

8. Now, use the pencil to draw the entranceway on the face of the building, tracing over the construction lines.

9. Finally, select the Push/Pull tool and use it to recess the entranceway by approximately 6 feet. Use the mouse to push the block inward, and watch the distance bar in the lower right corner.

10. Examine the building with Bing Maps' Bird's Eye imagery, and you'll see that Lincoln Building has several sets of windows that run around the top of the building—two sets of four windows on the front and back, and 18 windows on either side. Drawing all 52 of these identical windows individually would be a huge chore, so to save time and effort, you'll be designing the window only once and reusing it throughout the model.

11. One way of doing this is to set up multiple construction lines across the front of the building. Orbit around to see the top front of the building. Set up construction lines as follows (these are simplifications of the real measurements for purposes of this lab):

    a. The first window begins 6 feet from the edge of the building.

    b. The first window begins 3 feet from the top of the building.

    c. The window is 3 feet wide and 21 feet tall.

    d. There are five divisions on the window, and their measurements are (from the top):

Closing Time

This lab was involved, but it served to show off many of the 3D modeling features available within SketchUp and dealt with how to interface SketchUp together with GE. Exit SketchUp by selecting Exit from the File pull-down menu. Make sure you saved the final version of your SketchUp model as well as the KMZ version of it.

Final SketchUp Modeling Checklist

_____ Obtain overhead view of Lincoln Building and place it in SketchUp
_____ Digitize building footprint
_____ Extrude building to proper height
_____ Set building properly on the terrain
_____ Create entranceway opening on front of building
_____ Create components for windows
_____ Apply windows to all sides of building
_____ Apply appropriate colors and textures to all faces
_____ Create 3D text on front of building
_____ Add other features (front columns, windows, roof unit)
_____ Place 3D model in appropriate place in GE

15 Life in the Geospatial Cloud and Other Current Developments

Using the Cloud with Geospatial Technology, Web Maps, Story Maps, Who's Involved with Geospatial Technology, Geospatial Technologies in K–12 Education, and College and University Geospatial Educational Programs

It's an understatement to say that geospatial technology is a rapidly changing and advancing field. New Websites, imagery, tools, gadgets, and developments are coming at an incredible rate. GPS, digital maps, and satellite images are part of everyday life. When an app on your phone can provide you with the coordinates of your location, a map of what's nearby, a high-resolution remotely sensed image of your surroundings, and directions to wherever you want to go, you know that geospatial technologies have become an essential part of your life—and they're not going away. Or, as the Geospatial Revolution Project from Penn State University put it: "The location of anything is becoming everything." Back in Chapter 1, we discussed the incredibly wide variety of human activities that utilize some form of geospatial technology—if it involves location, it's going to involve geospatial technologies. We've covered a lot of ground over the last 14 chapters of this book, but every day there are new frontiers opening up, and every technological breakthrough brings the promise of new worlds to explore.

This chapter explores these new advances in the geospatial world and takes a look at who's making this progress possible through their participation, support, and sponsorship. How can you get involved with creating your own maps online? What kind of educational opportunities are there in higher education, and how is geospatial technology being used at the K–12 education levels? We'll look into all of these topics and review some examples of what's happening in geospatial technology (and, again, there's no way we're going to be able to cover everything or salute everybody—these pages aim to provide a sampling of what's out there).

The previous chapters have described numerous datasets and how to obtain them (either for free or for minimal cost), often through the

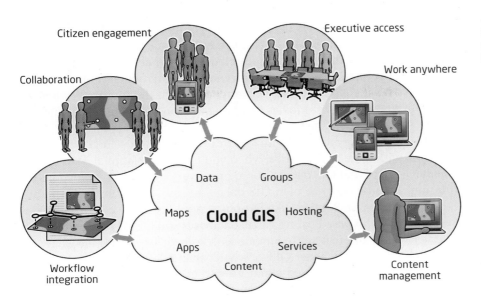

FIGURE 15.1 Using the cloud for geospatial technology applications.

hands-on applications in the chapters. Online resources are the quickest and easiest way to access geospatial data. Whether you want to acquire archived satellite imagery or get the newest updates for Google Earth, you can do everything via the Internet. The same holds true with distributing your own data and information—setting up a Web interface where your data can be accessed interactively is the way to go.

Geospatial technology is increasingly becoming part of a field referred to as the **cloud,** wherein resources are stored at another location but served to a user across the Internet (see **Figure 15.1**). The cloud is structured to let you store your data on other servers at other locations, but allows you to access it when you need to via the Internet. Similarly, other people would be able to access your data through these servers. However, there's much more to the cloud than simply serving datasets to users, as the cloud structure can be used for a wide variety of geospatial applications.

> **cloud** a computer structure wherein a user's data, resources, or applications are kept at another location and made available to the user over the Internet

How Is the Cloud Used with Geospatial Technology?

Chances are good that you use the cloud on a regular basis. For instance, if you take a photo with your phone, you can store your photos in the cloud rather than taking up space on your phone. Whenever you want to show your photos to a friend, your phone will retrieve your photos from the cloud to display on your phone. What this means is that your photos are stored somewhere else—on a data server at a remote location—and then you can access them on demand via the Internet to your phone. Similarly, when you want to watch a movie from a service like Netflix, you use the cloud to see it. The digital files for that movie are stored on a server at another location and streamed to your television via the Internet whenever you feel like watching *Sharknado 3*. The cloud is really synonymous with the Internet—whatever you want to use (photos, movies, etc.) is stored on a remote server and you can get access to it via the Internet.

If you can store and access movies or photos with the cloud, you can do the same with data. For instance, when you used The National Map back in Chapter 1, all of that GIS data is stored on servers that you download via the Internet. If you have your own GIS data, you can store it in the cloud and others can access it. Services like Dropbox, Apple's iCloud, or Amazon Web Services work this way. Posting parcel databases and DEMs for download on a server will be of great aid to anyone with some geospatial knowledge, but those who don't know what GIS is, what a shapefile is, or how to georeference imagery may be baffled about how to use the data. For instance, say you've used GIS to map the floodplain for your local area, and you've used that information to determine what land parcels are on the floodplain as well as the zoning type and property value of each parcel. If you posted the floodplain layer, the parcels layer, and any tabular data that would need to be joined to the parcels on a server for others to download, people with GIS knowledge may find this collection of data very useful. However, casual Internet users might be able to download these files, but probably would not know how to view them, overlay them, or query them to find out if their properties are on the floodplain or not.

It would be more helpful if you could construct an interactive map of this data and post it online. This type of map would allow casual users to see where their property boundaries are in relation to the floodplain and find out whether or not their house is likely to be flooded. They could also click on their parcel to see the property value and zoning information. Perhaps an aerial photo of their housing development might also be available, showing the locations of nearby rivers and other water bodies. In addition, the users of the map would have easy access to it through their Web browsers. Even though you may have created the map using ArcGIS, the users wouldn't need an expensive piece of software—they would need only a computer with a Web browser or a smartphone or tablet with access to the Web. Suddenly your online map has the potential to become a utility for realtors to use on-site as well as a mine of useful information for prospective home buyers.

The cloud makes an ideal platform for delivering these types of online maps to end users (and you've already used some examples of these cloud-based online geospatial tools in the *Hands-On Applications* of previous chapters). All of this geospatial information can be assembled into a **Web map** and placed onto a server in the cloud so that anyone can use and interact with its resources by simply using the Web.

Web map an interactive online representation of geospatial data, which can be accessed via a Web browser

When you're setting up (or "publishing") a Web map, you'll be using some type of software that will be hosted online in the cloud. There are several of these types of hosting services, but one of the most popular is Software as a Service (**SaaS**), in which the software you're using to run your application or map will be hosted in the cloud. With SaaS, you don't need to install software on your own computer—all of the software, tools, and accessories are stored in the cloud and you access them through a Web browser. For example, Google's Gmail is an example of SaaS. When you want to use your Gmail account, you log on to the Gmail Website using your browser and you have access to all of the Gmail features—you don't actually download a Gmail program on your computer.

In terms of geospatial technology, Esri's **ArcGIS Online** is a good example of SaaS wherein everything you need to create and publish Web maps is available to you through a Web browser. To make a Web map, sign in to the Website of ArcGIS Online and start using all of the available tools to create your Web map. ArcGIS Online allows users to share datasets with others by using the cloud, but it also gives users the tools to create Web maps and applications that can be served to others via the cloud.

ArcGIS Online comes in two versions. The first of these is free to use—all you need is to create a free Esri account to use when you log on to the ArcGIS Online Website and you can begin creating Web maps to share online. You can create maps from the huge amount of data made available by users of ArcGIS Online, or you can also add your own data to the map. This is what you'll do in *Geospatial Lab Application 15.1*—use the free resources and data to create and share a Web map.

The second version of ArcGIS Online enables you to use a lot more features, but also involves paying to use them. In this case, your free Esri account is tied to an organizational account that pays for the use of these added features. For instance, a university or a business may have an organizational account that your free account can be linked with. The organization has a certain number of **credits** that are consumed when you do things with ArcGIS Online. Think of credits like cell-phone minutes—you get a certain number of minutes to make phone calls with and when they're used up you need to purchase more. ArcGIS Online credits work the same way—whenever you use an ArcGIS Online feature, it deducts from the number of credits. For example, the organization version of ArcGIS Online allows you to do batch geocoding (see Chapter 8) and geoprocessing and spatial analysis (see Chapter 6), but these things use up credits from your organization.

The current version of ArcGIS for Desktop also directly interfaces with ArcGIS Online, allowing you to upload GIS data and maps as **services** to your ArcGIS Online account. You can then use these services to create Web maps or applications in ArcGIS Online (see **Figure 15.2** on page 524). For instance, you can add data and perform analysis in ArcGIS for Desktop, then take your final results, publish them as a service in ArcGIS Online, and create a Web map from them to share with the world.

SaaS Software as a Service—a method by which the software and data for an application are hosted in the cloud and accessed through a Web browser

ArcGIS Online Esri's cloud-based GIS platform where data and Web mapping software can be accessed via the Internet

credits A system used by Esri to control the amount of content that can be served by an organization in ArcGIS Online

service a format for GIS data and maps to be distributed (or "served") to others via the Internet

FIGURE 15.2 Somali
piracy trends for 2011 as
mapped in ArcGIS Online.
[Source: ArcGIS]

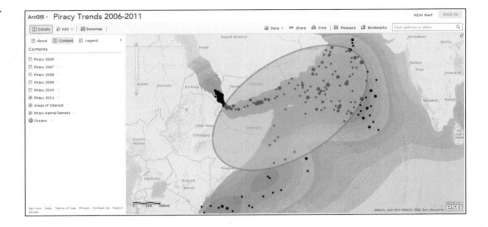

> **Story Map** an
> interactive Web map
> created using ArcGIS Online
> resources that can add
> descriptive text, photos, or
> videos to locations on a
> map to inform the user
> about the map's topic

FIGURE 15.3 A Story
Map showing the major
shale plays across the
United States. *[Source: ArcGIS]*

A story map is an example of a Web application that goes beyond a simple Web map in ArcGIS Online. In a **Story Map,** you can tailor your Web map to include pictures, videos, and text that help to communicate a certain theme or information (the "story" of the title) to the user. Story Maps can be freely created using the resources of ArcGIS Online. **Figure 15.3** shows an example of a Story Map created to examine the shale gas boom and where hydraulic fracturing (better known as "fracking") is taking place across the U.S. The Story Map contains three tabs, each showing a different map and containing different information—the first shows the location of various shale plays, the second shows the production rates per state, and the third shows the actual locations of fracking sites as points with information about who owns the well pad and when the location was fractured. In this case, the

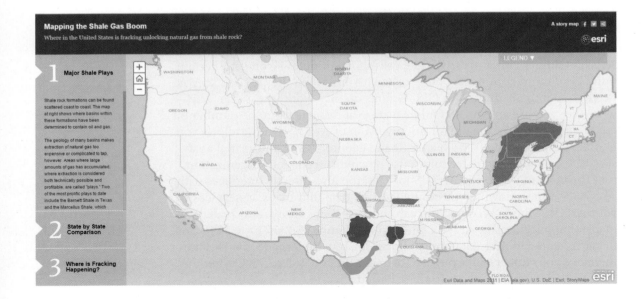

Story Map provides a lot of information about each development in a series of different interactive Web maps to tell the full "story" of the shale plays and fracking.

Figure 15.4 shows a different type of Story Map, referred to as a "Map Tour." In this type of Story Map, you can select different locations on a map and have descriptive text, photos, or videos linked to those places. In the Map Tour, clicking on a location or the photo thumbnail at the bottom of the screen will bring up the text and media for that place. For instance, in Figure 15.4, selecting a location in San Diego, CA, will give you information about the transportation innovations at that place, such as the San Diego Green Line Trolley, the Sprinter light rail, or electric car recharging stations. Text and photos for each location will also appear to help tell the "story" about transportation in San Diego. A Map Tour can also be used to give a traditional tour of areas (such as a walking tour of historic places in a city or a tour of highlights at a popular tourist destination). There are many different types and formats of Story Maps available beyond these two—for more about using Story Maps, see *Hands-On Application 15.1: Esri Story Maps.*

Tools like these are part of a larger initiative that Esri refers to as "Mapping for Everyone," which allows users to add to existing databases, use

FIGURE 15.4 A Map Tour showing various transportation innovations across the city of San Diego, CA. *[Source: ArcGIS]*

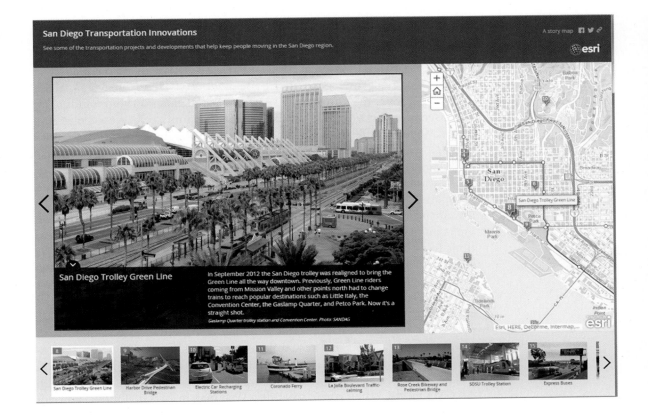

HANDS-ON APPLICATION 15.1

Esri Story Maps

To examine a bunch of different Story Maps, point your Web browser at **http://storymaps.arcgis.com.** This Website is the place to begin with Story Maps—if you want to create a certain type of Story Map, you would choose the Apps option and then select the type of Story Map you want to make. However, in order to view the different types of Story Maps available, choose the option for Gallery. In the Story Map Gallery, you'll see a wide variety of Story Maps created by Esri as well as people from the larger geospatial community. Under the choices for Story Map App you can view different kinds of Story Maps that use the various map types—this includes Map Tours as well as other Story Map types such as Map Journal, Shortlist, Swipe, and Spyglass. Story Maps are also grouped together by theme—by choosing one of the options under Collections, you can find Story Maps related to topics of interest to you, including those designed for Events and Disasters, Nature and Environment, and Sports and Entertainment. Investigate some of the different types of Story Maps as well as a sampling from various collections.

Expansion Questions

- Choose three collections related to your own interests. What topics (or "stories") are being presented with Story Maps in these collections?

- Look at Story Map examples from the different types available under Story Map Apps. How does each one present the topic (or "story") in a different fashion?

special software to distribute geospatial data, and use their own data (or data processed for them) to create and publish their own maps on their own Website. Like the user-generated content of VGI (see Chapter 1), these options put more mapping choices and geospatial data into your hands as an end user, allowing you to build, design, and customize your own mapping applications and then embed them onto a Website or e-mail a link to other people.

One service that's available from the cloud for use with Web maps or applications is a **basemap**—a pre-made layer over which you can lay other geospatial data. When you're using a utility like Google Maps (as in *Hands-On Application 8.4: Online Mapping and Routing Applications and Shortest Paths* on page 289), you can select a view of the shortest path between locations to be displayed over a street map or a satellite image of the area. The street map and the satellite image are examples of basemaps that other layers can be displayed on, and they can be streamed to you via the cloud and displayed in your Web browser. Geospatial data can be obtained through the cloud structure in the form of aerial or satellite imagery, digital topographic maps, and digital street maps, all of which can be used as basemaps with your own applications. Because this data is being served to you over the Internet, you don't have to download it and set it up on your own computer.

For instance, Esri makes a number of such basemaps available through ArcGIS Online. When you're creating a Web map in ArcGIS Online, you can

> **basemap** an image layer that serves as a backdrop for the other layers that are available in ArcGIS Online

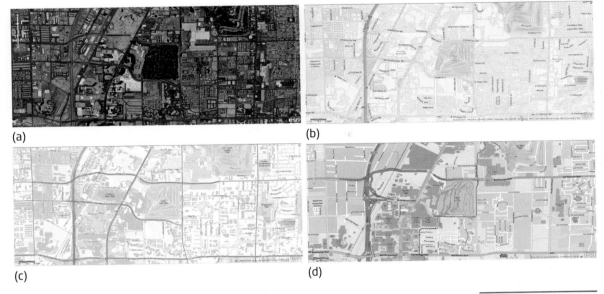

(a)

(b)

(c)

(d)

choose the type of basemap you want and display your data overlaid on them (see **Figure 15.5** for examples of the basemap data available through ArcGIS Online). In addition, these same basemaps can be streamed from the cloud into ArcGIS for Desktop as well, so that you can have on-demand access to georeferenced high-resolution satellite imagery or street maps to aid in your GIS analyses.

There are many programs available for taking geospatial data and setting it up online in an easy-to-use interactive format as well as creating an interactive map to incorporate into a Website. **ArcGIS for Server** is a good example—this Esri software is designed to facilitate the distribution of interactive maps and geospatial data across the Internet for viewing and analysis. (See **Figure 15.6**, which shows an online mapping and analysis application

FIGURE 15.5 Examples of basemaps available in ArcGIS Online: a) Imagery with Labels, b) Streets, c) Topographic, and d) OpenStreetMap. *[Source: ArcGIS]*

ArcGIS for Server an Esri utility for distribution of geospatial data in an interactive mapping format on the Internet

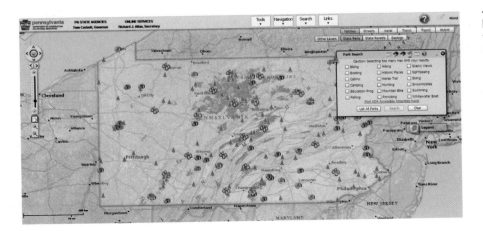

FIGURE 15.6 Online GIS for the Pennsylvania Department of Conservation and Natural Resources, created using ArcGIS for Server. *[Source: Pennsylvania Department of Conservation and Natural Resources, ArcGIS]*

created for the Pennsylvania Department of Conservation and Natural Resources using ArcGIS for Server.)

Setting up interactive online tools for managing and analyzing geospatial data is becoming increasingly common for a variety of utilities. Many other cloud-based geospatial resources are also available for creating Web maps. Products such as CartoDB or MapBox enable users to quickly and easily create powerful maps and geospatial applications that are stored in and served from the cloud.

Web applications are created with the help of an **API** (Application Programming Interface) that can be used for Web development. An API describes the available commands you can use and the format of those commands that allow two or more types of software or applications to interface or "talk" with one another. APIs are used by programmers when building software or applications. The W3C Geolocation API described on page 17 of Chapter 1 is an example of this—it allows programmers to obtain the location information of your computer or mobile device and plot that on a map.

For example, if you were running a bakery at a particular location, you'd have your shop's address on its Website and when people wanted to get to your shop, they'd have to leave your Website, go to Google Maps, type your address, and then juggle two Websites back and forth. What would be more beneficial is if you could embed an interactive version of Google Maps onto your shop's Website to show the bakery's location. To do this, you would use a special type of API (in this case, the Google Maps Embed API) to properly add this to your bakery's Website. This kind of technique was used by the Guggenheim Museum in New York City to add Google Maps interactivity to its Website (see **Figure 15.7**).

APIs are used for creating geospatial Web applications. For example, you would use the Javascript API for developing Web applications for ArcGIS. Other companies, like Google, Microsoft, and Yahoo!, provide APIs that allow you to use their mapping products (such as Google Maps, Bing Maps, or Yahoo! Maps) as functions on Websites. Also commonly used are open source products, such as Leaflet or OpenLayers, which provide libraries of Javascript functions for creating maps and geospatial applications for mobile devices and the Web. By combining several map layers together, a user can create a new map **mashup,** which uses different datasets together in one map. Check out *Hands-On Application 15.2: More Than a Map—The Google Maps API* for some examples of the Google Maps API in action as an agent for developing a wide variety of geospatial Web applications.

Users can create Web maps or applications that can be distributed to everyone via the cloud, thereby circulating geospatial knowledge and making geospatial tools freely available. Users of an interactive Web map of a local metropark may have never before seen the letters G, I, and S together in an acronym, but they can still access the Web map on their smartphone or tablet to locate service stations, places to eat, points of interest, and their

API Application Programming Interface—the functions and commands used when making requests of software

mashup the combination of two or more map layers into one new application

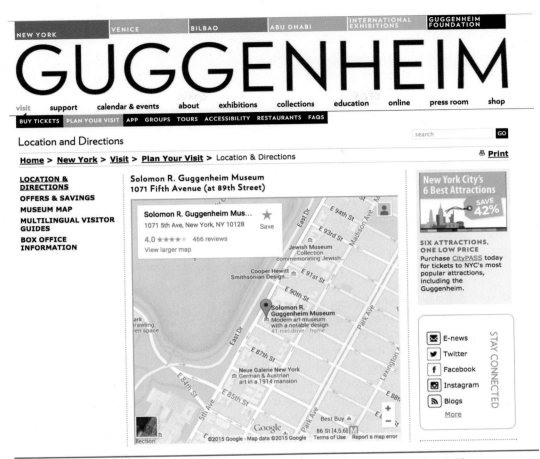

FIGURE 15.7 The Website of the Guggenheim Museum, which uses embedded interactive Google Maps. *[Source: Screenshot of Guggenheim's map from www.guggenheim.org. © SRGF, NY]*

HANDS-ON APPLICATION 15.2

More Than a Map—The Google Maps API

To examine some of the applications of the Google Maps API, point your Web browser to **https://developers.google.com/maps/.** Click on Get Started to see the various categories of Google APIs and select one to see a demo of it, including utilizing basemaps, satellite imagery, street views, routing, and data visualization. Within each one of these options, you can view different presentations of data—for instance, on the Satellite link you can examine high-resolution overhead views, as well as 45-degree oblique imagery.

Under the View Showcase link, you can read about different Google Maps API applications.

Expansion Questions

- What sort of real-world applications have been developed using the Google Maps API, and what have they been used for?

- How are the applications for satellite imagery, street views, and data visualization being used to visualize and communicate geospatial information?

THINKING CRITICALLY WITH GEOSPATIAL TECHNOLOGY 15.1

Who Owns Geospatial Data?

Here's something for you to think about: You start with a basemap sample (say, a world topographic map), and you obtain data from the USGS about earthquake locations and other natural hazards. You create a Web map using ArcGIS Online, and post it onto your Website. Who owns this geospatial resource, and who owns the geospatial dataset you've created? In essence, you've taken other freely available data and created your own unique geospatial resource—but can you really lay claim to it as something you own? Similarly, if you make changes or updates to a resource like OpenStreetMap, can you claim ownership of a shared online resource? With more and more geospatial data and customization functions becoming available via cloud resources (such as the high-resolution imagery that's available through ArcGIS Online), who really owns the end products created from them? Does anybody really own them?

location on a hiking trail. GIS analysts can carry out extensive studies of crime hotspots, and law enforcement officers can make use of this data via an updated Web application served to them from the cloud. In the aftermath of a disaster, rescue and relief teams can examine frequently updated maps showing blocked roads, locations of emergency calls, and remotely sensed imagery of the area, all available through Web maps or via applications shared in the cloud. All these new developments in the geospatial world open an increasingly wide variety of professional and educational opportunities to people who are interested or trained in geospatial technology.

 ## Who Is Involved with Geospatial Technology?

There are numerous "power players" in the geospatial world—we've already discussed the role of companies like Microsoft, Google, Esri, and Digital-Globe, and a growing number of government agencies, including NASA, NOAA, and the USGS, in generating and providing geospatial data and infrastructure. Other federal government agencies, including the EPA (Environmental Protection Agency) and the NGA (the National Geospatial-Intelligence Agency), play important roles in applications of geospatial technologies. Many state and local agencies utilize GIS, GPS, or remotely sensed data through city planning, county auditors' offices, or county engineers' offices. Also, the Open Geospatial Consortium (**OGC**) is dedicated to developing new standards of geospatial data interoperability.

There are also several prominent professional organizations in the geospatial field. The Association of American Geographers (**AAG**) is the

OGC the Open Geospatial Consortium, a group involved with developing new standards of geospatial data interoperability

AAG the Association of American Geographers; the national organization for the field of geography

national organization for geographers in all fields, whether physical, regional, cultural, or geospatial. Boasting several thousand members, AAG is a key organization for geographers, hosting a large annual national meeting, multiple regional branches, and publishing key geographic journals. The American Society for Photogrammetry and Remote Sensing (**ASPRS**) is a major professional organization dedicated to remote sensing and geospatial research. ASPRS publishes a prominent academic journal in the field and hosts national and regional meetings that are well attended by geospatial professionals. The Urban and Regional Information Systems Association (**URISA**) sponsors numerous conferences and publications dealing with the academic, professional, and practical aspects of GIS. Finally, but of equal importance, the National Council for Geographic Education (**NCGE**) is a nationally recognized organization of educators dedicated to promoting geographic education through conferences, research, and journal publications.

There are many ongoing research and educational initiatives dedicated to furthering geospatial technology and its applications, as well as to improving and fostering access to geospatial data and tools. For example, **OhioView,** created in 1996, is an initiative dedicated to remote sensing research and educational programs in Ohio. Part of OhioView's efforts led to the free distribution of Landsat imagery (previously, a single Landsat scene used to cost $4000). OhioView serves as a consortium for remote sensing and geospatial research, and is composed of members from each of the state universities in Ohio, along with other partners. With all of these statewide affiliations, OhioView serves as a source of many initiatives within the state, including educational efforts.

This type of statewide consortium may have originated with OhioView, but it has expanded to a national program called **AmericaView.** Administered in part by the USGS and set up as a 501(c)(3), AmericaView aims to spread remote sensing research and promote educational initiatives all over the United States (see **Figure 15.8** on page 532). Many states have developed a **StateView** program similar to OhioView (for instance, there's VirginiaView, WisconsinView, TexasView, and so on), with the eventual goal of having a complete set of 50 StateView programs. Each StateView is composed of state universities and related partners from that state and offers various geospatial services and data (such as Landsat imagery and other remotely sensed data) for that state. For instance, OhioView servers permit the free download of Ohio Landsat data from their multiyear archive. Similarly, WisconsinView provides fresh MODIS imagery of Wisconsin every day, and the other AmericaView member states make many data products (such as high-resolution aerial photography and Landsat imagery) freely available. In addition, TexasView offers free access to remotely sensed images and provides links to GIS data of Texas. See *Hands-On Application 15.3: AmericaView and the StateView Programs* to further investigate the resources available through the various StateView programs.

ASPRS the American Society for Photogrammetry and Remote Sensing; a professional organization in the geospatial and remote sensing field

URISA the Urban and Regional Information Systems Association; an association for GIS professionals

NCGE the National Council for Geographic Education; a professional organization dedicated to geographic education

OhioView a research and education consortium consisting of the state universities in Ohio, along with related partners

AmericaView a national organization dedicated to remote sensing research and education

StateView the term used to describe each of the programs affiliated with AmericaView in each state

FIGURE 15.8 The current membership composition of AmericaView. *[Source: AmericaView]*

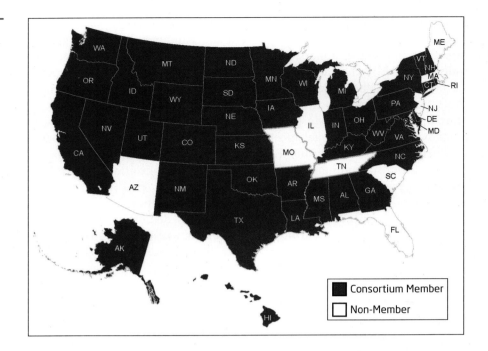

Consortium Member

Non-Member

How Is Geospatial Technology Used in K–12 Educational Efforts?

Groups such as NCGE, OhioView, and AmericaView are dedicated to promoting education in the ever-growing field of geospatial technology, but especially in K–12 classrooms. With geospatial technologies becoming

HANDS-ON APPLICATION 15.3

AmericaView and the StateView Programs

To see if your state is part of the AmericaView consortium (and to also see the current version of the program's membership), open your Web browser to the interactive map located at **www.americaview.org/membership-map**. If it is a full member, click on the state to bring up the contact info and check out the specific StateView's Website (that is, MichiganView's Website or VirginiaView's Website) to see who that state's members are and what type of data and services they provide. If your state's not yet a full member, take a look at some of your neighboring states and see what they offer as a StateView program.

Expansion Questions

- What is the status of your home state as a StateView program?

- What types of data and geospatial services does your StateView provide? (If your state is not yet a full member of AmericaView, look at the StateView Websites of neighboring states that are.)

increasingly utilized in today's society, introducing concepts early in the grammar-school classroom and using them for illustrating topics such as geography, spatial thinking, or earth science is becoming increasingly common. These efforts are aided by the increasing availability of free software programs (like those used in this book), and plenty of freely available geospatial data. Remotely sensed images from the Earth Observing System (EOS) can help to provide clear, graphic explanations of environmental or climatic conditions, such as changing global temperatures, while easily comprehensible Landsat imagery can be used to evaluate the health of vegetation or the loss of forested lands and calculate the subsequent effects. NASA and Google have released teacher kits and supplemental lesson plans to accompany their products, and many other examples of geospatial technology being integrated into K–12 curricula can be found on the Internet and in our schools (see *Hands-On Application 15.4: Educational Resources and Lesson Plans* for several examples).

Beyond using Google Earth (and other software) for interactive classroom investigations, many other programs are aimed at promoting geospatial technology at the K–12 level. The National Council for Geographic

HANDS-ON APPLICATION 15.4

Educational Resources and Lesson Plans

The widespread nature of geospatial technologies and the availability and access to geospatial software have allowed educators to integrate GIS, GPS, and remote sensing into lesson plans in multiple ways. Examine some of the following examples to see how geospatial technology is being incorporated into K–12 curricula.

1. Educaching (GPS lesson plans): **www.educaching.com**

2. GIS 2 GPS (GIS and GPS lesson plans): **http://gis2gps.com**

3. Google Maps Education: **http://maps.google.com/help/maps/education/learn/**

4. Integrated Geospatial Education and Technology Training (IGETT) (labs and other educational materials for GIS and remote sensing): **http://igett.delmar.edu/index.html**

5. Landsat Science Education (for use with Landsat imagery): **http://landsat.gsfc.nasa.gov/?page_id=11/teacherkit**

6. NASA Teacher's Guide (using the Image Composite Explorer in the classroom): **http://earthobservatory.nasa.gov/Experiments/ICE/ice_teacher_guide.php**

7. Sciencespot (GPS lesson plans): **http://sciencespot.net/Pages/classgpslsn.html**

8. VirginiaView Educational Resources (lesson plans incorporating remote sensing concepts and trackable geocoins): **http://virginiaview.cnre.vt.edu/education.html**

Expansion Questions

- What kinds of activities are teachers using in the classroom that involve geospatial technology?

- When you've examined these examples, locate some other examples of how geospatial technologies are being applied in the classroom. What other types of activities and lesson plans are being used in classroom settings?

Education (NCGE) is at the forefront of geographic education, supporting teachers and furthering the understanding of geographic concepts at the K–12 level. Another organization dedicated to promoting geographic education is the **Geographic Alliance Network,** a national organization of geography educators at the K–12 and higher education levels. Geographic Alliances are set up in each state as a way of hosting workshops, professional development, meetings, events, involvement with geography bees, lesson plans, and more at the statewide level.

Among OhioView's initiatives is **SATELLITES** (Students and Teachers Exploring Local Landscapes to Interpret the Earth from Space), a program designed to instruct K–12 teachers about geospatial technologies and to help them to introduce them into their classrooms. SATELLITES hosts week-long "teachers' institutes" during the summer in locations throughout Ohio. These institutes are free for teachers to attend, and teachers can earn free graduate credits, when available, through successful participation. During the week, teachers receive introductory instruction in several of this book's topics (including GIS, GPS, remote sensing, landscapes, and how to use remotely sensed environment and climate data) and participate in hands-on work with the technology itself. Teachers also receive free equipment to take back to their classrooms, including a GPS receiver and an infrared thermometer. The aim is for the teachers to develop a research project that will incorporate geospatial technologies and an environmental theme (previous SATELLITES institutes have engaged with global research initiatives like the International Polar Year).

The inquiry-based projects the K–12 students become involved with are part of **GLOBE** (Global Learning and Observations to Benefit the Environment), an international initiative sponsored by NASA and NOAA that incorporates over 20,000 schools from 114 countries. GLOBE is dedicated to the study of Earth's environments and ecosystems. Using specific data-collection protocols, students involved with GLOBE collect data related to land cover, temperature, water, soils, and the atmosphere. This data is shared with other students around the world via the GLOBE Website, creating rich and far-ranging datasets for use by everyone. Teachers attending the SATELLITES institutes learn GLOBE protocols and engage their students through field data collection coupled with geospatial technology applications.

An example of a SATELLITES success story is the "Satellite Girls" (a group of four middle school students from Akron, Ohio). Their teacher attended the SATELLITES program in the summer of 2007 and involved a group of his students with data collection and analysis through GLOBE. The girls competed with their project at state and national levels, and were selected to be one of five teams of students to represent the United States at the international GLOBE Learning Expedition in Cape Town, South Africa, in 2008. In 2010 and 2012, GLOBE projects developed by students as a result

---

Geographic Alliance Network a National Geographic-sponsored organization set up in each state to promote geographic education

SATELLITES Students and Teachers Exploring Local Landscapes to Interpret the Earth from Space—an OhioView initiative involving K–12 teachers and students with geospatial technologies

GLOBE Global Learning and Observations to Benefit the Environment—an education program aimed at incorporating user-generated environmental data observations from around the world

of their teachers' participation in the SATELLITES program were presented by the students at the White House Science Fair in Washington, D.C.

 ## What Types of Educational Opportunities Are Available with Geospatial Technology?

Beyond K–12 involvement, geospatial technology is integrated into the curricula of higher education programs internationally. The U.S. Department of Labor's Geospatial Technology Competency Model (**GTCM**) provides a framework for geospatial skills that are key for today's workforce and workforces of the future—a framework that has been used as the basis for educational curriculum development by the GeoTech Center. Geospatial technology course offerings are increasingly to be found in college and university geography departments, but also within civil engineering, natural resources, environmental sciences, and geology programs. Bachelor's degrees may offer some sort of concentration in geospatial technology, or at least provide the option to take several courses as part of the major. A minor in GIS or geospatial technologies is also a common option in geography departments. Graduate programs (both master's and doctorate degrees) in geography often have an option to focus on aspects of geospatial technology.

> **GTCM** the Geospatial Technology Competency Model set up by the U.S. Department of Labor

However, with the growth of the geospatial field, it's now possible to get a higher education degree in geospatial technology (or GIS or geographic information science, depending on the name of the program). Bachelor's, master's, and even doctorate programs have become available at several universities. Some of these are interdisciplinary programs, involving elements of various social sciences or computer science, and they often integrate elements of information technology like programming, creation of databases, or Web design. For instance, Youngstown State University offers a bachelor's degree in "Spatial Information Systems," which combines elements from the geospatial side of the geography discipline with computer-related coursework in programming and databases, classes in CAD and professional writing, and an option for the student and advisors to design a set of courses for specific applications (such as environmental science, biology, or archeology).

Certificate programs are also a common option for geospatial education. These programs vary from school to school, but typically involve taking a structured set of geospatial classes (perhaps four to seven classes total). These programs tend to be focused squarely on completing the requirements for a set of courses, and are often available at both the undergraduate and graduate levels. For example, the Geospatial Certificate at Youngstown State University involves students taking four required courses—an introductory mapping or geospatial class, an introductory GIS class, an introductory remote sensing course, and either an advanced

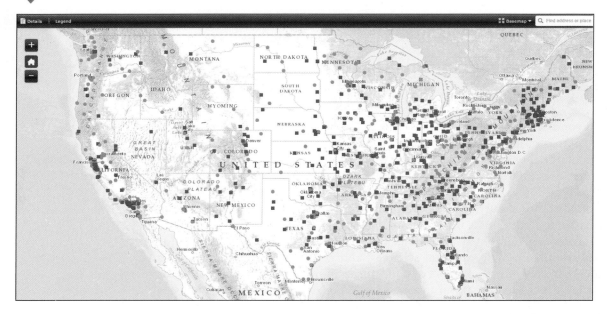

FIGURE 15.9 The GeoTechCenter's Web map of geospatial institutions showing two-year colleges (in green) and four-year colleges and universities (in blue).
[Source: GeoTechCenter.org]

GIS or advanced remote sensing class. Students take an additional two courses, chosen from a pool of options, including field methods, GPS and GIS, a geospatial-related internship, object-oriented programming, and database design. Certificates and degree programs are also becoming increasingly available at the community and technical college level (see **Figure 15.9** and *Hands-On Application 15.5: Degree Programs and Certificates for Geospatial Technology* for some further investigation).

HANDS-ON APPLICATION 15.5

Degree Programs and Certificates for Geospatial Technology

An online utility is available for examining the available geospatial programs at community colleges and technical colleges throughout the United States—open your Web browser and go to Change to: **www.geotechcenter.org/geospatial -national-map.html.** This map (part of the GeoTech Center Website at **www.geotechcenter.org**) shows those community and technical colleges that are involved with geospatial education. Schools are mapped by categories, such as whether they offer a degree in an aspect of geospatial technology, a certificate program, or have classes available. Check your local area to see which schools have geospatial courses, and find out what kinds of degree programs they offer (select the option for CC Info on GIS Offerings in the cube menu icon to get the names of the schools).

Expansion Questions

- What community or technical colleges in your local area offer some form of geospatial education?

- What kinds of programs are available at these schools—certificates, degrees, or classes?

Professional certification in geospatial technologies is also available. GIS Professional (**GISP**) certification is offered by the GIS Certification Institute to applicants who pass an exam as well as a portfolio review that includes contributions to the GIS field, evidence of candidates' professional experience, and their level of achievement in education. Esri offers technical certification based on an exam. ASPRS offers certification status to candidates in several geospatial fields, including Certified Photogrammetrist, Certified Mapping Scientist in GIS/LIS, and Certified Mapping Scientist in remote sensing.

GISP the title given for professional GIS certification

With geospatial technology being used in so many different fields, students from outside the discipline of geography may seek out a couple of courses or aim at certification to give them the kind of formal background in geospatial technology that will assist them when they apply for positions in their own field (be it business, real estate, archeology, geology, or human ecology). Online courses are offered by a variety of schools to allow students to complete a geospatial certificate or degree remotely, and "virtual universities" offer students at one school the option to take a geospatial course with a professor at another school.

Geospatial technology is playing an increasingly significant part in our daily lives, whether in our work, travel, or recreation. We have even established observance days dedicated to showcasing and celebrating these vital new technologies. The National Geographic Society kicks off Geography Awareness Week each November, and since 1998 the Wednesday of this week has been **GIS Day**. GIS Day is sponsored by a number of prominent agencies and groups (including the National Geographic Society, Esri, the USGS, and the Library of Congress). It is aimed at promoting GIS for students, educators, and professionals in the field. Each year on GIS Day, there are GIS-related events at schools, universities, and workplaces across the world. Similarly, 2006 saw the launch of **Earth Observation Day** (currently

GIS Day the Wednesday of Geography Awareness Week (observed in November), dedicated to GIS

Earth Observation Day a day dedicated to remote sensing research and education held in conjunction with Earth Sciences week (observed in October)

THINKING CRITICALLY WITH GEOSPATIAL TECHNOLOGY 15.2

What's Next for Geospatial Technologies?

Right now, you can run Google Earth on your phone, download satellite imagery for free, create and view interactive Web maps in no time flat, interact with landscapes and cities in immersive 3D, and earn a doctorate in the geospatial field. If that's what you can do today, what's going to happen tomorrow? Where do you see the geospatial field going? As access to information becomes faster and easier, as the power and importance of computers advances, and as our lives become more and more tied together in a global web, what's going to be the next stage for geospatial technology? What do you see the geospatial world looking like five years from now?

observed in conjunction with Earth Sciences week in October of each year) through AmericaView to showcase remote sensing research and education across the globe.

Chapter Wrapup

So this is it—the end of the last chapter. Over these 15 chapters, we've looked at many aspects of geospatial technologies, both from the theoretical side and also from working hands-on with various aspects of the technologies themselves. The geospatial world has changed a lot in the last few years (remember, Google Earth only debuted back in 2005), and it's going to keep changing. That next "killer app" for geospatial technology is probably being coded by someone out there as you read this. See *Geospatial Cloud and Organizational Apps* for a mobile app for ArcGIS Online, and also check out *Geospatial Organizations and the Geospatial Cloud in Social Media* for several blogs, Facebook pages, Twitter accounts, and YouTube videos related to using the cloud as well as this chapter's look at some of the people and groups out on the frontiers of geospatial technologies.

GIS, GPS, remote sensing, and all of their applications are only going to become more important in our lives, and even though many people may still not be familiar with the term "geospatial technology," by now you've probably got a pretty good idea of what it is and what it can do.

There's still one last Geospatial Lab Application. In *Geospatial Lab Application 15.1* you'll be using Esri's ArcGIS Online to create and publish your own Web maps in the cloud.

Important note: The references for this chapter are part of the online companion for this book and can be found at **www.macmillanhighered .com/shellito/catalog**.

Geospatial Cloud and Organizational Apps

Here's a sampling of available representative apps related to this chapter's contents for your phone or tablet. Note that some apps are for Android, some are for Apple iOS, and some may be available for both.

- **ArcGIS:** A mobile app version of ArcGIS, which allows access to many features of ArcGIS Online

- **ASPRS:** An app that gives you access to ASPRS events, publications, and conference information

- **Explorer for ArcGIS:** An app that allows you to access ArcGIS Online Web maps in the cloud (made by others or within your organization)

Geospatial Organizations and the Geospatial Cloud in Social Media

Here's a sampling of some representative Facebook and Twitter accounts, along with some YouTube videos and blogs related to this chapter's concepts.

f **On Facebook, check out the following:**

- **AmericaView:**
    www.facebook.com/AmericaView
- **Earth Observation Day:**
    www.facebook.com/EarthObsDay
- **Esri Geo Developers:**
    www.facebook.com/EsriGeoDev
- **GeoTech Center:**
    www.facebook.com/geotechcenter20
- **GIS Day:**
    www.facebook.com/gisday
- **The GLOBE Program:**
    www.facebook.com/TheGLOBEProgram

Become a Twitter follower of:

- **AAG:** @theAAG
- **AmericaView:** @AmericaView
- **ArcGIS Online:** @ArcGISOnline
- **ASPRS:** @ASPRSorg
- **CartoDB:** @cartoDB
- **Earth Observation Day:** @EarthObsDayAV
- **Esri GIS Education:** @GISEd
- **Esri Story Maps:** @EsriStoryMaps
- **Extension** (America's Research-Based Learning Network)—Geospatial Technology: @exGeospatial
- **GeoTechCenter:** @GeoTechCenter
- **GIS Day:** @gisday
- **Joseph Kerski** (geospatial educational news from Esri's Education program): @josephkerski
- **Leaflet:** @LeafletJS
- **Mapbox:** @Mapbox
- **NCGE:** @NCGE1915

- **OhioView:** @OHViewGeo
- **Open Geospatial Consortium:** @opengeospatial
- **The Globe Program:** @GLOBEProgram
- **URISA:** @URISA

 On YouTube, watch the following videos:

- **GIS Education** (an Esri YouTube channel devoted to educational concepts using GIS):
  www.youtube.com/user/ESRIEdTeam
- **The GLOBE Program** (The GLOBE Program's YouTube channel featuring videos of students' and teachers' research):
  www.youtube.com/user/globeprogram
- **Geographyuberalles** (a YouTube channel featuring all types of GIS and geographic videos):
  www.youtube.com/user/geographyuberalles

For further up-to-date info, read up on these blogs:

- **AmericaView Blog** (the latest news and geospatial applications from the members of AmericaView):
  http://blog.americaview.org
- **ArcGIS Resources—ArcGIS Online Category** (blog posts concerning topics related to working with ArcGIS Online resources):
  http://blogs.esri.com/esri/arcgis/category/arcgis-online
- **ArcGIS Resources—Esri Story Maps Category** (blog posts about using ArcGIS Online and Story Maps):
  http://blogs.esri.com/esri/arcgis/category/story-maps
- **Esri Applications Prototype Lab** (a blog for new developments coming out of Esri):
  http://blogs.esri.com/esri/apl
- **Esri GIS Education Community** (a blog about education applications of GIS):
  http://blogs.esri.com/esri/gisedcom
- **Google GeoDevelopers Blog** (a blog dealing with new developments and applications of Google's geospatial tools):
  http://googlegeodevelopers.blogspot.com
- **Letters from the SAL** (a blog about geospatial research from the Spatial Analysis Laboratory at the University of Vermont):
  http://letters-sal.blogspot.com
- **Spatial Reserves** (a blog about publically available geospatial data):
  http://spatialreserves.wordpress.com

Key Terms

AAG (p. 530)
AmericaView (p. 531)
API (p. 528)
ArcGIS for Server (p. 527)
ArcGIS Online (p. 523)
ASPRS (p. 531)
basemap (p. 526)
cloud (p. 521)
credits (p. 523)
Earth Observation
  Day (p. 537)
Geographic Alliance
  Network (p. 534)
GIS Day (p. 537)

GISP (p. 537)
GLOBE (p. 534)
GTCM (p. 535)
mashup (p. 528)
NCGE (p. 531)
OGC (p. 530)
OhioView (p. 531)
SaaS (p. 523)
SATELLITES (p. 534)
service (p. 523)
StateView (p. 531)
Story Map (p. 524)
URISA (p. 531)
Web map (p. 522)

15.1 Geospatial Lab Application

Creating Web Maps with ArcGIS Online

This chapter's lab will introduce you to the use of Esri's cloud-based GIS mapping and analysis package, ArcGIS Online. As noted in this chapter, there are two different versions of ArcGIS Online: one that requires you to be part of an organization and purchase credits for use, and a free version that doesn't use the credits system. This lab will use the free version, so all data, methods, and techniques you'll use in this lab are free of charge. During this lab, you'll use the cloud as your platform for working with GIS data, remotely sensed imagery, and geospatial applications. You'll create two Web maps—one from data available online and one from data within a spreadsheet—and submit the URLs for these Web maps to your instructor for grading.

Objectives

The goals of this lab are:

▶ To learn the basics of the ArcGIS Online environment

▶ To create a Web map from data available through ArcGIS Online

▶ To adjust the symbology of features on a Web map

▶ To save and share a Web map

▶ To create a Web application from a Web map

▶ To perform geocoding using ArcGIS Online

▶ To use different basemaps in creating a Web map

▶ To create a Web map and a Web application from your own data

Using Geospatial Technologies

The concepts you'll be working with in this lab are used in a variety of real-world applications, including:

▶ Engineering, where Web maps are utilized in the planning of new roads or highways by examining terrain conditions such as elevation, steepness, and land cover

▶ County auditing, where auditors examine Web maps and streaming high-resolution imagery of residential and commercial properties to assist them in assessing property taxes

[Source: Mint Images-Tim Pannell/Getty Images]

Obtaining Software

No software or data is necessary to download for this lab—everything you will be working with can be done through a Web browser.

Important note: Software and online resources can change fast. This lab was designed with the most recently available version of the software at the time of writing. However, if the software or Websites have significantly changed between then and now, an updated version of this lab (using the newest versions) will be available online at **www.macmillanhighered.com /shellito/catalog**.

Lab Data

Copy the folder Chapter15—it contains an Excel file (converted to CSV format) called OhioViewAddresses.csv, which contains the names and addresses of the colleges and universities that comprise the membership of OhioView.

Localizing This Lab

This lab starts at the broad scale of the entire United States and then narrows its focus onto areas within Ohio, specifically the members of the OhioView consortium. However, the same techniques used in this lab can be used for your own local area. When you're creating the Web map from your own data in Step 15.4, you can substitute a .csv file consisting of the names and addresses of the schools from your own StateView (such as the schools within the VirginiaView consortium, for instance). Alternatively, you could use a .csv file containing the names and addresses of the state universities for your state.

15.1 Getting Started With ArcGIS Online

1. Open your Web browser and go to **www.arcgis.com**. When you arrive at this page, click on Sign In.

2. In order to save and share your work with ArcGIS Online, you'll need to obtain a free Esri Global Account. You can do this by clicking on Create a Public Account. Follow the instructions to create your account, then sign in to ArcGIS Online using your new account. After you have signed in, you'll see your first name in the upper-right-hand corner of the screen where the words "Sign In" used to be—this indicates you're properly signed in to your Esri account.

3. After signing in, click on the Map link at the top of the Web page. The basic Web map interface for ArcGIS Online will appear. You'll see the controls to zoom in and out of the map in the upper-left-hand corner of the map view (you can also use the mouse for this). The "crosshairs" symbol below the zoom buttons will allow ArcGIS Online to focus in on your current location.

[Source: Esri]

15.2 Creating a Web Map From Online Data

1. We'll begin by making a Web map from data that's already available online. A dataset showing the 2015 AmericaView membership has already been created and shared through ArcGIS Online. To retrieve

this dataset and begin developing your Web map, click the Add button, and then select Search for Layers.

Search for Layers

Find:	Americaview	GO
In:	ArcGIS Online	▼
☐	Within map area	

[Source: Esri]

2. In the Find: box, type AmericaView.

3. In the In: box, select ArcGIS Online.

4. Remove the checkmark from the Within map area box.

5. Click GO to search for the AmericaView layer in ArcGIS Online.

6. ArcGIS Online will respond that it has found the result—the layer called AmericaView (by bradbook). Click the text that says Add next to this result. A new layer showing the current membership of states in AmericaView will be added to the Web map. Click on DONE ADDING LAYERS to continue making your map.

7. You'll see the AmericaView layer listed under Contents along with the current basemap being used (by default it should be the Topographic basemap). These are the two layers that your map will consist of. However, you'll see that the layer showing the AmericaView states is covering up the basemap. What you'll do next is change the color of the AmericaView states layer and also make it semi-transparent so that you can see both the states layer and the basemap underneath.

8. To begin, click on the small ellipsis symbol … next to AmericaView in the Contents box. A new group of options will appear.

[Source: Esri]

9. These options will allow you to affect the layer, rename it, remove it, or change some of its properties and appearance. To begin changing the appearance of the layer, select the Change Style icon.

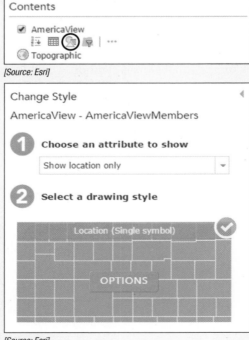

[Source: Esri]

10. As you want to show the locations of the states only, just use the Show location only option under the Choose an attribute to show menu. Under Select a drawing style, click on OPTIONS.

11. The next screen will allow you to change the transparency of the layer. The slider bar under Transparency will allow you to show the layer fully opaque (0%, on the far left) or fully transparent (100%, on the far right). Move the slider bar to about the 40% percent transparency range and you'll see the AmericaView layer become transparent enough to view the features and labels of the basemap underneath.

12. Next, click on the box next to the word Symbols. A new menu will appear and allow you to choose a different color from the default.

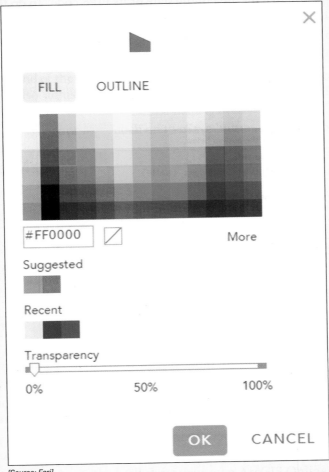

[Source: Esri]

13. Select FILL to change the color of the states and then click on one of the red color options.

14. Select OUTLINE to change the color of the states' borders and then click on one of the black color options.

15.3 Creating and Sharing a Web Application

Your Web map is now saved. You can reopen it at any time by selecting the ArcGIS pull-down menu in the upper left corner of the screen, choosing My Content. This will show you all of the Web maps you've created and any other services, Web applications, or anything else you've created using Arc-GIS Online. By selecting your Web map, you'll have another option to open the saved map. However, you're the only one who can access it until you share it with others.

1. To get started sharing your map, click on **Share** on the toolbar.

Share ✕

Choose who can view this map.
Your map is currently shared with these people.

☑ Everyone (public)

Link to this map
http://arcg.is/1VmuZAb   [f] Facebook   [t] Twitter
☑ Share current map extent

Embed this map

EMBED IN WEBSITE   CREATE A WEB APP

DONE

[Source: Esri]

2. In the Share dialog, put a checkmark in the box next to **Everyone (public)**. This will share your Web map with everyone out there. The line below that will give a shortened URL for the Web map—by using this URL, you or anyone who has it will be able to view and interact with your Web map.

3. ArcGIS Online gives you the ability to share this URL through Facebook or Twitter or embed your map on a Website. One thing to keep in mind is that any Web map created with ArcGIS Online will have the basic appearance of the ArcGIS Online interface. To distinguish it, change the overall appearance of the interface, and customize it further, you'd want to create a Web application. To get started with this, click on **CREATE A WEB APP**.

[Source: Esri]

4. There are numerous types of Web applications, each with their own distinct appearance and features. For the basic map you're creating, some of the Web application templates are not appropriate (such as the Elevation Profile or the Story Map options). Try out some of the Web application templates by clicking on the Create pull-down menu underneath each one and then selecting Preview to see what the final Web application will look like. Many of the templates would be appropriate for the AmericaView Membership map you've created, but in particular check out and evaluate the following templates: Basic Viewer, Information Lookup, Map Tools, Public Information, and Simple Map Viewer.

5. When you have settled on a template for what your final Web application will look like, click on the Create pull-down menu underneath that particular template, then select Create.

6. On the next screen, you'll see the Title, Tags, and Summary you added for the Web map. Adjust these as you wish, but make sure that there is a checkmark in the box next to Share this app in the same way as the map (Everyone). This will share your Web application with everyone. Lastly, click on DONE.

7. Your Web application will open in your Web browser. A new pane will open on the right-hand side, allowing you to make other changes or adjustments. Take a look at the many options available to you, including which widgets can be used on the Web application. However, for this basic Web application, just accept all of the defaults, scroll down to the bottom of the pane, click SAVE, and then click DONE.

8. The next screen will show you a summary of your Web application, as well as the URL assigned to it. Click on the URL to open your final Web application and examine it, then answer Question 15.1.

> **Question 15.1** Submit the URL for your final Web application to your instructor, who will check over your work for completeness and technical merit (do you have the pop-ups configured properly, do you have the map zoomed to the proper extent, etc.) as well as its appearance (did you choose an appropriate color scheme, did you select an appropriate Web application template, etc.) to make sure you get credit for this exercise.

15.4 Creating a Web Map From Your Own Data

1. Next, we'll make a second Web map, but using your own data (i.e., data not available through ArcGIS Online). Back in the main Web browser tab, select the option for MAP. When it returns to your basic Web map, select NEW MAP to effectively clear out your previous map and provide a fresh mapping canvas to work with.

2. Bring up Windows Explorer and navigate to the location where you copied the Chapter15 folder. Inside of it is a spreadsheet called OhioViewAddresses.csv. Open this file and you'll see the names of the 14 colleges and universities that make up the membership of OhioView. What we're going to do next is geocode these addresses (see Chapter 8) and turn them into a point layer on the map. In the end, you'll have a Web map of Ohio featuring these 14 different locations of the OhioView member schools.

3. Geocoding with ArcGIS Online is a snap. Resize your Windows Explorer window and your Web browser window so that you can see both of them side by side. Then simply drag and drop the OhioViewAddresses.csv file from Windows Explorer into the map shown in ArcGIS Online.

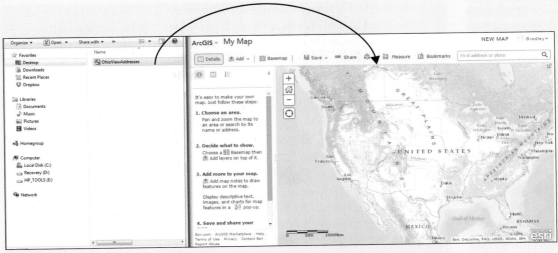

[Source: Esri]

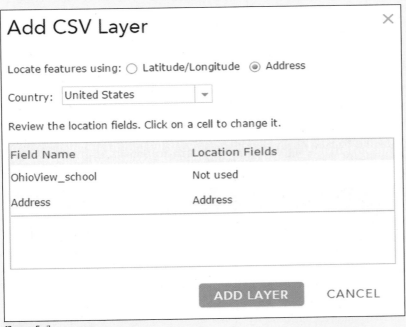

[Source: Esri]

4. The Add CSV Layer dialog box will appear. This lets you specify how you want to add this data to ArcGIS Online. Under Locate features using:, select the radio button for Address and make sure that the Field Name of Address is using the Location Field of Address. This lets ArcGIS Online know that you will be geocoding addresses (rather than plotting points based on coordinates) and that the attribute field called "Address" contains the necessary address information to be used in geocoding. Click ADD LAYER to proceed with geocoding your points.

5. The next screen will allow you to specify how you want the points to appear on the map. Make sure that the OhioView school field is selected for Choose an attribute to show (as you want to show the name of each school on the map), and under Select a drawing style, click on Options. You'll see that of the 14 schools, ArcGIS Online gives a unique color to 10 of them, then puts the last four in a category called "Others." To give each of the 14 schools its own unique color, scroll through the list of schools and click on the blue arrows symbol next to "Others." This will expand the option and show the other four schools as unique points (see the graphic below for a guide).

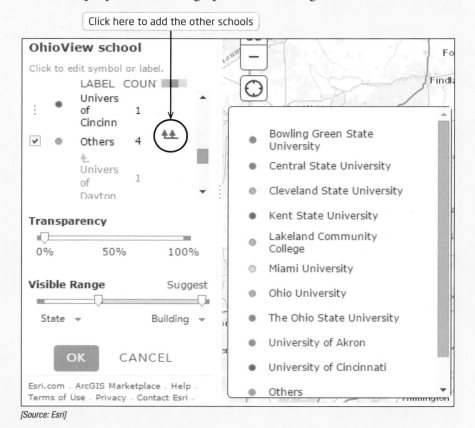

[Source: Esri]

6. Move the left-hand arrow of the slider bar to Country so that you'll be able to see the points when you're zoomed in to the level of the United States or closer.

7. Click on OK to close the dialog box and then click on DONE to complete the mapping of the schools as points on the map.

8. In the Web map, zoom out so you can see all of Ohio and your 14 geocoded schools.

9. Click on the Legend button and you'll see the names and colors that correspond with each point displayed. Click on a school on the map and you'll see a pop-up with its name and address.

15.5 Using Basemaps in Web Mapping

1. The next thing we'll do is select a different basemap for this Web map. To see your basemap options, click on the Basemap button.

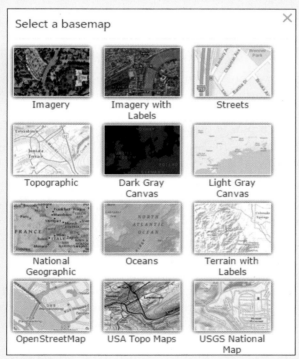

Select a basemap

Imagery   Imagery with Labels   Streets

Topographic   Dark Gray Canvas   Light Gray Canvas

National Geographic   Oceans   Terrain with Labels

OpenStreetMap   USA Topo Maps   USGS National Map

[Source: Esri]

2. In the Select a basemap dialog box, you'll see several different options for the basemap (which is the graphics layer underneath your data). By default, ArcGIS Online will start you off with the Topographic basemap.

3. To switch to the Imagery basemap, click on its icon in the Select a basemap dialog box. The basemap will switch to high-resolution satellite imagery of Ohio (and the surrounding areas).

4. While this looks nice, it doesn't put your geocoded points in much context. Click on Basemap again, and this time choose Imagery with Labels. The map will change to now have borders for the state as well as some labeled cities. As you zoom in on the map, you'll see other cities appear as well as county borders.

5. Your final map will have a single basemap underlying it, so try out all of the other choices for basemaps. You'll find that some of them are less appropriate for the kind of map you're making (such as Imagery, Dark Grey Canvas, Light Grey Canvas, and Oceans), but other choices will be more appropriate. Keep in mind, you'll want to select a basemap that looks good while examining Ohio at the state scale as well as when zooming in on particular school points. Move on to the next step when you've chosen which basemap you want to use in your final map (just make sure it's not Topographic).

15.6 Creating a Web Application From Your Own Data

1. When you have your Web map looking the way you want it, the last step will be to turn it into a Web application (like you did to the AmericaView map back in Step 15.3—you may want to refer back to that section for how to do each of the following things).

2. Save your Web map (giving it an appropriate title and summary, and using IGT3e for a tag).

3. Share your Web map with Everyone.

4. Make a Web application from your Web map, choosing an appropriate Web application template.

5. When you have created your final Web application and have a URL assigned to it, open that URL in a new tab, examine your final Web application, and then answer Question 15.2.

> **Question 15.2** Submit the URL for your final OhioView Web application to your instructor, who will check over your work for completeness and technical merit (do you have all 14 schools labeled properly, do you have the map zoomed to the proper extent, etc.) as well as its appearance (did you choose an appropriate basemap, did you select an appropriate Web application template, etc.) to make sure you get credit for this exercise.

Closing Time

ArcGIS Online is a great tool for creating Web maps from your own data or pre-existing data online. There are a number of other types of data that can be used in ArcGIS Online, including GPX files (see Chapter 4) and a shape-file within a zipped file (see Chapter 5), as well as numerous types of adjust-ments and visualization options for your Web maps.

In addition, there are numerous types of Web applications that can be created from the templates, including Story Maps. While the organizational version of ArcGIS Online adds many other features, you can create profes-sional looking Web maps quickly and easily with only using the free version. When you're finished, simply exit your Web browser—your Web maps and Web apps will remain safely stored in your space in the cloud that came with your free Esri account. Then you can say good-bye to lab applications for the rest of the book.

3D modeling (p. 481) designing and visualizing data that contains a third dimension (a z-value) in addition to x- and y-values.

3D trilateration (p. 95) finding a location on Earth's surface in relation to the positions of three satellites.

3D Warehouse (p. 490) Trimble's online repository of 3D objects created using SketchUp.

3DEP (p. 458) the 3D Elevation Program, a U.S. government program that provides digital elevation data for the entire United States.

8-bit imagery (p. 353) a digital image that carries a range of brightness values from 0 to 255.

AAG (p. 530) the Association of American Geographers; the national organization for the field of geography.

absorption (p. 347) when light is trapped and held by a target.

across-track scanning (p. 377) a scanning method using a rotating mirror to collect data by moving the device back and forth across the width of the satellite's swath.

address matching (p. 281) another term for geocoding.

address standardization (p. 282) setting up the components of an address in a regular format.

aerial photography (pp. 2, 310) acquisition of imagery of the ground taken from an airborne platform.

affine transformation (p. 74) a linear mathematical process by which data can be altered to align with another data source.

Afternoon Constellation (p. 423) a set of satellites (including Aqua and Aura) that pass the Equator in the afternoon during their orbits.

AIRS (p. 421) the Advanced Infrared Sounder instrument onboard Aqua.

algorithm (p. 287) a set of steps used in a process designed to solve a particular type of problem (for example, the steps used in computing a shortest path).

ALI (p. 391) the multispectral sensor onboard EO-1.

almanac (p. 93) data concerning the status of a GPS satellite, which is included in the information being transmitted by the satellite.

along-track scanning (p. 376) a scanning method using a linear array to collect data directly on a satellite's path.

AmericaView (p. 531) a national organization dedicated to remote sensing research and education.

AMSR-E (p. 422) the Advanced Microwave Scanning Radiometer for EOS instrument onboard Aqua.

AMSU-A (p. 421) the Advanced Microwave Sounding Unit instrument onboard Aqua.

anaglyph (p. 495) a new image created by slightly offsetting the features in an existing image from each other and highlighting them in two different colors.

AND (p. 180) the Boolean operation that corresponds with an intersection operation.

API (p. 528) Application Programming Interface—the functions and commands used when making requests of software.

Aqua (p. 421) an EOS satellite whose mission is to monitor Earth's water cycle.

ArcGIS (p. 138) Esri's GIS platform.

ArcGIS for Desktop (p. 138) the component of ArcGIS that runs on a desktop computer.

ArcGIS for Server (p. 527) an Esri utility for distribution of geospatial data in an interactive mapping format on the Internet.

ArcGIS Online (pp. 139, 523) Esri's cloud-based GIS platform where data and Web mapping software can be accessed via the Internet.

ArcGIS Pro (p. 139) a new stand-alone GIS program that comes as part of ArcGIS for Desktop 10.3.

ArcGlobe (p. 486) a 3D visualization tool used to display data in a global environment, which is part of Esri's ArcGIS for Desktop program.

ArcMap (p. 138) the component of ArcGIS for Desktop used for viewing and analyzing data.

ArcScene (p. 486) the 3D visualization and design component of Esri's ArcGIS for Desktop program.

ASPRS (p. 531) the American Society for Photogrammetry and Remote Sensing; a professional organization in the geospatial and remote sensing field.

ASTER (p. 419) the Advanced Spaceborne Thermal Emission and Reflection Radiometer instrument onboard Terra.

atmospheric windows (p. 346) those wavelengths of electromagnetic energy in which most of the energy passes through Earth's atmosphere.

ATMS (p. 426) the Advanced Technology Microwave Sounder instrument onboard Suomi NPP.

A-Train (p. 423) another term for the Afternoon Constellation.

attribute table (p. 134) a spreadsheet-style form where the rows consist of individual objects and the columns are the attributes associated with those objects.

attributes (p. 133) the non-spatial data that can be associated with a spatial location.

Aura (p. 423) an EOS satellite dedicated to monitoring Earth's atmospheric chemistry.

AVHRR (p. 431) Advanced Very High Resolution Radiometer; the six-band sensor onboard NOAA satellites.

band (p. 344) a narrow range of wavelengths that may be measured by a remote sensing device.

base heights (pp. 462, 483) the elevation values assigned to the terrain upon which the objects will be placed.

basemap (p. 526) an image layer that serves as a backdrop for the other layers that are available in ArcGIS Online.

batch geocoding (p. 284) matching a group of addresses together at once.

block (p. 481) a flat polygon that has been extruded to transform it into an object with a z-value.

blue band (p. 344) the range of wavelengths between 0.4 and 0.5 micrometers.

Blue Marble (p. 430) a composite MODIS image of the entire Earth from space.

Boolean operator (p. 180) one of the four connectors (AND, OR, NOT, XOR) used in building a compound query.

brightness values (p. 352) the energy measured at a single pixel according to a predetermined scale; also referred to as digital numbers (DNs).

buffer (p. 182) a polygon of spatial proximity built around a feature.

C/A code (p. 93) the digital code broadcast on the L1 frequency, which is accessible by all GPS receivers.

cartographic generalization (p. 226) the simplification of representing items on a map.

cartography (p. 223) the art and science of creating and designing maps.

Catalog (p. 139) the component of ArcGIS for Desktop used for managing data (which contains the functionality of the previous ArcCatalog).

CERES (p. 415) the Clouds and Earth's Radiant Energy System instruments onboard Terra and Aqua.

CGIS (p. 125) the Canadian Geographic Information System—a large land inventory system developed in Canada, and the first system of this type to use the name "GIS."

channels (p. 93) the number of satellite signals a GPS unit can receive.

choropleth map (p. 231) a type of thematic map in which data is displayed according to one of several different classifications.

CIR photo (p. 317) color infrared photo—a photo where infrared reflection is shown in shades of red, red reflection is shown in shades of green, and green reflection is shown in shades of blue.

citizen science (p. 16) the activities of untrained volunteers conducting science.

CityEngine (p. 492) a powerful rule-based 3D visualization program designed for quickly modeling 3D cities.

cloud (p. 521) a computer structure wherein a user's data, resources, or applications are kept at another location and made available to the user over the Internet.

CMYK (p. 236) a color scheme based on using the colors cyan, magenta, yellow, and black.

collar (p. 453) the white, information-filled border around a US Topo.

color composite (p. 354) an image formed by placing a band of imagery into each of the three color guns (red, green, and blue) to view an image in color rather than grayscale.

color gun (p. 354) equipment used to display a color pixel on a screen through the use of the colors red, green, and blue.

color ramp (p. 237) a particular range of colors applied to the thematic data on a map.

Compass (p. 103) China's GNSS, currently in development.

components (p. 487) predesigned (or user-designed) objects in SketchUp.

compound query (p. 180) a query that contains more than one operator.

connectivity (p. 277) the linkages between edges and junctions of a network.

constellation (p. 91) the full complement of satellites comprising a GNSS.

continuous field view (p. 129) a conceptualization of the world in which all items vary across Earth's surface as constant fields, and values are available at all locations along the field.

contour interval (p. 452) the vertical difference between two adjacent contour lines drawn on a map.

contour line (p. 451) an imaginary line drawn on a map to connect points of common elevation.

control points (p. 70) point locations where the coordinates are known, used to align the unreferenced image to the source.

control segment (p. 92) one of the three segments of GPS, consisting of the control stations that monitor the signals from the GPS satellites.

Corona (p. 373) a U.S. government remote sensing program, which utilized film-based camera equipment mounted in a satellite in Earth's orbit; Corona was in operation from 1960 to 1972.

CORS (p. 101) Continuously Operating Reference Stations; a system operated by the National Geodetic Survey to provide a ground-based method of obtaining more accurate GPS positioning.

coverage (p. 141) a data layer represented by a group of files in a workspace consisting of a directory structure filled with files and also associated files in an INFO directory.

credits (p. 523) a system used by Esri to control the amount of content that can be served by an organization in ArcGIS Online.

CrIS (p. 426) the Cross-track Infrared Sounder instrument onboard Suomi NPP.

crowdsourcing (p. 16) the activities of untrained volunteers to create content and resources that can be utilized by others.

Cubesat (p. 394) a type of satellite with a weight of less than three pounds and a length of about four inches.

data classification (p. 232) various methods used for grouping together (and displaying) values on a choropleth map.

datum (p. 40) a reference surface of Earth.

datum transformation (p. 41) changing measurements from one datum to measurements in another datum.

decimal degrees (DD) (p. 43) the fractional decimal equivalent to coordinates found using degrees, minutes, and seconds.

degrees, minutes, and seconds (DMS) (p. 42) the measurement system used in GCS.

DEM (p. 456) digital elevation model—a representation of the terrain surface, created by measuring a set of equally spaced elevation values.

DGPS (p. 100) differential GPS; a method of using a ground-based correction in addition to the satellite signals to determine position.

digitizing (p. 127) the creation of vector objects through sketching or tracing representations from a map or image source.

Dijkstra's Algorithm (p. 287) an algorithm used in calculating the shortest path between an origin node and other destination nodes in a network.

discrete object view (p. 126) a conceptualization of the world in which all reality can be represented by a series of separate objects.

dissolve (p. 182) the ability of the GIS to combine polygons with the same features together.

DLG (p. 451) Digital Line Graph—the features (such as roads, rivers, or boundaries) digitized from USGS maps.

DOQ (p. 320) a Digital Orthophoto Quad—orthophotos that cover an area of 3.75 minutes of latitude by 3.75 minutes of longitude, or one-fourth of a 7.5 minute USGS quad.

DPI (p. 238) dots per inch—a measure of how coarse (lower values) or sharp (higher values) an image or map resolution will be when exported to a graphical format.

draping (p. 462) a process in which an image is given z-values to match the heights in a digital terrain model.

DRG (p. 450) Digital Raster Graphic—a scanned version of a USGS topographic map.

DSM (p. 457) digital surface model; a measurement of the heights of ground elevations as well as the objects on top of the ground as captured by lidar.

DTM (p. 455) a representation of a terrain surface calculated by measuring elevation values at a series of locations.

dual frequency (p. 93) a GPS receiver that can pick up both the L1 and L2 frequency.

Earth Observation Day (p. 537) a day dedicated to remote sensing research and education held in conjunction with Earth Sciences week (observed in October).

Earth Observatory (p. 430) a Website operated by NASA that details how EOS is utilized in conjunction with numerous global environmental issues and concerns.

easting (p. 50) a measurement of so many units east (or west) of some principal meridian.

edge (p. 277) a term used for the links of a network.

EGNOS (p. 103) an SBAS that covers Europe.

electromagnetic spectrum (p. 343) the light energy wavelengths and the properties associated with them.

ellipsoid (p. 40) a model of the rounded shape of Earth.

EO-1 (p. 389) a satellite launched in 2000 as part of NASA's New Millennium Program, which orbits 1 minute behind Landsat 7.

EOS (p. 413) NASA's Earth Observing System mission program.

ephemeris (p. 93) data referring to the GPS satellite's position in orbit.

equal intervals (p. 233) a data classification method that selects class break levels by taking the total span of values (from highest to lowest) and dividing by the number of desired classes.

Equator (p. 42) the line of latitude that runs around the center of Earth and serves as the 0 degree line from which to make latitude measurements.

EROS (p. 377) the Earth Resources Observation Science Center, located outside Sioux Falls, South Dakota, which serves (among many other things) as a downlink station for satellite imagery.

Esri (p. 138) a key developer and industry leader of GIS products.

ETM+ (p. 382) the Enhanced Thematic Mapper sensor onboard Landsat 7.

exclusive or (p. 181) the operation wherein the chosen features are all of those that meet the first criterion as well as all of those that meet the second criterion, except for the features that have both criteria in common in the query.

extrusion (p. 481) the extending of a flat object to have a z-value.

face (p. 483) one side of a block.

false color composite (p. 357) an image arranged when the distribution of bands differs from placing the red band in the red color gun, the green band in the green color gun, and the blue band in the blue color gun.

false easting (p. 51) a measurement made east (or west) of an imaginary meridian set up for a particular zone.

false northing (p. 51) a measurement made north (or south) of an imaginary line such as is used in measuring UTM northings in the southern hemisphere.

feature class (p. 141) a single data layer in a geodatabase.

fields (p. 134) the columns of an attribute table.

file geodatabase (p. 141) a single folder that can hold multiple files, with nearly unlimited storage space.

Flock-1 (p. 394) a constellation of high-resolution remote sensing Cubesats operated by PlanetLabs.

fonts (p. 229) various styles of lettering used on maps.

Galileo (p. 103) the European Union's GNSS, currently in development.

geocaching (p. 105) using GPS and a set of coordinates to find locations of hidden objects.

geocoding (p. 281) the process of using the text of an address to plot a point at that location on a map.

geocoin (p. 106) a small coin used in geocaching with a unique ID number that allows its changing location to be tracked and mapped.

geodatabase (p. 141) a single item that can contain multiple layers, each as its own feature class.

geodesy (p. 40) the science of measuring Earth's shape.

GeoEye-1 (p. 392) a satellite launched in 2008 that features a panchromatic sensor with a spatial resolution of 0.46 meters.

Geographic Alliance Network (p. 534) a National Geographic-sponsored organization set up in each state to promote geographic education.

geographic coordinate system (GCS) (p. 41) a set of global latitude and longitude measurements used as a reference system for finding locations.

geographic information systems (GIS) (p. 123) a computer-based set of hardware and software used to capture, analyze, manipulate, and visualize geospatial information.

geographic scale (p. 224) the real-world size or extent of an area.

Geographic Information System (GIS) (p. 2) computer-based mapping, analysis, and retrieval of location-based data.

geoid (p. 40) a model of Earth using mean sea level as a base.

GEOINT (p. 8) Geospatial Intelligence.

geolocation (p. 16) the technique of determining where something is in the real world.

GeoPDF (p. 239) a format that allows maps to be exported to a PDF format, yet contain multiple layers of geographic information.

geoprocessing (p. 184) the term that describes when an action is taken to a dataset that results in a new dataset being created.

georeferencing (p. 68) a process of aligning an unreferenced dataset with one that has spatial reference information.

geospatial data (p. 8) items that are tied to specific real-world locations.

geospatial technology (p. 2) a number of different high-tech systems that acquire, analyze, manage, store, or visualize various types of location-based data.

geostationary orbit (p. 374) an orbit in which an object follows precisely the direction and speed of Earth's rotation and is therefore always directly above the same point on Earth's surface.

geotag (p. 17) connecting real-world location information to an item.

GeoTIFF (p. 451) a graphical file format that can also carry spatial referencing.

GeoWall (p. 495) a powerful computer tool for displaying data and imagery in stereoscopic 3D.

GIS Day (p. 537) the Wednesday of Geography Awareness Week (observed in November), dedicated to GIS.

GIS model (p. 192) a representation of the factors used for explaining the processes that underlie an event or for predicting results.

GISP (p. 537) the title given for professional GIS certification.

GLOBE (p. 534) Global Learning and Observations to Benefit the Environment—an education program aimed at incorporating user-generated environmental data observations from around the world.

Global Positioning System (GPS) (p. 2) acquisition of real-time location information from a series of satellites in Earth's orbit.

GLONASS (p. 102) the former USSR's (now Russia's) GNSS.

GloVis (p. 386) the Global Visualization Viewer set up by the USGS for viewing and distribution of satellite imagery.

GNSS (p. 90) the global navigation satellite system, an overall term for the technologies that use signals from satellites to find locations on Earth's surface.

GOES (p. 431) Geostationary Operational Environmental Satellite; a series of geostationary satellites tasked with monitoring weather conditions.

Google Earth (p. 18) a freely available virtual globe program first released in 2005 by Google.

GPS (p. 89) the Global Positioning System, a technology using signals broadcast from satellites for navigation and position determination on Earth.

GPX (p. 104) a standard file format for working with GPS data collected by GPS receivers.

graduated symbols (p. 231) the use of different-sized symbology to convey thematic information on a map.

great circle distance (p. 43) the shortest distance between two points on a spherical surface.

green band (p. 344) the range of wavelengths between 0.5 and 0.6 micrometers.

grid cell (p. 131) a square unit, representing some real-world size, which contains a single value.

GTCM (p. 535) the Geospatial Technology Competency Model set up by the U.S. Department of Labor.

hillshade (p. 461) a shaded relief map of the terrain created by modeling the position of the Sun in the sky relative to the landscape.

HIRDLS (p. 424) the High Resolution Dynamics Limb Sounder instrument onboard Aura.

HSB (p. 421) the Humidity Sounder for Brazil instrument onboard Aqua (which ceased operation in 2003).

Hyperion (p. 391) the hyperspectral sensor onboard EO-1.

hyperspectral imagery (p. 354) remotely sensed imagery created from the bands collected by a sensor capable of sensing hundreds of bands of energy at once.

hyperspectral sensor (p. 391) a sensor that can measure hundreds of different wavelength bands simultaneously.

identity (p. 185) a type of GIS overlay that retains all features from the first layer along with the features it has in common with a second layer.

IKONOS (p. 392) a satellite launched in 1999 that features sensors with a multispectral spatial resolution of 3.2 meters and panchromatic resolution of 0.8 meters.

incident energy (p. 348) the total amount of energy (per wavelength) that interacts with an object.

International Charter (p. 397) an agreement between multiple international space agencies and companies to provide satellite imagery and resources to help in times of disasters.

International Date Line (p. 47) a line of longitude that uses the 180th meridian as a basis (but changes away from a straight line to accommodate geography).

intersect (p. 185) a type of GIS overlay that retains the features that are common to two layers.

intersection (p. 180) the operation wherein the chosen features are those that meet both criteria in the query.

interval data (p. 134) a type of numerical data in which the difference between numbers is significant, but there is no fixed non-arbitrary zero point associated with the data.

IR (p. 344) infrared; the portion of the electromagnetic spectrum with wavelengths between 0.7 and 100 micrometers.

join (p. 135) a method of linking two (or more) tables together.

JPEG (p. 238) the Joint Photographic Experts Group image or graphic file format.

junction (p. 277) a term used for the nodes (or places where edges come together) in a network.

key (p. 136) the field that two tables have to have in common with each other in order for the tables to be joined.

KML (p. 488) Keyhole Markup Language—the file format used for Google Earth data.

KMZ (p. 488) the compressed version of a KML file.

label (p. 229) text placed on a map to identify features.

Landsat (p. 381) a long-running and ongoing U.S. remote sensing project that launched its first satellite in 1972.

Landsat 7 (p. 382) the seventh Landsat mission, launched in 1999, which carries the ETM+ sensor.

Landsat 8 (p. 382) the latest Landsat mission, launched in 2013.

Landsat scene (p. 385) a single image obtained by a Landsat satellite sensor.

LandsatLook (p. 386) an online utility used for the viewing and downloading of Landsat imagery.

large-scale map (p. 225) a map with a higher value for its representative fraction; such maps will usually show a small amount of geographic area.

LAS (p. 458) the industry standard data format used for lidar data.

latitude (p. 41) imaginary lines on a globe north and south of the Equator that serve as a basis of measurement in GCS.

layout (p. 229) the assemblage and placement of various map elements used in constructing a map.

LDCM (p. 382) the Landsat Data Continuity Mission—a previous name for the Landsat 8 mission.

legend (p. 228) a graphical device used on a map that explains what the various map symbols and colors represent.

lidar (p. 457) a process in which a series of laser beams fired at the ground from an aircraft is used both to create highly accurate elevation models and also to measure the height of objects from the ground.

line segment (p. 278) a single edge of a network that corresponds to one portion of a street (for instance, the edge between two junctions).

linear interpolation (p. 283) a method used in geocoding to place an address location among a range of addresses along a segment.

lines (p. 127) one-dimensional vector objects.

longitude (p. 42) imaginary lines on a globe east and west of the Prime Meridian that serve as a basis of measurement in GCS.

map (p. 223) a representation of geographic data.

map algebra (p. 186) combining datasets together using simple mathematical operators.

map projection (p. 47) the translation of locations on the three-dimensional (3D) Earth to a two-dimensional (2D) surface.

map scale (p. 225) a metric used to determine the relationship between measurements made on a map and their real-world equivalents.

map templates (p. 230) premade arrangements of items in a map layout.

mashup (p. 528) the combination of two or more map layers into one new application.

metadata (p. 137) descriptive information about geospatial data.

MCE (p. 192) multi-criteria evaluation—the use of several factors (weighted and combined) to determine the suitability of a site for a particular purpose.

micrometer (p. 343) a unit of measurement equal to one-millionth of a meter, abbreviated μm.

Mie scattering (p. 347) scattering of light caused by atmospheric particles the same size as the wavelength being scattered.

MISR (p. 415) the Multi-angle Imaging SpectroRadiometer instrument aboard Terra.

MLS (p. 424) the Microwave Limb Sounder instrument onboard Aura.

MODIS (p. 417) the Moderate Resolution Imaging Spectroradiometer instrument onboard Terra and Aqua.

MOPITT (p. 417) the Measurements of Pollution in the Troposphere instrument onboard Terra.

Morning Constellation (p. 415) a set of satellites (including Terra, Landsat 7, SAC-C, and EO-1), each one passing the Equator in the morning during its own orbit.

MSAS (p. 103) an SBAS that covers Japan and nearby regions.

MSI (p. 389) Multispectral Instrument, the multispectral sensor onboard the Sentinel-2 satellites.

MSS (p. 382) the Multispectral Scanner aboard Landsat 1 through 5.

multipath (p. 99) an error caused by a delay in the signal due to reflecting from surfaces before reaching the receiver.

multispectral imagery (p. 354) remotely sensed imagery created from the bands collected by a sensor capable of sensing several bands of energy at once.

multispectral sensor (p. 382) a sensor that can measure multiple wavelength bands simultaneously.

NAD27 (p. 41) the North American Datum of 1927.

NAD83 (p. 41) the North American Datum of 1983.

nadir (p. 317) the location on the ground that lies directly below the camera in aerial photography.

NAIP (p. 321) the National Agriculture Imagery Program, which produces orthophotos of the United States.

nanometer (p. 343) a unit of measurement equal to one-billionth of a meter, abbreviated nm.

nanosatellite (p. 394) a type of satellite that weighs between 2.2 and 22 pounds.

NASA (p. 373) the National Aeronautics and Space Administration, established in 1958; it is the U.S.

government's space exploration and aerospace development branch.

NASA NEO (p. 430) the NASA Earth Observations Website, which allows users to view or download processed EOS imagery in a variety of formats, including a version compatible for viewing in Google Earth.

The National Map (p. 10) an online basemap and data viewer with downloadable geospatial data maintained and operated by the USGS and part of the National Geospatial Program.

natural breaks (p. 232) a data classification method that selects class break levels by searching for spaces in the data values.

NAVD88 (p. 450) the North American Vertical Datum of 1988, the vertical datum used for much geospatial data in North America.

NAVSTAR GPS (p. 90) the U.S. Global Positioning System.

NCGE (p. 531) the National Council for Geographic Education; a professional organization dedicated to geographic education.

NDGPS (p. 101) National Differential GPS, which consists of ground-based DGPS locations around the United States.

NDVI (p. 350) Normalized Difference Vegetation Index; a method of measuring the health of vegetation using near-infrared and red energy measurements.

near-polar orbit (p. 374) an orbital path close to the North and South Poles that carries an object around Earth at an unvarying elevation.

NED (p. 459) the National Elevation Dataset, which provided digital elevation coverage of the entire United States, and is now part of 3DEP.

negation (p. 180) the operation wherein the chosen features are those that meet all of the first criteria and none of the second criteria (including where the two criteria overlap) in the query.

network (p. 277) a series of junctions and edges connected together for modeling concepts such as streets.

NHD (p. 129) The National Hydrography Dataset—a collection of GIS data of water resources for the United States.

NIR (p. 344) near infrared; the portion of the electromagnetic spectrum with wavelengths between 0.7 and 1.3 micrometers.

NLCD (p. 131) the National Land Cover Database is a raster-based GIS dataset that maps the land-cover types for the entire United States at 30-meter resolution.

NOAA (p. 426) the National Oceanic and Atmospheric Administration, a U.S. federal agency focused on weather, oceans, and the atmosphere.

nominal data (p. 133) a type of data that is a unique identifier of some kind—if numerical, the differences between numbers are not significant.

non-selective scattering (p. 347) scattering of light caused by atmospheric particles larger than the wavelength being scattered.

non-spatial data (p. 9) data that is not directly linked to a geospatial location (such as tabular data).

normalized (p. 235) altering count data values so that they are at the same level of representation (such as using them as percentages instead of regular count values).

north arrow (p. 228) a graphical device on a map used to show the orientation of the map.

northing (p. 50) a measurement of so many units north (or south) of a baseline.

NOT (p. 180) the Boolean operator that corresponds with a negation operation.

oblique photo (p. 321) an aerial photo taken at an angle.

off-nadir viewing (p. 380) the capability of a satellite to observe areas other than the ground directly below it.

offset (p. 482) a value applied to an object that is not attached to the ground, but is entirely above ground level.

OGC (p. 530) the Open Geospatial Consortium, a group involved with developing new standards of geospatial data interoperability.

OhioView (p. 531) a research and education consortium consisting of the state universities in Ohio, along with related partners.

OLI (p. 383) the Operational Land Imager, the multispectral sensor onboard Landsat 8.

OMI (p. 424) the Ozone Monitoring Instrument onboard Aura.

OMPS (p. 426) the Ozone Mapping Profiler Suite instrument onboard Suomi NPP.

OR (p. 180) the Boolean operator that corresponds with a union operation.

ordinal data (p. 134) a type of data that refers solely to a ranking of some kind.

orthophoto (p. 319) an aerial photo with uniform scale.

orthorectification (p. 319) a process used on aerial photos to remove the effects of relief displacement and give the image uniform scale.

overlay (p. 185) the combining of two or more layers in the GIS.

P code (p. 93) the digital code broadcast on the L1 and L2 frequencies, which is accessible by the military.

panchromatic (p. 317) black-and-white aerial imagery.

panchromatic imagery (p. 353) black-and-white imagery formed by viewing the entire visible portion of the electromagnetic spectrum.

panchromatic sensor (p. 378) a sensor that can measure one range of wavelengths.

pan-sharpening (p. 378) the technique of fusing a higher-resolution panchromatic band with lower-resolution multispectral bands to improve the clarity and detail seen in the image.

parsing (p. 282) breaking an address up into its component parts.

pattern (p. 324) the arrangement of objects in an image, used as an element of image interpretation.

PDOP (p. 98) the position dilution of precision; describes the amount of error due to the geometric position of the GPS satellites.

personal geodatabase (p. 141) a single file that can contain a maximum of 2 GB of data.

perspective view (p. 460) the oblique angle view of a digital terrain model from which the model takes on a "three-dimensional" appearance.

photo scale (p. 328) the representation used to determine how many units of measurement in the real world are equivalent to one unit of measurement on an aerial photo.

photogrammetry (p. 327) the process of making measurements using aerial photos.

Pléiades (p. 395) a series of satellites operated by Airbus Defence and Space, which feature sensors with a multispectral spatial resolution of 2 meters and panchromatic resolution of 0.50 meters.

POES (p. 431) Polar Orbiting Environmental Satellites; a series of polar-orbiting NOAA satellites.

point cloud (p. 457) the name given to the massive number of elevation data measurements collected by lidar.

points (p. 127) zero-dimensional vector objects.

polygons (p. 127) two-dimensional vector objects.

Prime Meridian (p. 42) the line of longitude that runs through Greenwich, England, and serves as the 0 degree line of longitude from which to base measurements.

principal point (p. 319) the center point of an aerial photo.

prism map (p. 485) a 3D thematic map from whose surface shapes (like the polygons that define states or nations) are extruded to different heights that reflect different values for the same type of data.

prisms (p. 485) the extruded shapes on a prism map.

pseudo-3D (p. 460) a term often used to describe the perspective view of a terrain model since it is often a 2.5D model, not a full 3D model.

pseudorange (p. 94) the calculated distance between a GPS satellite and a GPS receiver.

Python (p. 139) a free computer programming language used for writing scripts in ArcGIS.

QGIS (p. 142) an open source GIS program.

quantile (p. 232) a data classification method that attempts to place an equal number of data values in each class.

query (p. 178) the conditions used to retrieve data from a database.

QuickBird (p. 392) a satellite launched in 2001 by DigitalGlobe whose sensors have 2.4-meter multispectral resolution and 0.61-meter panchromatic resolution.

radiometric resolution (p. 379) the degree of a sensor's ability to determine fine differences in a band of energy measurements.

Rapideye (p. 396) a constellation of five satellites operated by Blackbridge that feature sensors with 5-meter multispectral sensors and a daily global revisit.

raster data model (p. 131) a way of representing spatial data that utilizes a series of equally spaced and sized grid cells.

ratio data (p. 134) a type of numerical data in which the difference between numbers is significant, but there is a fixed non-arbitrary zero point associated with the data.

Rayleigh scattering (p. 346) scattering of light caused by atmospheric particles smaller than the wavelength being scattered.

records (p. 134) the rows of an attribute table.

red band (p. 344) the range of wavelengths between 0.6 and 0.7 micrometers.

reference database (p. 282) the base network data used as a source for geocoding.

reference map (p. 230) a map that serves to show the location of features, rather than thematic information.

relational operator (p. 179) one of the six connectors (=, <>, <, >, >=, or <=) used to build a query.

relief displacement (p. 319) the effect seen in aerial imagery where tall items appear to "bend" outward from the photo's center toward the edges.

remote sensing (pp. 2, 309, 340) the process of collecting information related to the electromagnetic energy reflected or emitted by a target on the ground, using a device a considerable distance away from the target on board an aircraft or spacecraft.

reproject (p. 66) changing a dataset from one map projection (or measurement system) to another.

RF (p. 225) representative fraction—a value indicating how many units of measurement in the real world are equivalent to how many of the same units of measurement on a map.

RGB (p. 236) a color scheme based on using the three primary colors red, green, and blue.

root mean square error (RMSE) (p. 75) an error measure used in determining the accuracy of the overall transformation of the unreferenced data.

rotated (p. 74) when the unreferenced image is turned during the transformation.

SaaS (p. 523) Software as a Service—a method by which the software and data for an application are hosted in the cloud and accessed through a Web browser.

satellite imagery (p. 2) digital images of Earth acquired by sensors onboard orbiting spaceborne platforms.

satellite navigation system (p. 275) a device used to plot the user's position on a map, using GPS technology to obtain the location.

SATELLITES (p. 534) Students and Teachers Exploring Local Landscapes to Interpret the Earth from Space—an OhioView initiative involving K–12 teachers and students with geospatial technologies.

SBAS (p. 101) Satellite-based augmentation system; a method of using correction information sent from an additional satellite to improve GPS position determination.

scale bar (p. 227) a graphical device used on a map to represent map scale.

scaled (p. 74) when the unreferenced image is altered during the transformation.

SDTS (p. 141) the Spatial Data Transfer Standard, a "neutral" file format for geospatial data, which makes the data capable of being imported into various geospatial software programs.

selective availability (p. 97) the intentional degradation of the timing and position information transmitted by a GPS satellite.

Sentinel (p. 388) a satellite program operated by the European Space Agency.

Sentinel-2 (p. 389) a Sentinel mission, consisting of two satellites with characteristics similar to Landsat.

service (p. 523) a format for GIS data and maps to be distributed (or "served") to others via the Internet.

shadow (p. 325) the dark shapes in an image caused by a light source shining on an object, used as an element of image interpretation.

shape (p. 325) the distinctive form of an object, used as an element of image interpretation.

shapefile (p. 141) a series of files (with extensions such as .shp, .shx, and .dbf) that make up one vector data layer.

shortest path (p. 287) the route that corresponds to the lowest cumulative transit cost between stops in a network.

single frequency (p. 93) a GPS receiver that can pick up only the L1 frequency.

site and association (p. 325) the information referring the location of objects and their related attributes in an image, used as elements of image interpretation.

site suitability (p. 188) the determination of the "useful" or "non-useful" locations based on a set of criteria.

size (p. 325) the physical dimensions (length, width, and area on the ground) of objects, used as an element of image interpretation.

SketchUp (p. 487) a 3D modeling and design software program distributed online by Trimble.

skewed (p. 74) when the unreferenced image is distorted or slanted during the transformation.

SKP file (p. 488) the file format and extension used by an object created in SketchUp.

Skysat (p. 394) a high-resolution, small satellite program operated by Skybox Imaging and Google.

SLC (p. 386) the Scan Line Corrector in the ETM+ sensor; its failure in 2003 caused Landsat 7 ETM+ imagery to not contain all data from a scene.

slope (p. 459) a measurement of the rate of elevation change at a location, found by dividing the vertical height (the rise) by the horizontal length (the run).

slope aspect (p. 460) the direction that a slope is facing.

small satellite (p. 394) a type of satellite that weighs between 220 and 1100 pounds.

small-scale map (p. 225) a map with a lower value for its representative fraction; such maps will usually show a large amount of geographic area.

SOS (p. 432) Science on a Sphere, a NOAA initiative used in projecting images of Earth onto a large sphere.

space segment (p. 90) one of the three segments of GPS, consisting of the satellites and the signals they broadcast from space.

spatial analysis (p. 177) examining the characteristics or features of spatial data, or how features spatially relate to one another.

spatial query (p. 184) selecting records or objects from one layer based upon their spatial relationships with other layers (rather than using attributes).

spatial reference (p. 68) the use of a real-world coordinate system for identifying locations.

spatial resolution (pp. 352, 378) the size of the area on the ground represented by one pixel's worth of energy measurement.

SPCS zone (p. 54) one of the divisions of the United States set up by the SPCS.

spectral reflectance (p. 348) the percentage of the total incident energy that was reflected from that surface.

spectral resolution (p. 380) the bands and wavelengths measured by a sensor.

spectral signature (p. 349) a unique identifier for a particular item, generated by charting the percentage of reflected energy per wavelength against a value for that wavelength.

SPOT (p. 395) a high-resolution imagery satellite program operated by Airbus Defence and Space.

SQL (p. 179) Structured Query Language—a formal setup for building queries.

SRTM (p. 456) the Shuttle Radar Topography Mission, flown in February 2000, which mapped Earth's surface from orbit for the purpose of constructing digital elevation models of the planet.

standard deviation (p. 234) a data classification method that computes class break values by using the mean of the data values and the average distance a value is away from the mean.

standard false color composite (p. 357) an image arranged by placing the near-infrared band in the red color gun, the red band in the green color gun, and the green band in the blue color gun.

State Plane Coordinate System (SPCS) (p. 53) a grid-based system for determining coordinates of locations within the United States.

StateView (p. 531) the term used to describe each of the programs affiliated with AmericaView in each state.

stereoscopic 3D (p. 495) an effect that creates the illusion of depth or immersion in an image by utilizing stereo imaging techniques.

stops (p. 288) destinations to visit on a network.

Story Map (p. 524) an interactive Web map created using ArcGIS Online resources that can add descriptive text, photos, or videos to locations on a map to inform the user about the map's topic.

street centerline (p. 278) a file containing line segments representing roads.

Street View (p. 292) a component of Google Maps and Google Earth that allows the viewer to see 360-degree imagery around an area on a road.

suitability index (p. 190) a system whereby locations are ranked according to how well they fit a set of criteria.

Sun altitude (p. 461) the value between 0 and 90 used in constructing a hillshade to model the Sun's elevation above the terrain.

Sun azimuth (p. 461) the value between 0 and 360 used in constructing a hillshade to model the Sun's position in the sky to show the direction of the Sun's rays striking the surface.

sun-synchronous orbit (p. 376) an orbital path set up so that the satellite always crosses the same areas at the same local time.

Suomi NPP (p. 426) a joint satellite mission of NASA, NOAA, and the U.S. Department of Defense.

swath width (p. 374) the width of the ground area the satellite is imaging as it passes over Earth's surface.

SWIR (p. 344) shortwave infrared; the portion of the electromagnetic spectrum with wavelengths between 1.3 and 3.0 micrometers.

symmetrical difference (p. 185) a type of GIS overlay that retains all features from both layers except for the features that they have in common.

tasking (p. 391) the ability to direct the sensors of a satellite on demand to collect imagery of a specific area.

temporal resolution (p. 380) the length of time a sensor takes to come back and image the same location on the ground.

Terra (p. 414) the flagship satellite of the EOS program.

TES (p. 424) the Tropospheric Emission Spectrometer instrument onboard Aura.

texture (p. 326) repeated shadings or colors in an image, used as an element of image interpretation.

textures (p. 484) graphics applied to 3D objects to give them a more realistic appearance.

thematic map (p. 231) a map that displays a particular theme or feature.

three-dimensional (3D) model (p. 455) a model of the terrain that allows for multiple z-values to be assigned to each x/y coordinate location.

TIFF (p. 238) the Tagged Image File Format used for graphics or images.

TIGER/Line (p. 279) a file produced by the U.S. Census Bureau that contains (among other items) the line segments that correspond with roads all over the United States.

time zones (p. 43) a method of measuring time around the world, created by dividing the world into subdivisions of longitude and relating the time in that subdivision to the time in Greenwich, England.

TIN (p. 456) Triangulated Irregular Network—a terrain model that allows for non-equally spaced elevation points to be used in the creation of the surface.

TIR (p. 345) thermal infrared; the portion of the electromagnetic spectrum with wavelengths between 3.0 and 14.0 micrometers.

TIRS (p. 383) the Thermal Infrared Sensor, the instrument that acquires thermal imagery onboard Landsat 8.

TM (p. 382) the Thematic Mapper sensor onboard Landsat 4 and 5.

Tomnod (p. 397) an online crowdsourcing initiative that applies satellite imagery for a variety of real-world issues.

tone (p. 326) the grayscale levels (from black to white), or range of intensity of a particular color discerned as a characteristic of particular features present in an image, used as an element of image interpretation.

topographic map (p. 450) map created by the USGS to show landscape and terrain as well as the location of features on the land.

topology (p. 128) how vector objects relate to each other (in terms of their adjacency, connectivity, and containment) independently of the objects' coordinates.

transit cost (p. 287) a value that represents how many units (of time or distance, for example) are used in moving along a network edge.

translated (p. 74) when the unreferenced image is shifted during the transformation.

transmission (p. 347) when light passes through a target.

trilateration (p. 94) finding a location in relation to three points of reference.

true color composite (p. 357) an image arranged by placing the red band in the red color gun, the green band in the green color gun, and the blue band in the blue color gun.

true orthophoto (p. 320) an orthophoto where all objects look as if they're being seen from directly above.

two-and-a-half-dimensional (2.5D) model (p. 455) a model of the terrain that allows for a single z-value to be assigned to each x/y coordinate location.

type (p. 228) the lettering used on a map.

UAS (p. 314) unmanned aircraft systems—a reconnaissance aircraft that is piloted from the ground via remote control.

union (p. 180) the operation wherein the chosen features are all that meet the first criterion as well as all that meet the second criterion in the query.

union (overlay) (p. 186) a type of GIS overlay that combines all features from both layers.

United States National Grid (USNG) (p. 52) a grid system of identifying locations in the United States.

Universal Transverse Mercator (UTM) (p. 50) the grid system of locating coordinates across the globe.

UN-SPIDER (p. 397) the United Nations Space-Based Information for Disaster Emergency and Response Program.

URISA (p. 531) the Urban and Regional Information Systems Association; an association for GIS professionals.

US Topo (p. 452) a digital topographic map series created by the USGS to allow multiple layers of data to be used on a map in GeoPDF file format.

user segment (p. 93) one of the three segments of GPS, consisting of the GPS receivers on the ground that pick up the signals from the satellites.

UTM zone (p. 50) one of the 60 divisions of the world set up by the UTM system, each zone being 6 degrees of longitude wide.

UV (p. 344) ultraviolet; the portion of the electro-magnetic spectrum with wavelengths between 0.01 and 0.4 micrometers.

vector data model (p. 127) a conceptualization of the world that represents spatial data as a series of vector objects (points, line, and polygons).

vector objects (p. 127) points, lines, and polygons that are used to model real-world phenomena using the vector data model.

vertical datum (p. 449) a baseline used as a starting point in measuring elevation values (which are either above or below this value).

vertical exaggeration (p. 464) a process whereby the z-values are artificially enhanced for purposes of terrain visualization.

vertical photo (p. 317) an aerial photo in which the camera is looking directly down at a landscape.

VGI (p. 14) volunteered geographic information, a term used to describe user-generated geospatial content and data.

viewshed (p. 459) a data layer that indicates what an observer can and cannot see from a particular location due to terrain.

VIIRS (p. 426) the Visible Infrared Imaging Radiometer Suite instrument onboard Suomi NPP.

virtual globe (p. 18) a software program that provides an interactive three-dimensional map of Earth.

Visible Earth (p. 430) a Website operated by NASA to distribute EOS images and animations of EOS satellites or datasets.

visible light spectrum (p. 344) the portion of the electromagnetic spectrum with wavelengths between 0.4 and 0.7 micrometers.

visual hierarchy (p. 224) how features are displayed on a map to emphasize their level of prominence.

visual image interpretation (p. 324) the process of examining information to identify objects in an aerial (or other remotely sensed) image.

W3C Geolocation API (p. 17) the commands and techniques used in programming to obtain the geolocation of a computer or mobile device.

WAAS (p. 101) Wide Area Augmentation System; a satellite-based augmentation system that covers the United States and other portions of North America.

wavelength (p. 342) the distance between the crests of two waves.

Web map (p. 522) an interactive online representation of geospatial data, which can be accessed via a Web browser.

WGS84 (p. 41) the World Geodetic System of 1984 datum (used by the Global Positioning System).

wiki (p. 14) a database available for everyone to utilize and edit.

WorldView-1 (p. 392) a satellite launched in 2007 by DigitalGlobe whose panchromatic sensor has 0.5-meter spatial resolution.

WorldView-2 (p. 393) a satellite launched in 2009 by DigitalGlobe that features an 8-band multispectral sensor with 1.84-meter resolution and a panchromatic sensor of 0.46-meter resolution.

WorldView-3 (p. 393) a satellite launched in 2014 by DigitalGlobe that features shortwave infrared and CAVIS bands, plus a panchromatic sensor of 0.31-meter resolution and 8 multispectral bands with 1.24-meter resolution.

WorldView-4 (p. 393) a satellite planned for a future launch that is expected to feature a panchromatic sensor with a spatial resolution of 0.34 meters.

Worldwide Reference System (p. 385) the global system of Paths and Rows that is used to identify what area on Earth's surface is present in which Landsat scene.

XOR (p. 180) the Boolean operator that corresponds with an exclusive or operation.

Y code (p. 93) an encrypted version of the P code.

z-value (pp. 449, 481) the elevation assigned to an x/y coordinate.

Index

Note: Page numbers with f indicate figures; those with t indicate tables.